Sustainable Production Technologies for Horticultural Crops

NIPA® GENX ELECTRONIC RESOURCES & SOLUTIONS P. LTD.
New Delhi-110 034

About the Editors

Prof. Balraj Singh is currently serving as Vice Chancellor at S.K.N. Agriculture University in Jobner, Rajasthan. He has held various leadership positions in the past, including as Vice-Chancellor at Agriculture University in Jodhpur, Director of ICAR-NRC on Seed Spices in Ajmer, and Project Coordinator of ICAR-AICRP on Honey Bees and Pollinators at ICAR-IARI in New Delhi. Throughout his 33-year career, Prof. Singh has gained extensive experience in research, teaching, HRD, and horticultural crop extension. He has also undergone international training in "Leadership and Decision Making" at the prestigious Harvard Kennedy School at Harvard University in Cambridge, USA, and in "Horticultural Crop Production under Climatic Constraints" at the Faculty of Agriculture, Hebrew University in Rehovot, Israel. Prof. Singh was a member of the pioneering team that introduced Protected Cultivation Technology to India as part of an Indo-Israeli collaborative project. He has developed numerous low-cost and energy-efficient protected cultivation techniques that are suitable for various agro-climatic conditions, with the goal of assisting Indian farmers. Prof. Singh was also the Consortia Principal Investigator for a NAIP project funded by the World Bank, which was worth approximately 1.0 million USD and focused on "Protected Cultivation of High Quality Vegetables and Cut Flowers: A Value Chain Approach." He has served as a resource person on various working groups formed by Indian states to implement Protected Cultivation and Hi-Tech Vegetable Farming. Prof. Singh's contributions have been widely recognized both nationally and internationally, and he has received several awards and honors. He is the founder secretary of the Indian Society for Protected Cultivation and the Vice President of the Indian Society of Seed Technology. Additionally, he is a Fellow member of the Academy of Sciences, Engineering and Technology; the Indian Society of Seed Technology; the Horticulture Society of India; the Indian Society of Seed Spices; and the President of the Society for Integrated Development of Agriculture at AU, Jodhpur, and SHRD in Ghaziabad.

He has provided guidance to numerous postgraduate and Ph.D. students at ICAR-IARI New Delhi. His extensive research experience is reflected in the publication of over 145 research papers in reputable national and international journals, as well as more than 36 book chapters published both nationally and internationally. Additionally, he has written over 125 technical and popular articles to disseminate technology among farmers and stakeholders. Prof. Singh is the author of the first Indian textbook on "Protected Cultivation of Vegetable Crops" and has also edited the book "Advances in Protected Cultivation," which was forwarded by the former President of India, Dr. A.P.J. Abdul Kalam. He has conducted over 50 national and international training programs for scientists, professionals, and farmers on various topics such as high-tech nursery raising in vegetables, advances in protected cultivation of vegetables, plant protection of horticultural crops under protected conditions, and seed production in vegetables. Prof. Singh's contributions include the identification and release of eight varieties of different vegetable and seed spice crops at the state and national levels, as well as the standardization of vegetable crop and seed production technologies under protected and open field conditions. Under his leadership, the ICAR-NRC on Seed Spices, Ajmer, achieved new heights in R&D and received ISO 9001:2008 accreditation for laboratory analysis, seed spice research and development, consultancy, training, and technology transfer to stakeholders. Prof. Singh's accomplishments have been recognized with the Dr. Kirti Singh Gold Medal by HSI and the Life Time Achievement Award by SHRD for his outstanding contributions to the field of Vegetable Sciences.

Dr. A.K. Singh, the Principal Scientist and Incharge of the Regional Station at the Central Horticultural Experiment Station (ICAR-CIAH) in Godhra, Gujarat, has dedicated over 24 years to the advancement of dryland fruits, including both major and minor varieties that are grown in hot semi-arid and arid conditions. He has demonstrated exceptional skill in the ex-situ establishment and conservation of the genetic diversity of dryland underutilized fruits in field gene banks, particularly bael, where he has developed 217 clonal and 121 seedling varieties. Additionally, Dr. Singh has developed 30 varieties of fruit and vegetable crops, including 8 varieties of bael (Goma Yashi, Thar Divya, Thar Neelkanth, Thar Srishti, Thar Prakriti, Thar Shivangi, Thar Gauri, and Thar Bhavya) and 12 varieties of other underutilized semi-arid fruit crops, such as jamun (Goma Priyanka and Thar Kranti), tamarind (Goma Prateek and Thar Rashmi), chironji (Thar Priya), khirni (Thar Rituraj), phalsa (Thar Pragati), karonda (Thar Kamal), wood apple (Thar Gaurav and Thar Prabha), custard apple (Thar Amrit), and mahua (Thar Madhu). Additionally, he has co-developed 10 varieties of vegetable crops, including dolichos bean (Thar Ganga, Thar Lakshmi, Thar Vinaya, Thar Kiran), yard long bean (Thar Surya, Thar Deeksha, Thar Prateeksha), vegetable cow pea (Thar Jyothi), ivy gourd (Thar Sadabahar), and spine gourd (Thar Varsha). Dr. Singh has specialized in dryland fruit culture and has developed various production technologies, such as propagation techniques, high-density planting, orchard floor management, organic farming, intercropping, and cropping models, which have been widely adopted by farmers, including resource-poor and tribal farmers, in India. He has also reported the first instances of morphological variations in bael, including metaxenia, cauliflory, and vivipary.

He has published an impressive number of research papers, including 147 in total, with 50 articles having a NAAS rating of over 6. He has also written 11 review papers, 124 popular articles, 59 book chapters, 17 leaflets, 13 technical bulletins, 3 training manuals, 60 full-length papers, and 163 abstracts in conference proceedings. Additionally, he has contributed to 14 books, two of which are e-books. He has played a significant role in generating valuable scientific information from 12 in-house and 10 externally funded projects, which have helped to disseminate technologies among farmers and stakeholders. Dr. Singh has received numerous awards for his work, including best oral and poster presentations, research paper awards, and recognition from scientific professional societies. He is a recipient of national fellowships from CHAI, HSI, and ISNS and has served as the Editor-in-Chief of Current Horticulture and the editor of Indian Horticulture, ICAR, New Delhi. Dr. Singh is also recognized as a Nodal Officer for DUS testing centers on bael, chironji, and tamarind, and as a Co-Nodal Officer for the DUS testing center on aonla, as recognized by PPV&FRA, New Delhi. He is widely respected as a scientist who is dedicated to advancing dryland horticulture research and development for the benefit of the country's stakeholders.

Dr. B.S. Tomar has consistently been involved in the development of seed production technology for vegetable crops under both open field and protected conditions. For over two-and-a-half decades, he has focused on the genetic improvement of brinjal, onion, okra, and cucumber. Dr. Tomar has been instrumental in the development of 30 different vegetable varieties, including brinjal (6), onion (3), okra (1), cucumber (3), musk melon (2), garden pea (1), bitter gourd (2), velayti palak (1), and cow pea (1), among others. Many of these varieties have been commercialized through MoUs, the National Seeds Corporation, and the NHRDF, as well as IARI and its regional station in Karnal. Dr. Tomar has also organized live vegetable demonstrations on vegetable technologies at the UVRD farm and mela site. He has been teaching Seed Science and Technology and the Division of Vegetable Science for the past 24 years and has guided 8 Ph.D. students and 6 M.Sc. students. One of his students, Dr. R Y Vishwanath, won the IARI gold medal for his outstanding contribution to his Ph.D. program. Dr. Tomar has published 101 peer-reviewed research papers on vegetable crops and over 100 popular articles in Indian Horticulture, Phal-Phool, and Khad Patrika, among other publications.

Dr. B.S. Tomar has demonstrated a significant track record of accomplishments in his career as an agricultural expert. He has authored 11 training manuals, 2 practical manuals, and over 30 book chapters. In addition, he has organized a short training course and winter school sponsored by the Agricultural Education Division of the Indian Council of Agricultural Research, two Model Trainings sponsored by the Directorate of Agricultural Extension, Ministry of Agriculture and Farmers Welfare, and other sponsored farmers training programs in vegetable and seed production. Dr. Tomar has also organized the Sabzi Pathshala program at All India Radio, sponsored by M/s. Dayal Seed, Meerut, and delivered more than 500 talks on All India Radio, Delhi Doordarshan, and DD Kisan program for the benefit of the farming community. Moreover, Dr. Tomar has served as an expert member in various organizations, including the Union Public Service Commission, the Agricultural Scientists Recruitment Board, the National Fertilizers Limited, the National Seed Corporation, the Central Horticultural Experiment Station, the National Dairy Development Board, the Indian Council of Forestry Research and Education, and the Indian Veterinary Research Institute. He has also served as the Incharge of the Seed Production Unit from 2011-2013.

Dr. Tomar is recognized as a Fellow of the Indian Society of Seed Technology, the Indian Society of Seed Spices, the Indian Society of Vegetable Science, and the National Academy of Horticulture Sciences. He has received several awards for his contributions to the field of agriculture, including the Best Teacher Award-2014 at the Indian Agricultural Research Institute, New Delhi, and the Dr. G. Kalloo Outstanding Horticulture Scientist Award-2018 by the Society for Horticulture Research and Development, Ghaziabad, Uttar Pradesh.

Dr. J.K. Ranjan was born on October 7th, 1974, in the Madhubani district of Bihar. He completed his 10th grade from MRG High School in Andhra Tharhi, Madhubani, Bihar, and his 12th grade from Marwari College, LN Mithila University, Darbhanga, Bihar. In 1998, he earned his B.Sc. in Agriculture from Banaras Hindu University in Varanasi. He also holds an M.Sc. in Horticulture (2000) and a Ph.D. in Horticulture (2005) from ICAR-IARI in New Delhi. Dr. Ranjan received the ICAR-Junior Research Fellowship for his Master's program and the CSIR-Senior Research Fellowship for his Doctoral program.

Dr. Ranjan joined the Agricultural Research Service of ICAR as a Scientist in Horticulture (Vegetable Science) in 2005. He was later selected as a Senior Scientist in Horticulture for ICAR-NRCSS in Ajmer. Currently, he works as a Principal Scientist in Vegetable Science at ICAR-IARI in New Delhi. In 2012, he visited Clemson University in South Carolina as a Visiting Scientist. Dr. Ranjan has been recognized as a Fellow of the Indian Society of Seed Spices (2013, Ajmer), the Society for Horticultural Research and Development (2021, Uttar Pradesh), the Indian Seed Vegetable Society (ISVS, Varanasi), and the Indian Seed and Nursery Association (ISNS, Chennai). He has also received the Outstanding Horticultural Scientist Award from the Society for Horticultural Research and Development, Uttar Pradesh.

Dr. J.K. Ranjan has dedicated the past 14 years to the study and improvement of various crops, including temperate fruits and vegetables, seed spices, Solanaceous, and aquatic vegetables. Currently, he is concentrating on the genetic advancement of bottle gourd and brinjal. Throughout his career, Dr. Ranjan has developed two garlic varieties, one apple variety, two Indian spinach varieties, one Ipomea aquatica variety, and a hybrid chili variety, as well as several production technologies and genetic stocks for vegetable crops.

Dr. Ranjan has published an impressive 41 research papers in peer-reviewed national and international journals, five books, 24 book chapters, four training manuals, eight conference proceedings papers, 20 extension folders/leaflets, and over 50 popular articles. He is also a member of several professional societies, including the Society for Horticulture Research and Development, Indian Society of Vegetable Science, Indian Academy of Horticulture Sciences, Indian Society of Allium, Indian Society of Seed Spices, Association for Promotion of Innovations in Vegetables, and the International Society for Noni Sciences. Dr. Ranjan currently holds the position of Joint Secretary in the Society for Horticulture Research and Development and the Indian Society of Seed Spices.

Dr. Som Dutt obtained his M.Sc. and Ph.D. in Botany from Chhatrapati Shahu Ji Maharaj University, Meerut. He commenced his professional career as a Research Scholar at the Indian Agricultural Research Institute (IARI), New Delhi in 1982. Due to his keen interest in popular science writing, he joined the Publications and Information Division of the Indian Council of Agricultural Research (ICAR) at their headquarters in New Delhi in 1984. As an editor, he oversaw the Indian Journal of Agricultural Sciences and Indian Horticulture, a national semi-technical magazine that disseminated information on all research and developmental activities in horticultural crops across the country. He also edited the Handbook of Horticulture, the first-ever published comprehensive book on horticulture by the ICAR, and other important horticultural publications.

Dr. Dutt has written over 500 articles on agricultural sciences, book chapters, and one book titled 'Phoolon Ki Vyavsayik Kheti' in Hindi. He has made a significant impact on farmers by giving radio talks on agriculture in various programs of All India Radio. He has delivered numerous lectures to farmers, extension specialists, young scientists, and Horticulture Development Officers of various states, which has contributed to increasing agricultural production significantly. Dr. Dutt has been honored with the Chaudhary Charan Singh Award for Excellence in Journalism in Agricultural Research and Development 2008 (ICAR National Award), Raizada Memorial Award for Young Information Scientist 1990, and the Best Editor Award 2008, for his efforts in popularizing agricultural sciences through various media outlets. Currently, he is the Managing Editor of Current Horticulture, a peer-reviewed research journal published by the Society for Horticultural Research and Development (SHRD), Ghaziabad, Uttar Pradesh. He has also organized mega events such as the Indian Horticulture Summit-2020 and 2022 and a Webinar on Boosting Immunity through Horticulture in 2020, as the Secretary of SHRD, Ghaziabad, Uttar Pradesh.

Sustainable Production Technologies for Horticultural Crops

Balraj Singh
A.K. Singh
B.S. Tomar
J.K. Ranjan
Som Dutt

NIPA® GENX ELECTRONIC RESOURCES & SOLUTIONS P. LTD.
New Delhi-110 034

NIPA® GENX ELECTRONIC RESOURCES & SOLUTIONS P. LTD.

101,103, Vikas Surya Plaza, CU Block
L.S.C.Market, Pitam Pura, New Delhi-110 034
Ph : +91 11 27341616, 27341717, 27341718
E-mail: newindiapublishingagency@gmail.com
www: www.nipabooks.com

For customer assistance, please contact
Phone: + 91-11-27 34 17 17
Fax: + 91-11- 27 34 16 16

ISBN: 978-93-58874-77-8

Composed and Designed by NIPA®.

Prof. G. Kalloo
Former Deputy Director General (Horticulture & Crop Science)
Indian Council of Agricultural Research, New Delhi
&
Ex-Vice Chancellor, Jawaharlal Nehru Krishi Vishwavidyalaya
Adhartal, Jabalpur, Madhya Pradesh

Foreword

In recent years, there has been a growing concern due to ill effects of food adulteration on health and well being. Food adulteration and contamination are ever-present challenges that threaten our health as well as the very essence of our food heritage. It brings together leading experts and scholars from diverse fields to dissect the multifaceted challenges of food safety and authenticity. This book is a compendium of the collaborative efforts of experts from diverse fields who have come together to provide insights into this multifaceted issue. With each turn of the page, readers will gain a deeper understanding of the challenges we face and the innovative solutions that are being offered to address them. From the fundamental principles of food toxicology to the intricacies of detecting adulterants in products as diverse as coffee, spices, and dietary supplements, this book provides a comprehensive and up-to-date overview of the field. The threat of food bioterrorism looms large in our modern world, and its implications for human physiology are dissected. The book takes us on a journey through food processing, additives, and packaging materials, all of which can be sources of adulterants and contaminants. The field of foodomics provides a powerful tool for the detection of adulterants, while nanotechnology offers promising avenues for enhancing food quality and safety. Chapters dedicated to specific food categories, such as cereals, oils, fruits, and meats, shed light on the unique challenges each faces in terms of adulteration. Moreover, it does not stop at detection but also delves into prevention strategies, tracing the evolving landscape of food fraud prevention and the dynamic relationship between authenticity, traceability, branding, and consumer preference. They also culminate in a discussion of strategies to prevent food fraud, addressing the ever-evolving challenges in this domain. This book is more than just a compilation of scientific knowledge; it is a call to action. It challenges us to be vigilant consumers, informed policymakers, and proactive researchers. It urges us to prioritize food safety and authenticity in a world where the global food supply chain is more interconnected than ever before. I commend the dedication and expertise of the editors and authors who have contributed to this book and am confident that their work will inspire further research, innovation, and, most importantly, a renewed commitment to safeguarding the purity and trustworthiness of our food. May this book serve as a beacon of knowledge and a catalyst for positive change in the world of food safety and authenticity.

Wishing the book very best.

Thanking You

(G. Kalloo)

Preface

The horticultural science has emerged as the leading discipline in total agricultural development in India. Playing a pivotal role, horticultural science has surpassed total food grain production in our country. Besides, horticultural science has proven facts of protecting mankind from so many dreaded diseases, making food and nutritionally secured country in the world.

A number of ICAR institutes and State Agricultural Universities are devoted in the field of research and development of horticultural crops. The large network of National Agricultural Research System consisting of huge manpower is credited for such fruitful research and developmental activities across the country. Though, we are at comfortable place, still we need further research in certain fields. A number of new varieties, genetic resources, technological interventions and certain breakthroughs need to be updated, compiled and disseminated to the target audience and stakeholders. So many crops having enormous potential need to be tapped. A number of challenges, problems and biotic and abiotic stresses have challenged our research system. Collection of adequate scientific information and their timely dissemination on challenging topics has become the need of the hour.

The Society for Horticultural Research and Development (SHRD) in Uttar Pradesh, India, has compiled a comprehensive collection of research findings and outcomes from the Indian Horticulture Summit-2020 on Mitigating Climate Change and Doubling Farmers Income through Crop Diversification, the Webinar on Boosting Immunity through Horticulture, and the 2nd Indian Horticulture Summit-2022 on Horticulture for Prosperity and Health Security. The publication brings together valuable research contributions from horticultural scientists across various disciplines, as well as invited chapters from other innovative researchers. The book includes a number of lead lectures from senior subject matter specialists and provides the most current information on the topic.

The book covers a range of topics, including genetic resources, varietal wealth, crop improvement, plant health management, insect, pest, and disease management, and value-added horticultural products. Additionally, the book also delves into general horticulture.

The editors would like to extend their gratitude to Drs G. Kalloo, T. Janakiram, Z.P. Patel, S. Rajan, P.L. Saroj, Vishal Nath, and T.R. Ahlawat for their guidance and support during the compilation of the book. The Society is grateful to NIPA, New Delhi for their prompt publication of the book.

This publication is a valuable resource for students, scientists, researchers, policymakers, and other stakeholders in the field of horticultural science.

Editors

Contents

Section 02: Vegetables

Section 03: Floriculture and Landscaping

Section 04: Tuber and Plantation Crops

Section 05: Crop Protection

Section 06: Institution

1

Opportunities and Potential of Expansion of Protected Cultivation in India

Balraj Singh

Vice Chancellor, S.K.N. Agriculture University, Jobner, Rajasthan

With time it has been well proved that protected cultivation is a better technology to enhance productivity and quality of the crops by providing a logical and technical solution to manage the major and minor biotic and abiotic stresses encountered under open field cultivation of major horticultural crops. The effectiveness of the technology has also been observed in many parts of the world (Singh, 2013). Presently, the area under different forms of protected cultivation is around 4.5 million ha in the world. During the last two and half decades, the area under protected cultivation has increased exponentially in various countries and around Mediterranean sea. In China, various kind of protected cultivation technologies like mulching, use of temporary plastic walls in open fields, plastic low tunnels, plastic covered walk-in-tunnels, high tunnels, temporary and permanent insect-proof net houses, shade net houses, rain shelters and different kinds of greenhouses etc are being adopted. Presently, China is the world leader in cultivating horticultural crops under different protected conditions, total area under protected cultivation being around 3.5 million ha. Out of this area, nearly 96 percent is being used for cultivation of commercial fresh vegetables and their hybrid seed production. Simultaneously, growth like China has also been observed in India and African continent. But the success rate varies significantly because of poor correlation between design of various protected structures and prevailing agroclimatic conditions of a region. It has been clearly established and proven that mere transportation or adoption of technology as such from Northern Europe to other parts of the world, irrespective of prevailing agroclimatic conditions was not a valid decision and step. Each region and area actually require further research, development, extension, training and new norms, procedures and methods of application of technology to meet the region-specific requirements for protected cultivation of horticultural crops (Singh, 2019).

Opportunities of Protected Cultivation in India

India, with its vast and diverse agricultural landscape, has always been a significant contributor to the global agricultural market. The challenges of increasing population and changing dietary preferences have placed immense pressure on our country's food production systems. To meet the growing demands for fresh and nutritious produce, India must innovate and adopt modern techniques. Protected cultivation methods, including greenhouses, net-houses, high tunnels, and low tunnels, hold the key to increasing vegetable production and securing India's food future. India being a country with diverse climatic regions has shown an overall growth of around 2.0 lakh ha area under protected cultivation in different states during the last two decades. The success rate of these technologies varied significantly depending upon the climatic conditions of various regions in India. In plains of Northern India, these technologies faced high challenges for making them successful against the harsh climatic conditions, whereas in mild climatic areas like Bengaluru and Pune, the success rate achieved has always been high. Basically, the dismal growth of protected cultivation technology in the country happened mainly due to Government policies providing handsome subsidies under various schemes launched by the government of India under MIDH (NHM), TM, NHB, RKVY etc., and up to some extent due to the technical beauty of technology. The technical know-how for adoption of protected cultivation technology under Indian conditions was not to the level at the time of inception. With passage of time, research and developmental work carried out by various public sector institutions in collaboration with developed countries gradually reflected that technical designs of different protected structures needs modification suitable to the region-specific needs of prevailing climatic conditions and problems under open field cultivation. During last few years, varieties & hybrids and production technology suitable for protected cultivation have been developed by various public sector institutions in India. Further there are some most successful examples and models for adoption of this technology even under harsh climatic conditions like Bassi-Jhajhraand Basari village, just 15 km away from Jobner (Jaipur), but this seems only in cluster approach. This most successful example in cluster approach clearly shows the potential of the technology expansion up to 15.0-17.0 lakh ha for growing horticultural crops in different states in India by the year 2050. Presently, the area under protected cultivation in India stands at a mere 2.0 lakh ha. In contrast our country produces 207.00 million tonnes of vegetables annually. Looking ahead to 2050, our nations vegetable requirement is expected to surge to 450.00 million tonnes. To bridge this massive gap, protected cultivation has the potential to play a pivotal role, offering a promising solution (Tuzel and Kacira, 2021).

Expansions of Protected Cultivation

Promotion and expansion of protected cultivation technology will not only be going to help for creation of huge self-employment for unemployed educated youths, but it will also increase the national economy by sale of quality healthy planting material, fresh horticultural produce for domestic and international markets. Under the era of FDI (Foreign Direct Investment) in retail, these kinds of models possess high potential for enhancing the income of farmers opting for quality and off-season vegetables and cut flowers cultivation through various protected cultivation technologies. On one hand, production of vegetable and cut flower crops under protected conditions provides high-water and nutrient-use efficiency under varied agroclimatic conditions of the country. On the other, this technology has very good potential especially in peri-urban areas in cluster approach adjoining to major cities, a fast-growing market for fresh quality produce. Thus, it can be profitably used for growing high-value vegetables like, tomato, cherry tomato, coloured peppers, parthenocarpic cucumber and brinjal, cut flowers like roses, gerbera, carnation, chrysanthemum etc. and virus-free healthy seedlings and planting material of horticultural crops in agro-entrepreneurial models. But protected cultivation technologies require careful planning, attention, adequate timing of production and moreover, harvesting time to coincide with high market prices, choice of varieties adopted to off-season environments, and able to produce economical yield of high-quality produce etc. Even though application of chemicals for controlling biotic stresses is also low under protected structures which gives a high-quality safe vegetables for human consumption. By using protected structures, it is also possible to raise off-season and long duration vegetable crops with high yield (3-5 fold as compared to open field cultivation) withquality. Vegetables and cut flowers farming in agro-entrepreneurial models targeting various niche markets of big cities is inviting regular attention of vegetable and flower growers for diversification from traditional ways of crop cultivation to the modern methods like protected cultivation of horticultural crops. (Singh *et al.* 2010).

Important Tips for Protected Cultivation

The basis behind the successful implementation of protected cultivation technology largely depends on selection of suitable designs with proper, crops,their varieties and adoption of successful cultivation are fundamental variables that may significantly affect success and economic return of the entire. A most successful example of protected cultivation in cluster approach is Bassi-Jhajhraand Basari village 15 km away from Jobner (Singh, 2023). The success of protected cultivation technology entirely depends upon four basic concepts,*viz.* what to produce, when to produce, how to produce and where to sell and where the export the high-quality produce. The growers must know two basic options, i.e. choice of crop or variety for its high economic potential/return and to develop most suitable production system in cluster approach. Crop should be selected based on existing structures, wide consumption and good adoption to unsteady climatic conditions and suitable for long cultivation cycles. While adopting the protected cultivation technology, following most important points, viz. market requirement of the produce, distance from the market for the fresh produce, climatic conditions of the area, soil characteristics and quality of water, economic convenience, crop requirement, labour and skilled manpower requirement should be well considered in advance.(Singh, 2005).

Major Challenges in India

Major Challenges are

- Non-availability of region-specific designs of green houses and insect-proof net houses for varied agroclimatic conditions and regions.
- Fabrication of protected structure has come up as a big business, taking an opportunity small industry are sacrificing with quality of material to be used to gain more profit and also lack of understanding of designs of the protected structures for different regions and quality of basic steel and cladding material used for fabrication and installation of structures.
- Limited traineds killed and professional manpower for designing, fabrication of protected structures and thereafter, maintenance of the structures and for protected cultivation of various high-value crops.
- Lack practical training institutions and advisory services in the area of protected cultivation of horticultural crops.
- Limited of availability of suitable crop varieties and planting material specific to protected cultivation, especifically with public sector institutions, its management practices etc. asplanting material/seeds, cladding material, water-soluble fertilizers, etc. available with private sector companies andare too costly.
- Lack of demand-driven cultivation without proper marketing strategies several times creates problem for proper disposal of quality produce and sometime farmers are even not getting reasonable price of their produce.
- Increasing threat of soil-born problems like root-knot nematodes and Fusarium for protected cultivation of horticultural crops and more specifically vegetable crops.
- Non-availability of virus-free healthy planting material in vegetables, fruits and ornamental crops.
- Poor power supply for drip irrigation is major challenge in protected cultivation.

Potential and Strategies for Adoption of Protected Cultivation Technologies

Protected cultivation technology for horticultural crops has tremendous potential for adoption under varied agroclimatic conditions of the country. The most potential areas where high scale interventions are required to be promoted are as under (Report of the Working Group, 2013).

- Use of plug tray nursery technology on commercial scale for raising virus free healthy vegetable and flower seedlings as Agri- business models.
- Large scale use of different designs of insect poof net houses with different designs for commercial vegetable cultivation and for hybrid seed production of vegetable crops.
- Use of modified naturally ventilated greenhouses equipped with mini sprinklers on the roof top area and zero energy consuming exhausts on gutters for continuous removal of greenhouse hot air for cultivation of vegetable and flower crops cultivation under harsh conditions of arid and semi-arid regions of the country.
- Use of naturally ventilated green house for hybrid seed production in vegetables for increasing the overall profitability of the farmers.
- Large scale use of plastic mulches for commercial vegetable cultivation under open fields and also under green houses and even under insect proof net houses under varied agro-climatic conditions but now the use of synthetic plastic needs replacement with biodegradable or organic mulch.
- Commercial application of micro-grafting technology for developing resistant plant material against soil borne problems.
- To promote cultivation of fruit crops like papaya, pomegranate under insect proof net houses for protection against biotic and abiotic stresses.
- Large scale application of low-pressure drip irrigation system for managing an area of 1000-2000 sq. mts. area for cultivation of different horticultural crops under protected structures.
- Commercial use of open roof type high tunnels in northern parts of the country and even under temperate conditions can be more rewarding.
- Large scale skill development of youth manpower in two ways following the Chinese model i.e., designing, fabrication and maintenance of protected structures and secondly the production management of crops under protected conditions
- Application of GAP procedures and standards for protected cultivation of high value vegetablecrops will help to catch international trade.
- Government support should be extended for self-fabrication mode of temporary low-cost structures like insect proof net houses, shade net houses, walk-in-tunnels and plastic low tunnels for the cultivation of suitable vegetable and flower crops.
- Promotion of protected cultivation technology for its sustainability in cluster approach especially in peri-urban areas of the country by following Bassi-Jhajhra village in Jobner.

- Government should also promote to develop input hubs for protected cultivation in multi-locations in PPP mode.
- Use of solar energy for running drip system and up to some extent for running heating and cooling devices of the protected structures should also be promoted.
- All the protected cultivation clusters must be mandatorily clubbed with rain water harvesting infrastructure facilities.
- Promotion of most suitable crop sequences for different protected structures and seasons based on research data.
- Promotion of large-scale mechanization in vegetable and flower cultivation by using raised bed makers, plastic laying machines, plastic low tunnel making machines, pipe bending machines for making walk-in-tunnels, drip lateral laying and binding machines
- To establish convergence and synergy among various on-going and planned government programmes in the field of protected cultivation development.
- To ensure adequate, appropriate, time bound and concurrent attention to all links in production under protected conditions, post production on-farm value addition, processing,consumption chainsand for export.
- Cluster approach for taking up protected cultivation as a whole is required and may be clubbed with processing industriesand export along with national high ways and dedicated freight corridors.

Table 1: Expansion potential of protected cultivation technologies in different Indian states by 2050

Type of Protected Structure	Suitable crops	States	Pace expected expansion
Low-cost polyhouse up to $50m^2$-$250m^2$ area	Several vegetable crops for protection against frost and pests and viruses	Uttarakhand, Himachal Pradesh, Tamil Nadu,Jammu and Kashmir and all North East states	8-10% growth every year over the present level of this technology
Rain shelters-cum-insect proof net-houses	Several vegetables for protection against insect pests and viruses and continuous rains	All North East States, Bay Islands, Eastern Coastal Region, Western Coastal Region, Kerala etc	6-8% growth over the present level of use of the technology
Plastic low tunnel technology (but now needs replacement of Synthetic plastic with Non-woven material for covering the tunnels)	Mainly for off season cultivation of cucurbits during winter months	Punjab, Haryana, Rajasthan (parts adjoining to Haryana, Punjab), Uttar Pradesh	15-20% over the level of use in around 20,000-22,000 acres

Type of Protected Structure	Suitable crops	States	Pace expected expansion
Walk-in-tunnels	For off-season vegetable cultivation during winter months	Foot Hills Uttarakhand, Himachal Pradesh, Jammu and Kashmir and North Eastern	8-10% over the present use of the technology
Insect-proof net houses	For protection of vegetable and fruit crops like Papaya against viruses and Insect pests and up to some extent from frost	North Indian Plains,central India, western Parts like Rajasthan, Gujarat	15-20% over the present status of use in around 8000-10000 ha
Naturally ventilated greenhouses	For large scale and year-round cultivation vegetables viz. tomato, cherry tomato, Capsicum, seedless cucumber, eggplant, etc. and hybrid seed production of vegetable Large scale	Punjab, Haryana, Rajasthan, Uttar Pradesh, Madhya Pradesh, Gujarat, Maharashtra, Kerela, Telengana, Orissa, Bihar, Jharkhand, Tamil Nadu, Andhra Pradesh	15-20% over and above the present use in around 1.20-1.30 lakh ha.
Semi-climate-controlled greenhouses	Nursery production and use for commercial vegetable cultivation and seed production of high value vegetable crops.	Rajasthan, Gujarat, Madhya Pradesh, Maharashtra, Arid Region and Semi-arid regions	8-10% over the present status of use in around area of 1500-2000 ha.
Shade net houses	For cultivation of leafy. Vegetables & herbs during peak summer months.	Arid and semi-arid regions of Madhya Pradesh, Rajasthan, Gujarat, Maharashtra, Andhra Pradesh., Tamil Nadu,etc.	10-12% over the present area in use around 12,000-14,000 ha
High tunnels	For off-season cultivation of cucurbits and other vegetables during winter months + large-scale drying of chilli and pan methi	Rajasthan, Andhara Pradesh, Maharashtra, Madhya Pradesh, Gujarat and other arid and semi-arid regions	8-10% over and above the present level of use
Retractable greenhouses	For cultivation of black berry and blue berries	Uttar Pradesh, Uttrakhand, Trait Region Madhya Pradesh, etc.	5-8% over the leaf level 20 -30 ha

Type of Protected Structure	Suitable crops	States	Pace expected expansion
Climate controlled greenhouse/glass-houses	For quality planting material production of Potato under aeroponics systems and hydroponic system	Haryana, Punjab, Uttar Pradesh and Rajasthan	3-5% over the leaf level
Mulching (replace-ment of system-atic plastic with biodegradable or compostable plastic is needed or by organic mulch in fruit orchards	For cultivation of major vegetables, fruit crops, spices ornamental crops.	Across the country clubbed with raised beds and drip irrigation system	15-20% over the plant level of 50,000-60,000 ha

Benefits of Protected Cultivation

Protected cultivation provides a barrier against pests and diseases. The enclosed environment helps prevent the entry of harmful insects and pathogens, reducing the need for chemical pesticides. This contributes to a more environment-friendly and sustainable crop production systems.

The increased efficiency in resource utilization is another advantage. Controlled environments enable precise application of fertilizers and nutrients, minimizing waste and optimizing plant nutrition. This efficiency not only benefits the environment but also contributes to cost savings for farmers.

Protected cultivation also offers a range of benefits, including extended growing seasons, enhanced control over environmental factors, protection from adverse weather conditions, water conservation and pest and disease management. These advantages collectively contribute to higher agricultural productivity with better food quality, and increased sustainability in the face of evolving challenges in agriculture sector (Singh,2005).

Its versatility in adapting to diverse agroclimatic conditions. This technology allows year-round cultivation.It is beneficial for having with extreme climates. It supports sustainable crop diversification,enabling the cultivation of high-value crops such as fruits, vegetables, and flowers. Furthermore, it facilitates crop intensification by maximizing yields per unit area (Singh, 2011).

Addressing Water Scarcity

Water scarcity is a significant concern in many parts of India. Protected cultivation provides a means to optimize water-use efficiency by controlling the micro environment within the structure. Reduced water wastage and precise irrigation management are critical benefits, especially in areas struggling with water availability.

Ensuring Product Quality and Safety

In an era of increasingly stringent market demands, standards, and regulations, product quality and safety are paramount. Protected cultivation allows for meticulous control of environmental factors such as temperature, humidity, and light, resulting in consistent and high-quality produce. Moreover, protection from external contaminants and pests minimizes the need for chemical interventions, aligning with the growing consumer preference for organic and safe products.

Cluster Approach for Success

The cluster approach has been instrumental in successful expansion of protected cultivation. Cluster of farmers adopt this technology in proximity benefit from shared knowledge, resources, and market access. This approach fosters collaboration, enhances economies of scale, and minimizes risks associated with technology adoption (Singh, 2023).

Multiple Applications of Protected Cultivation

Protected cultivation technology finds diverse applications in Indian agriculture. Firstly, it plays a pivotal role in raising disease and virus-free healthy planting material. The controlled environment of greenhouses and polyhouses ensures that the initial crop material is of highest quality, contributing to the success of subsequent cultivation.

Secondly, it is instrumental in hybrid seed production, especially for vegetables. This technology enables the isolation and controlled pollination of different varieties, ensuring the purity and quality of hybrid seeds (Singh and Tomar, 2015).

Thirdly, protected cultivation is central to the production of fresh food for better economic viability. The extended growing seasons and higher yields empower farmers to meet market demands consistently, enhancing their income.

Role of Skilled Manpower

For the sustained growth of protected cultivation, developing a skilled workforce is paramount. This involves training rural youth in two key areas: designing, fabricating, installing, and maintaining protected structures, and managing the entire crop production process under protected conditions. Skilled labour not only ensures the efficient operation of these systems but also opens up opportunities for employment and entrepreneurship in agricultural sector.

Expansion of Protected Cultivation in Different States

The future of protected cultivation technology in India is promising. It is envisioned that by 2050, the technology could be expanded to cover an estimated 15.0-17.0 lakh ha of agricultural land across the country. This expansion would offer solutions to the multifaceted challenges facing Indian agriculture, making it more resilient and sustainable in the face of changing climate patterns.

Gujarat

Gujarat has been a pioneer in greenhouse farming and is known for its successful adoption of protected cultivation techniques. The state's moderate climate and progressive government policies make it conducive for expansion. Vegetables, flowers, and export-oriented crops like capsicum, tomato, and seedless cucumbers are successfully being grown under protected conditions. The two sea ports for fast export of value-added options are the major ladder for the State of Gujarat for making fast progress in protected cultivation:

Maharashtra

Maharashtra, with its diverse agroclimatic zones, offers ample opportunities for protected cultivation. The state has seen increased adoption, especially in the Nashik region for grapes, Pune for floriculture, and areas around Mumbai for vegetables. The state government has also been encouraging farmers to invest in protected cultivation. Some of the areas have already come up as protected cultivation hubs around Pusa for cultivation of vegetable and ornamental crops. More areas in the State have potential for protected cultivation like Konkan region and areas which are boarding with Karnataka State.

Protected cultivation can be taken up on a large scale-seed production of vegetable crops.

Himachal Pradesh

The hilly terrain and cold climate in Himachal Pradesh make it suitable for high-value crops like strawberries, capsicum, cucumbers, tomatoes and cut flowers under protected cultivation. The state government has been promoting polyhouses for extending growing season and increasing yield. Seed production of vegetable crops can also be taken up in foothills of the State under low cost protected structures.

Karnataka

Karnataka, with its varied agroclimatic zones, is another state with immense potential for protected cultivation. Regions like Bengaluru, Mysore, and parts of North Karnataka have seen growth in greenhouse farming for vegetables, while the Coorg region is known for its floriculture. Seed production of vegetable crops can also be taken up in different regions of the state. Dharwad, Rani Benuer etc. are ideal for cut flower crops like gerbera, carnation or even roses. Protected cultivation of flowers can be taken up in large scale in different regions of State like Oaty hills, Coimbatore, Solem area, etc.

Tamil Nadu

Tamil Nadu's favourable climate allows for year-round cultivation of various crops. Protected cultivation is extensively used for vegetables, flowers, and even fruits. The state government has provided subsidies to encourage farmers to adopt these techniques. Protected cultivation of flowers can be taken up in large scale in different regions of State like Oaty Hills, Coimbatore, Salem areas, etc.

Punjab and Haryana

These states in northern India have adopted protected cultivation for vegetables and fruits, especially during the off-season. The region's extreme weather conditions make greenhouses and polyhouses vital for year-round production of high value vegetables, seed production of selected vegetables and large scale healthy nursery production. Protected cultivation can be taken up in cluster approach along with mega highways like NH 152D KMP, NH-1, etc. for cultivation of high value vegetables and nursery hubs can also be developed in these areas for fast transportation of the produce and planting material to far-flung areas of the country.

Rajasthan

Rajasthan's arid climate and extreme temperatures can be challenging, but protected cultivation, particularly shade net houses and polyhouses, have been adopted and can be further expended for growing fruits and vegetables like tomatoes, capsicum, cucumbers, and papaya. In Rajasthan, insect-proof net houses and shade-net houses can be used at large scale and Bassi-Jhajhara village model can be replicated in several other parts of Rajasthan and other states.

Andhra Pradesh and Telangana

These states have seen an increase in protected cultivation for export-oriented crops like flowers, vegetables, and fruits. The government has introduced subsidies to promote these practices. Seed production of vegetable crops can also be taken up to large scale in different regions mainly around Hyderabad.

Kerala

In Kerala, due to its humid tropical climate, protected cultivation is used for growing orchids, vanilla, and spices. Polyhouses and shade net houses have gained popularity and up to some extend vegetables like cucumbers and capsicum are being grown. There is a need to use protected conditions for large-scale healthy planting material production of spices like black pepper, cardamom, ginger turmeric etc.

Uttarakhand

Uttarakhand's hilly terrain and cold climate provide opportunities for protected cultivation of high-value crops such as organic vegetables, medicinal herbs, and strawberries. Seed production of vegetable crops can also be taken up in foothills of the state.

Bihar and Jharkhand

Greenhouses and other low-cost protected structures in like insect-proof net houses, high tunnels, walking tunnels and shade net houses can be used at large-scale vegetable cultivation and also ten Gerbera, Carnation etc. Protected structures like greenhouses are required to be used at large scale for disease and virus-free planting material of different horticultural crops.

North-Eastern States

Low-cost protected structures like poly houses and rain shelter-cum insect-proof net houses can be used for cultivation of vegetable and ornamental crops. There is a need to use protected conditions for production of healthy planting material of different horticultural crops. In these states protected cultivation can be well linked with low-pressure drip irrigation system.

Orissa and West Bengal

Protected cultivation like greenhouses or net houses can be used for a large number of vegetable cultivation but in cyclone-prone coastal regions only low-cost protected structures are required to be used for cultivation of different horticultural crops.

Uttar Pradesh

In Uttar Pradesh, protected cultivation of vegetable and flower crops have very good potential. Same time, large scale planting material production can also be expanded along with national highways and in NCR region including Peri-urban areas of Lucknow, Prayagraj, Banaras, etc.

The expansion of protected cultivation in India also depends on market demand, access to technology, infrastructure development, and government incentives. Many state governments offer subsidies, technical assistance, and training programmes to encourage farmers to adopt these practices. With climate change affecting traditional agriculture, protected cultivation can play a crucial role in ensuring food security and income generation for farmers.

Diversified Use of Protected Structures

Post-harvest management

Plastic high tunnel can be used for better post-harvest drying, as it acts as a protected site under plastic and operates like a solar dryer with ventilation. The open sun drying under field reduces the face value of the produce by loss of excess moisture and discolouration besides chances of microbial contaminated if done on *Kachcha* platforms. Looking to the benefits of high tunnel, following are potential uses in arid/semi-arid regions (Singh, 2019).

Drying of Nagaurimethi: For drying of *Nagaurimethi* use of high tunnels can be very quick and clean process wherein the leaves can be dried 2-3 days earlier than sun drying.It is also safe from damage caused due to unusual weather situations like winter rains. This technology is more effective to have a clean and green produce of *Nagaurimethi* leaves especially for export purposes.

Drying of vegetable crops: Introduction of high tunnels is a cheap and clean process for drying vegetable like chilli, mint, spine gourd, ashwagandha etc. with maintaining quality for better consumer acceptability.

Drying of specific commodities: Some products like *Panchkuta, kachri* etc. which are unique to arid ecologies are sun dried by farmers. The use of high tunnel can be innovative approach in drying of these commodities for better quality and fast drying.

Drying of plantation crop products: In the region of Sojat and Pali region cultivation of heena leaves holds special identity as a Geopgraphical Indicator. The drying of henna leaves under high tunnels will be very useful in producing better heena powder for local and international market.

Conclusion

Protected cultivation has become an integral part of modern agriculture, significantly contributing to global food security and horticultural production. The diverse array of structures and crops grown under protection underscores the adaptability and versatility of these systems. As the world grapples with the challenges of feeding a growing population, protected cultivation is set to play an increasingly vital role in the future of agriculture, offering solutions to the problems of climate change, resource scarcity, and food quality. To realize the full potential of protected cultivation, governments, agricultural institutions, and private sector must work together to support and promote sustainable practices, ensure training and knowledge sharing, and create favourable policies for its continued growth. Looking to the future, the potential of protected cultivation is enormous. As global population growth and climate change continue to exert pressure on traditional agriculture, protected cultivation offers a sustainable solution to increase food production, ensure crop quality, and reduce the dependency on seasonal variations. Additionally, technological advancements in energy-efficient systems, renewable energy adoption, and integrated pest management are expected to make protected cultivation even more sustainable and economically viable.

There is a huge potential of expansion of the protected cultivation technology in India,in different states huge but highly dependent upon the technicality and recommendation of the technology implementation.The technology can be expended up to 15.0- 17.0 lakh by the year 2050. The cluster approachin it adoption can play crucial role in it successful expansion.The protected cultivation technology has to play a significant role under varied agroclimatic conditions of our country as a means for sustainable crop diversification, intensification, and for vertical growth of productivity of horticultural crops leading to optimization of water and fertilizer-use efficiency. In the near future, first and most important requirement of its is raising disease and virus-free healthy planting material of horticultural crops. The use of technology for hybrid seed production of vegetables and production of fresh food for better economic-viability are other uses. Further, sustainability of the technology is utmost important to develop a large skilled manpower in form of rural areas. There is a need to replace use of synthetic plastic mainly used for mulching purposes or as a covering plastic low tunnels either with the use of biodegradable or non-woven material.

References

Report of the Haryana Kisan Ayog (Working group on protected cultivation) Panchkula, Haryana 2013.

Singh, Balraj and Sirobi, N.P.S. 2006. Protected cultivation of vegetable crops in India: Problems and future prospects. Acta Horticulture 710:339-42.

Singh, Balraj and Tomar, B.S. 2015. Vegetable seed production under protected and open field conditions in India: A review. Indian J. of Agricultural Sciences 85(9):86-89.

Singh, Balraj, 2005. Protected cultivation of Vegetable Crops, Kalyani Publication, New Delhi, 1-188.

Singh, Balraj, 2011. Protected cultivation technologies for diversification and livelihood security. Progressive Agriculture 11:112-117.

Singh, Balraj, 2013. Protected Cultivation in India: Challenges and Strategies. Current Horticulture 1(2):3-6.

Singh, Balraj, 2019. Prospects of Protected Horticulture in Arid and Semi-Arid Regions of India. Acta Scientific Agriculture 3(3):93-99.

Singh, Balraj, 2023. Opportunities for expansion of protected cultivation in different states in India. Paper presented as key note speaker in the 10th Indian Horticultural Congress at AAU, Guwati.

Singh, Balraj, Singh, A.K. and Tomar, B.S. 2010. In peri urban areas: protected cultivation technology brings prosperty. Indian Horticulture 55(4): 31-32

Tuzel, Y and Kacira, M. 2021. Recent developments in protected cultivation. Acta Horticulture 1320(1):1-14.

Section 01: Fruits

2

Exchange and Utilization of Fruit Genetic Resources in India

Pragya Ranjan*[1], *S.K. Singh*[2], *J.K. Ranjan*[2], *Vartika Srivastava*[1] *Vandana Tyagi*[1], *Pratibha Brahmi*[1]

[1]ICAR- National Bureau of Plant Genetic Resources, New Delhi
[2]ICAR-Indian Agricultural Research Institute, New Delhi

India is the home to a number of fruit species including mango, banana citrus etc. which have global importance. The North-Eastern (NE) region is particularly endowed with tremendous diversity of banana and citrus and a number of wild/ minor species. However, India has been and continues to be highly interdependent for germplasm sharing with many countries. In the early times, development in Indian fruit industry brought about largely by making selections from indigenous variability for large scale production. Selection from natural or introduced diversity was performed by the local growers, often patronized by rulers, continued over the years and resulted in the development of several popular types of different fruits. Generally, a large number of exotic germplasms were introduced, adapted and utilized by breeders and adopted by farmers for extensive cultivation, which helped in increasing the horticultural production in our country during 1990's.

In India, there is a single window system operated by ICAR-NBPGR, New Delhi for germplasm exchange, which includes import, export and quarantine of small samples, including those of transgenics, meant for research use. India has a good experience of introducing quite a number of Fruit crops Genetic Resources (FGR), which has boosted the fruit production in our country. Classical use of delicious group of apple, low-chilling peaches and introduction of excellent varieties in grapes, *viz.,* Perlette, Thomson Seedless and Beauty Seedless have been instrumental in gaining production jumps, thereby increasing export share of fruit crops from India. Kiwi fruit introduced from New Zealand during 1960s, has played a very important role in enriching our fruit basket. The crops like blueberry, longan, mangosteen, rambutan, avocado, persimmon, macadamia nut, pineapple guava, opuntia pear, dragon fruit, monk fruit etc. are being adapted and awaiting commercialization. On the other hand, Indian mango, citrus and banana germplasm finds an important place in reaching the breeding goals in several countries. Development of Floridian mangoes of the USA and use of Indian citrus rootstocks worldwide are the classical examples of utilization of Indian fruit germplasm outside the world.

Domestication/Introduction of Fruit Crops in India: Historical Perspective

Vedic Period

Fruit culture in India is as old as Aryan civilization and dates back to Pre-Vedic period. *Shruti, Smriti* and *Samhita* of the Vedic age bear a clear testimony to the fact that the inhabitants of the towns of Mohanjodaro and Harappa used to cultivate fruits. The *Vriksh* Ayurveda, written by Kautilya in 300 B.C is one of the oldest known text on fruit culture (shodhganga.inflibnet.ac.in). Medicinal uses of fruits like *aonla*, *bael*, citrus, grapes, hog plum, jackfruit, wild fig, monkey jack, *jamun*, ber, karonda, khirni, lemon, lime, mango, mulberry, orange, *phalsa*, banana, pomegranate, wood apple etc. have been mentioned in *Charaka Samhita* and *Sushrut Samhita* (Chadha and Pareekh, 1993). References regarding *bael* is abundantly seen in *Yajurveda*, *Atharvaveda*, *Brahmanas*, *Kalpasutras* and *Puranas*; mangoes and date palms in *'Ramayana'* and *'Mahabharata'* (Singh 1969; Shah 2014).

Pre-British Era

Many new fruits were introduced through Mughals, Portuguese, Spanish and Dutch. Grape was introduced in tropical India during 620 BC (Olmo 1976) and subsequently by invaders from Afghanistan and Persia in 1300 AD (Chadha and Pareek 1993). Pomegranate, sapota and loquat reached India so early that their exact period of introduction is difficult to trace. Sixteenth century was in fact remarkable for plant introduction. Sultan Zain-ul-Abidin "Budshah" (1420-1470) imported many fruit crops from Central Asia. Almond was introduced by Persians in 16th century, pineapple by Portuguese in 1548 AD, guava in 1300 AD by Muslim invaders from Iran and Afghanistan, persimmon by European settlers in early 20th century and cashew in Malabar Coast of India in 16thcentury by the Portuguese. Earliest in-depth description about mango, pineapple and custard apple are recorded in *Ain-I-Akbari*, an encyclopaedia written during 1590 AD (Chadha and Pareek 1993; Rajan and Hudedamani 2019). Mughal emperor Akbar who ruled northern India from 1556 to 1605 planted an orchard of hundred thousand mango trees near Darbhanga the Lakhi Bagh (Mehta 2017) and many of these trees were found to be vigorous even after 300 years. Vavilov (1926) has suggested Indo-Burma region as the centre of origin of mango based on the observed level of genetic diversity. Scientists of BSIP, Lucknow, have traced the origin of genus *Mangifera* from 60 million years old fossil compressions of carbonized mango leaves in the Palaeocene sediments near Damalgiri, West Garo Hills, Meghalaya (Mehrotra *et al.* 1998).

British Era

The period of British colonization (AD 1757 - AD 1947) in India is also repleted with many references on fruits. This is the period when many exotic fruits like apple, pear, grapes, strawberry got impetus. British brought strawberry to India in the early 19th century. Lord Auckland, the Governor General during 1836-42, is said to have grown the first modern strawberries in India, but others were also trying across the subcontinent. Mr W.G. Mc Ivor, began large scale cultivation of fruit plants in South India in 1855 and by 1859, he claimed to have 178 species and varieties, most of which were imported from England (Krishnamurthi 1953). Apples were introduced by Europeans in Ootacamund (Nilgris, Tamil Nadu) in 1850s. Captain R C Lee, Mr. Wilson, M. Ermens, Alexander Coutts, M Pychard are the pioneer in introducing apple to different parts of India. However, commercial apple production was given an impetus by Satya Nand (Samual Nicholas) Stokes, who introduced 'Delicious' group of apples in the first quarter of the 20th century at Kotgarh, Shimla Hills.

Introduction of Arabian date palms into India perhaps began at the time of the first Mohammadan invasion of the Sind in the early eighth century when the date was believed to have been brought in for cultivation in Western India (Blatter 1926; Cooke 1908). This is the era when several promising varieties of grape were introduced which gave impetus to make grape cultivation highly profitable and successful venture. Though Muhammed Bin Tughlaq introduced, Bhokri, Fakhri and Sahebi cultivars in Aurangabad (Daulatabad) as early as 1338, Punjab was perhaps the first state to take up comprehensive introduction and trials for testing the adaptability of grape varieties. The work started at Lyallpur (now in West Pakistan) as early as 1928 and the work initiated by Randhawa and his co-workers during late fifties enriched Indian grape germplasm wealth. A number of cultivars like Ruby Seedless, Gordo Blano from Australia; Totlocha from Brazil, Dogridge, Pride, Dixie, Wedor and Black Cornith-2 from USA; Surnak Kitabiskij, Pozdrijwir and Shirajx-6 from USSR, Malvasiafina (Douro), Boal De Alicante, Tinta Deira Preta, Jampal, Tinta Roriz from Portugal was introduced. A number of species resistant to biotic and abiotic stresses, e.g., *Vitis gigas, V. caribbeana, V. munsoniana, V. smalliana, V. cineraria, V. shuttleworthi, V. arizonica* and *V. monticola* were introduced from USA (Singh and Rana 1993).

Though exchange of germplasm was a part of fruit research in India since long ago, systematic efforts were started only after establishment of Botany Division, Imperial Agricultural Research Institute in 1905. Further work was strengthened by opening of a separate Plant Introduction Station/ Service, Division of Botany at IARI in 1946. In 1956, Plant Introduction and Exploration Organization (Botany Division, IARI) was operational. An independent "Plant Introduction Division" was created at ICAR-IARI, New Delhi in 1961 followed by establishment of National Bureau of Plant Introduction in 1976 which further christened to National Bureau of Plant Genetic Resources in 1977.

Post-Independence (1947-1974): At the beginning of 20th century, much emphasis was given to introduction of horticultural crops. In 1950s, British introduced green English varieties of apple followed by coloured delicious varieties which were used to improve spur types (Singh *et al.*, 2006). The research on temperate fruits were further strengthened by establishment of research stations at Shalimar in Jammu and Kashmir; Boktu, Kandaghat, Kothkai, Kullu, Mashobra, Palampur and Shimla in Himachal Pradesh, Chaubattia in Uttarakhand, Chhindwara in Madhya Pradesh, Shillong in Meghalaya and Kodaikanal in Tamil Nadu during 1950-62. Large number of exotic collections were established at IARI, Pusa; FRS, Abohar; CES, Chethalli; and CES at Srirampur and Tirupati during the fifties and sixties. Kinnow mandarin was introduced from USA by Dr J.C. Bakhshi which has become one of the most promising introductions in India. Kinnow is now cultivated throughout North India and even in other citrus growing states.

Blueberry (EC12779) was first introduced during 1957 from USA, then from Holland (EC22244), USA (EC25191-93, EC25219), followed by several species and varieties from USA. This is the period when several improved varieties of grape (EC28389, EC44472) and kiwifruit (EC44764) were imported from USA, date palm from Egypt (EC28034) and Iran (EC111282-90), coconut (EC31103) from Spain, Cleopatra mandarin (EC18989) from USA. Avocado was initially introduced from Italy (EC26237) and Srilanka (EC31529), but improved varieties were imported during 80's. The Purple and Green varieties were introduced into India from Ceylon in 1941. Date palm was imported from different date-growing countries through the UNDP and their evaluation was undertaken in Kutch.

Post-Establishment of NBPGR, New Delhi

Since 1976, National Bureau of Plant Genetic Resources (NBPGR) looks after the activities of exchange of germplasm for research purpose within country as well as beyond the national territory. Since late 70's, planned introduction of fruit germplasm has been undertaken through ICAR-NBPGR under specific set of procedures. Due to increased emphasis on use of imported germplasm under on-going crop improvement programmes in the country, a separate unit of Germplasm Exchange was incepted. GEPU, ICAR-NBPGR has been instrumental in introducing a number of promising varieties/germplasm of fruit crops for research use in the country and till date a total of 9684 accessions/varieties of fruit crops from over 40 countries have been imported under strict phytosanitary conditions. Some of the notable introduction are Thompson Seedless, Perlette, New Perlette, Beauty Seedless, Delight, Himrod and Early Muscat grape cultivars from USA; Kishmish Charni and Kishmish Beli grape from USSR; Kinnow mandarin and low chilling peach varieties Flordasun and Sunred from USA, apple cvs Fuji from Japan and Majda from Yugoslavia, West Germany and USA.

Exotic Germplasm in Crop Improvement

Introduction is perhaps the easiest way to enhance crops with new alleles from exotic sources and wild species through introgression or incorporation. Introgression involves backcrossing a few chromosome segments with easily identifiable effects (often disease resistances) into elite cultivars while incorporation aims to create populations that are adapted to a breeder's target set of environments but that are also genetically distinct from elite cultivars (Simmonds, 1993). The introduced germplasm after evaluation and testing at different environments are used in the improvement programmes in either of these ways.

Selection and Release as Variety

When a new germplasm is introduced, it is tested and if adapted well and found suitable, it is directly released as variety so that farmers can access them. This is known as Primary introduction. Some of the outstanding introductions in fruit crops directly released as variety (Table 1) or released after selection (Table 2) are presented in tabular form. Owing to perennial nature of the most of the fruit crops, exotic introductions have been and will continue to be preferred over selection and hybridization programmes. This way new avenues are opened to the farmers to grow a new crop/variety with enhanced market potential.

Table 1: Outstanding examples of primary introduction in fruit crops

Crop	Varieties	Source country	Year of release
Banana	Lady Finger	Australia	-
	Grand Nain M.S.	France	-
	Valery	West Indies	-
Papaya	Sunrise Solo	USA	-
Pomegranate	Shirin Anar	USSR	-
	Wonderful	USA	-
Jacoticaba (*Myrciaria cauliflora*)	Sahara Blanc M 8962	USA	-
Apple	Vered	Israel	-
	Delicious-II	USA	-
	Red Baron	USA	-
	Mollies Delicious	USA	-
	Skyline Supreme Red Delicious	USA	-
Pear	Flemish Beauty	USA	-
	Max Red Bartlett	Italy	-
	Devoe	USA	-
	Manning Elizabeth	USA	-

Crop	Varieties	Source country	Year of release
Peach	Stark Early Glo	USA	-
	Candor	USA	-
	Flordasun	USA	1976
	Earli Granade	USA	1976
Plum	Methley	Kenya	-
	Kanto-5	USA	-
Apricot	Nugget	USA	-
	Coninos	Italy	-
Almond	Non Pareil	USA	-
Walnut	Lake English	USA	-
	Hansen	USA	-
	Payne	USA	-
	Tutle 31	USA	-
Pecan nut (*Carya illinoensis*)	Mahan	USA	1932
Kiwi fruit	Allison, Bruno, Abbott, Monty, Hayward	New Zealand	-
Pineapple guava (*Feijoa sellowiana*)	Nikitsky	USSR	-
Pineapple	Kew	USA	-
Grape	Perlette	USA	1958
	Thompson Seedless	Australia	1936
	Muscat of Hamburg (Gulabi)	Australia	1936
	Kishmish Chorni	USSR (Uzbekistan)	1964
	Beauty Seedless	USA	1964
	Flame Seedless	Australia	1993
	Red Globe	USA	1985
	Crimson Seedless	USA	2002
	Italia	Originally from Italy but introduced from USA	2002
	Cabernet Sauvignon	France	1982
	Chenin Blanc	France	1987
	Sauvignon Blanc	Blanc	1987
	Shiraz	France and Australia	1993
	Zinfandel	France	1995
	Malbec	France	1995
	Ugni Blanc	France	1995
	Grenache	France	1995
	Centennial Seedless	USA and Australia	1985
	Fantasy Seedless	USA	2001

Source: Chadha and Pareek 1993; Singh *et al*. 2009; Mitra and Dinesh 2019; Ranjan *et al*., 2022

Table 2: Utilization of exotic germplasm in varietal development through selection/ mutation

Crop	Variety	Exotic germplasm utilized	Method of breeding	Releasing institute/ Farmer's name	Year of release
Apple	Chaubattia Agrim	Early Shanburry	Mutant of Early Shanburry	HETC, Ranikhet, Uttarakhand	1984
Apricot	Chaubattia Alankar	(Kaisha x Charmagz) budwood irradiation with gamma rays	Induced mutation	HETC, Ranikhet, Uttarakhand	1984
	Chaubattia Madhu	(Turkey x Charmagz) budwood irradiation with gamma rays	Induced mutation	HETC, Ranikhet, Uttarakhand	1984
	Chaubattia Kesri	(St Ambroise x Charmagz) budwood irradiation with gamma rays	Induced mutation	HETC, Ranikhet, Uttarakhand	1984
Mango	Pusa Surya	Eldon	Selection	IARI, New Delhi	2002
Grape	New Perlette (HS 37-6)	Perlette	Natural sport	HAU, Hisar	1976
	Pusa Seedless	Thompson Seedless	Clonal selection	IARI, New Delhi	1970
	Sonaka	Thompson Seedless	Clonal selection	Late Nanasaheb Kale, Nanaj, Solapur, Maharashtra	1977
	Manik Chaman	Thompson Seedless	Clonal selection	Mr. T.R. Dabade, Nannaj, Solapur, Maharashtra	1982
	Tas-a-Ganesh	Thompson Seedless	Clonal selection	Mr. Vasant Rao Arve, Tasgaon, Sangli, Maharashtra	1976
	Monika	Thaompson Seedless	Clonal selection	Mr. A.G. Meher, Narayangaon, Pune, Maharashtra	1980
	Maruti Seedless	Thompson Seedless	Clonal Selection	Mr. Maruti Ramchandra Mali, isal, Miraj, Sangli, Maharashtra	1993
	Manjari Naveen	Centennial Seedless	Clonal Selection	NRCG, Pune	2008
	Manjari Kishmish	A mutant selection from Kishmish Rozavis	Mutant selection	NRCG, Pune	2017

Crop	Variety	Exotic germplasm utilized	Method of breeding	Releasing institute/ Farmer's name	Year of release
	Madhu Angoor	Carolina Black Rose	Clonal selection	ANGRAU, Hyderabad	1998
	Sharad Seedless	Kishmish Chorni	Bud sport of Kishmish Chorni	Nanasaheb Kale, Nanaj, Solapur	1980
	Mahadev Seedless	Kishmish Chorni	Bud sport of Kishmish Chorni	Mr. Gausmohammed Saipan Shaikh, Boramani, Solapur	2006
	Sarita Seedless	Sharad Seedless	Clonal Selection	Mr. Dattatraya Nanasaheb Kale, Nanaj, Solapur	1996
	Nanasaheb Purple	Sharad Seedless	Clonal Selection	Mr. Dattatraya Nanasaheb Kale, Murajipeth, Nanaj, Solapur	1998
	Nath Jambo	Sharad Seedless	Clonal Selection	Mr. Vithal Nivrutti Thorat, Kalamb, Tal. Ambegaon	2006
	Krishna Seedless	Sharad Seedless	Clonal Selection	Mr. Narayan Sangapa Mali, Sangli	2006
Papaya	CO5	Selection from Washington	Selection	TNAU, Tamil Nadu	1985
	Punjab Sweet	Selection from germplasm from Kenya	Selection after sib-mating	PAU, Ludhiana	-
	Coorg Honey Dew	Selection from Honey Dew	Selection	IIHR, Bengaluru	-

Source: Singh *et al.* 2009; Mitra and Dinesh 2019; Ranjan *et al.*, 2022

Hybridization for Specific Traits

Exotic germplasm in fruit crops are best utilized by the Indian breeders and more than 60 varieties have been released (Table 3) through hybridization followed by selection. The extent of reliance on exotic sources varies from crop to crop. Most of the commercial cultivars of apple, pear, cherry, plum, peach, apricot, persimmon and strawberry are of exotic origin; whereas some cultivars of walnut, sand pear and almond are selections from indigenous germplasm. However, in indigenous crops such as mango, banana, jackfruit, there is less reliance on foreign germplasm and most of the cultivars are of indigenous origin. Moreover, in citrus and banana a number of interspecific hybrids have been developed using various exotic species.

Table 3: Utilization of exotic germplasm of fruit crops in varietal development through hybridization in India

Crop	Variety/Hybrid	Exotic germplasm utilized	Releasing institute	Year
Apple	Lal Ambri	Red Delicious x Ambri	SKUAST (K), Srinagar, J&K	1973
	Sunehari	Ambri x Golden Delicious	SKUAST (K), Srinagar, J&K	1973
	Akbar	(Golden Del. v Rome Beauty) x *M. floribunda*	SKUAST (K), Srinagar, J&K	1987
	Firdous	Ambri x Cox's Orange Pippin	SKUAST (K), Srinagar, J&K	1984
	Shireen	Lord Lamborne x Melba XR-12740-7A	SKUAST (K), Srinagar, J&K	1984
	Ambred	Red Delicious x Ambri	YSPUH&F, Nauni (Mashobra), HP	1978
	Ambrich	Richard x Ambri	YSPUH&F, Nauni (Mashobra), HP	1978
	Ambstarking	Starkign Delicious x Ambri	YSPUH&F, Nauni (Mashobra), HP	1978
	Ambroyal	Starking Delicious x Ambri	YSPUH&F, Nauni (Mashobra), HP	1986
	Chaubattia Anupam	Early Shanburry x Red Delicious	HETC, Ranikhet, Uttarakhand	1984
	Chaubattia Princess	Early Shanburry x Red Delicious	HETC, Ranikhet, Uttarakhand	1978
	Chaubattia Swarnima	Benoni x Red Delicious	HETC, Ranikhet, Uttarakhand	1978
	Chaubattia Anurag	Esopus Spitzenberg x Red Delicious	HETC, Ranikhet, Uttarakhand	1984
	Chaubattia Alankar	Fanny x Red Delicious	HETC, Ranikhet, Uttarakhand	1986
	Pusa Gold	Golden Del. X TEW	IARI, RS, Shimla	-
	Pusa Amartara Pride	Royal Delicious x Prima	IARI, RS, Shimla	-
Custard apple	Arka Sahan	Island Gem (*Anonna atemoya*) x Mammoth (*A. squomosa*)	IIHR , Bengaluru	1995
Peach	Saharanpur Prabhat	Sharbati x Flordasun	FRS, Saharanpur	1988
	Shan-i-Punjab	(South Island) F_3 x Springtimes	PAU, Ludhiana	
Mango	Pusa Arunima	Amrapali X Sensation	IARI, New Delhi	2002
	Pusa Pratibha	Amrapali x Sensation	IARI, New Delhi	2011
	Pusa Shreshth	Amrapali x Sensation	IARI, New Delhi	2011
	Pusa Lalima	Dashehari x Sensation	IARI, New Delhi	2011
	Pusa Deepshikha	Amrapali x Sensation	IARI, New Delhi	2020
	Konkan Samrat	Alponso x TommyAtkins	RFRC, Vengurle	2014

Crop	Variety/Hybrid	Exotic germplasm utilized	Releasing institute	Year
Papaya	Co3	Co2 x Sunrise Solo	TNAU, Tamil Nadu	1983
	Co4	Co1 x Washington	TNAU, Tamil Nadu	1983
	Arka Surya	Sunrise Solo x Pink Flesh Sweet	IIHR, Bengaluru	1994
	Arka Prabhath	(Surya x Tainung-1) x Local Dwarf	IIHR, Bengaluru	-
	HPSC-3	Tripura Local x Honey Dew	ICAR RC-Tripura Centre	1995
	RCTP-1	Tripura Local with different genotypes including Honey Dew	ICAR	2014
Pomegranate	Solapur Lal	Bhagawa x[(Ganesh x Nana) x Daru]	NRC, Pomegranate, Solapur	2017
	Solapur Anardana	Bhagawa x[(Ganesh x Nana) x Daru]	NRC, Pomegranate, Solapur	2017
Grape	Pusa Navrang	Madeleine Angevine x Rubi Red	IARI, New Delhi	1997
	Pusa Urvashi	Hur x Beauty Seedless	IARI, New Delhi	1997
	Pusa Aditi	Banqui Abyad x Perlette	IARI, New Delhi	2018
	Pusa Trishar	(Hur x Bharat Early) x Beauty Seedless	IARI, New Delhi	2018
	Pusa Swarnika	Hur x Cardinal	IARI, New Delhi	2018
	Arkavati	Black Champa X Thomson Seedless	IIHR, Bengaluru	1980
	Arka Kanchan	Anab-e-Shahi X Queen of Vineyards	IIHR, Bengaluru	1980
	Arka Chitra	Angur Kalan and Anab-e-Shahi	IIHR, Bengaluru	1994
	Arka Majestic	Angur Kalan and Black Champa	IIHR, Bengaluru	1994
	Arka Neelmani	Black Champa and Thompson Seedless.	IIHR, Bengaluru	1991
	Arka Soma	Anab-e-Shahi x Queen of Vineyards	IIHR, Bengaluru	1994
	Arka Trishna	Bangalore Blue x Convent Large Black	IIHR, Bengaluru	1994
	Arka Shyam	Bangalore Blue x Black Champa	IIHR, Bengaluru	1980
	Arka Shweta	Anab-e-Shahi and Thompson Seedless	IIHR, Bengaluru	1994
	Arka Hans	Bangalore Blue x Anab-e-Shahi.	IIHR, Bengaluru	1980
	Arka Krishna	Black Champa x Thompson Seedless	IIHR, Bengaluru	1994
	Manjari Medika	Pusa Navrang x Flame Seedless	NRCG, Pune	2017

Source: Singh *et al*. 2009; Singh and Sharma 1996; Chadha and Pareek 1993; Ranjan *et al*., 2022

Impact of Exotic Germplasm in Fruit Production

New crops introduced from various countries enriched our fruit basket with more diversity, which in turn opened new avenues for the growers. Now we have all the seven colours of rainbow in our fruit basket ranging from orange oranges, red apples, green grapes, yellow bananas, blue blueberries, brown sapota, green avocado, black grapes, golden apricots, crimson red peaches, dark red plums, yellow persimmons and so on. Earlier eating fruits was supposed to be a status symbol; even a middle class family cannot afford to relish fruits due to very low *per capita* availability, *i.e.* 82 g during 1960. Only mango, banana, citrus and minor fruits like *bael*, *aonla* and *karonda* were available in the market for common people. Fruits like apple and pomegra1nate was used only during sickness. The holistic growth of horticulture sector due to availability of improved varieties and innovative technologies and their large scale adoption has made India a leader in fruit production with per capita availability reaching to 200 g.

The trend of productivity witnessed a significant rise over the years due to continuous improvement work attempted by various institutions under NARS and their adoption by the farmers. However, introduction of promising germplasm has certainly played a significant role in crafting horticulture industry to shower rainbow revolution. There is no doubt about the fact that the present day temperate fruit industry is based on efforts made by western settlers/ missionaries in the 19th century, who introduced exotic temperate fruit cultivars on a large scale. Though, the records indicate that temperate fruit growing in India dates back to some 2000 years ago (Walter 1904), yet commercial cultivation picked up sometime in the middle of the 19th century. In the process, several outstanding introductions of temperate fruits were made. Farmers are the pillars of any agricultural system. In India, a number of grape varieties has been developed by farmers through selection from exotic collections (Table 2).

The period during 1930-1970 was very important for horticulture sector when a number of promising varieties were introduced and appropriate technologies were adopted which altogether made the path to reach "Golden Revolution" reflecting its effect through remarkable jump in production, productivity, availability, and export of fruit crops. India has witnessed a sharp increase in area and production of banana, apple, grape, papaya and pineapple since last 57 years. Germplasm imported from other countries certainly had played great role to achieve this position. In apple old Delicious group were found promising which later on replaced by spur type, cultivars and some scab resistant cultivars. Besides Barlett pears, New Castle and Royal apricots, Alberta peaches and Santa Rosa plums are leading the market. In citrus, several exotic collections have been utilized as commercial cultivars. Kinnow is a novel introduction which has revolutionized the citrus production in north western region (Chadha 1995) and has replaced local cultivars. Tahiti lime, cultivars of grapefruit and sweet oranges were successfully adopted. Introduced germplasm have a significant impact on export market of India.

Papaya, a native to tropics of America and introduced to India from Philippines through Malaysia, has become a very important crop in India which is now leading the world in papaya production with an annual output of about 3 million tonnes. The most important commercial varieties grown in India are Washington Solo, Sunrise Solo, Honey Dew, Red Lady and varieties released through their selection or hybridization. Red Lady is the leading variety of papaya grown under 80% of the area. Dwarf Cavendish and Robusta are among the leading banana varieties of India. Grand Naine introduced from Israel is gaining popularity and may soon become the most preferred variety due to its tolerance to abiotic stresses and good quality bunches.

Some of the promising introduction in rootstocks particularly of apple, citrus and grape has revolutionized the fruit production in India. Clonal rootstocks of M and MM series were introduced from EMRS, U.K. and M-9, M-26, M-4, M-7, MM-106 and MM-111 were identified as promising (Ghosh 2012). Merton 779 was recommended as commercial rootstock for apple in Kumaon hills, EMLA-111 for drought prone areas, EMLA-7 for sloppy land and EMLA-106 for sloppy and less clayey soil and EMLA-9 for high density planting on irrigated deep soils (Ghosh, 2012). In grape, rootstocks like Dog Ridge from USA, 110 R from France Salt Creek from Australia introduced during 1966-86 are very promising. Dog Ridge has been identified as the best rootstock for drought and salinity resistance in table grape varieties particularly for Maharashtra, Andhra Pradesh and Northern Karnataka areas and is reported to help in mitigating the problems of salinity, drought and nematodes (www.iihr.res.in).

Diffusion/ outflow of Indian Fruit Germplasm around the World: Historical Perspective

The Chinese traveller Hiuen Tsang, who visited India between 632-645 AD, was the first to bring mango to the notice of the outside world (Rajan and Hudedamani, 2019). Several centuries later in 1328, Friar Jordanus visited Konkan region and appreciated the progenitors of Goa and Bombay mangoes. Today mango is relished throughout the world and may be considered as one of the choicest Indian gift to the world. It is believed that King Rana Bahadur Shah established a well-designed garden at Sera Phant (*Sera Bagaincha*) of Nuwakot focusing mango orchard for the first time in Nepal (Gotame *et al.* 2020). The earliest cultivation of banana is reported from India (Reynolds, 1951), however, it was Alexander's delectable dessert in 327 B.C. during his invasion of India that led to the fruit's widespread migration (www.news.un.org/en/story/2006/05/177262). From India it travelled to the Middle East, where it acquired its current name from the Arabic banan, or finger, and from there Arab traders took it to Africa, where the Portuguese transported it to the Caribbean and Latin America (Reynolds, 1951). This reflects the old history of outflow of fruit germplasm from India.

Though in early days germplasm outflow through travellers, pilgrims etc. was a common phenomenon, formal outflow started with the inception of ICAR-NBPGR under a set procedure for research use. As per the reports of GEPU, since 1976 a total of 579 accessions/ varieties of fruit crops were exported to countries namely Austria, Australia, Bangladesh, Brazil, Bulgaria, Canada, Cuba, Egypt, Ethiopia, Germany, Israel, Ivory Coast, Korea, Mauritius, Netherlands, Niger, Oman, Pakistan, Philippines, Senegal, Somalia, Spain, Trinidad & Tobago, UAE, UK, USA, Vietnam, West Indies, Zimbabwe. The data from 1995 to 2019 was studied to compile the information on major varieties which were exported to different countries on request or as per the collaborative research programmes or under special schemes such as PL 480.

Impact of Indian Germplasm in World Fruit Production

Indian FGR is being utilized well in several countries either directly as a variety or as a parents in hybridization programme. Some of the important fruit germplasm from India being maintained/utilized in the breeding programme worldwide are presented in Table 4. The Indian subcontinent has made an enormous contribution to the global genetic base of mangoes and bananas. The impact of Indian mango 'Mulgoba' in Florida mango industry is the best example. Indian mango varieties are very well placed in the commercial growing of several countries including Nepal, Pakistan, Bangladesh, Brazil and Egypt. Out of the 260 mango varieties reported in Pakistan, Indian varieties Samar Bahisht Chaunsa, Sufaid Chuansa, Kala Chaunsa, Anwar Ratole, Fajri, Sindhri, Dushehri, Faiz Kareem and Langra are reported to be commercially important (Amin and Hanif, 2002) and Sindhri, Samar Bahisht Chaunsa and Sufaid Chaunsa are leading varieties exported to Middle East, Southeast Asia, EU and USA (Nafees *et al.* 2013). Indian varieties Fazli, Gopal Bhog, Himsagar, Khirsapati, Langra, Kishan Bhog, Kohinoor, Mohan Bhog are commercially grown in Bangladesh, Alphonso, Bullock's Heart, Hindi Be Sennara, Langra, Mabrouka, Taimour, Zebda in Egypt and Anwar Ratol, Baganapalli, Chausa, Dashehari, Gulab Khas, Langra, Siroli, Sindhri, Suvarnarekha, and Zafran are common in Pakistan (www.mangifera.res.in).

Most of the cultivars of tropical fruits grown in Nepal are introduced from India. A number of accessions/varieties of citrus, *aonla*, apple and kiwi fruit were supplied from India and being maintained in Nepal (Gautam and Gotame 2020). Indian papaya cv. Pusa Dwarf and Farm Selection-1 are reported to be most preferred genotypes in Nepal due to excellent fruit bearing, dwarfness and yield (Chaudhary *et al.* 2012). Farmers are particularly interested in Indian and Thai varieties that produce very large and tasty fruits (Kalinganire *et al.* 2008; Kone *et al.* 2009).

In Florida, Indian mango varieties Mulgoba and Sandersha were utilized well for development of several mango varieties. In USA and Brazil, varieties like Embrapa Alfa, Embrapa Lita, BRS Beta, BRS Omega, Bailey's Marvel, Haden JIRCAS etc are released by using Indian mango varieties in the breeding programme (Table

5). Banana variety 'Malbhog' which is one of the most preferred indigenous table banana variety of Assam has been selected by participatory varietal selection approach and registered in 2019 in Nepal (Gautam and Gotame 2020). In Mali, Indian *ber* germplasm has been introduced and are being utilized in their breeding programme for development of pest resistant locally adapted varieties (Kalinganire *et al.* 2012).

Table 4: Fruit germplasm from India being maintained/ utilized in different countries

Crop	Indian germplasm/variety	Country	Reference
Mango	Bombay Green, Bombay Yellow, Maldahiya, Dashehari, Calcuttia, Mallika, Amrapali, Chausa, Gola, Jarda, Shilhat, Alphonso, Anupati, Bhadaiya Maima, Fazli, Gulab Khas, Jardalu, Sukul, Sukhtara, Anarbahartal, Bathuwa, Kishna Bhog, Langra, Ladbi Midhuwa, Lal Maldahiya, Maghukupia, Sipiya, Safeda Maldahiya, Radi, Neelam, Gulab Bhog, Hattijhul, Suvarnarekha, Langra, Kapari, Gurkha, Kalepad, Late Green, Malewa Safeda, Mombasa, Sukhia, Totapuri	Nepal	Kaini 1994; Gotame *et al.* 20140, ARS-Hort 2014
	Anwar Ratole, Dushehri, Samar Bahisht Chaunsa, Langra, Neelam, Fajri, Ratole No. 3	Pakistan	Riaz *et al.* 2018
	Mulgoba, Alphonso, SuvarnRekha, Sendura, Galour, Kesar, Pairi, Ratna, Sindhu, Dolores, Brindibani, Himasagar	Israel	Sherman *et al.* 2015
	Alphonso	Japan	Yamanaka *et al.* 2019
	Mallika, Langra Benarasi, Alphonso, Royal Special and Totapuri	Florida	Warschefsky and Wettberg 2019
	Amrapali, Mallika	Brazil	Pinto *et al.* 2004
Banana	Harichhal, Chinia Champa	Nepal	Kaini 1994
	Calcutta-4	IITA, Nigeria	Khokher 2015
	Poovan, Calcutta	EMBRAPA-CNPMF, Brazil	Khokher 2015
Guava	Lucknow-49 and Alahabadi Safeda	Nepal	Kaini 1994
Litchi	Early Seedless, Early Large Red, Late Large Red, Rose Scented, Majfpuri, Raja Saheb, Dehraduni, China, Calcuttia, Patharia Red, Shahi, Bambai, Kalkattiya, Early Large Red, Muzaffarpur, Maclean, Seedless, Early Green, Koshelia, Rose Scented, Dehara Rose, Bombay Red, Bedana, Deharadun, Late Large, Phash, Ujali, Purbi, Deshi,	Nepal	Kaini 1994; Bose 2001; Gautam *et al.*, 2020

Crop	Indian germplasm/variety	Country	Reference
	Bombai, Muzaffarpuri, Bedana and China Number Three	Bangladesh	Bose 2001
Citrus	Muntala and sweet orange	NCRP	NCRP 2018
	Citrus × *paradisi* 'Chakotra' Lour., *Jatti khatti Citrus jambhiri* Lush. cv '*Jatti khatti*', *Citrus* x *reticulata* 'Nagpuri santra	Pakistan	Shahzadi *et al.*, 2016
	Rangpur lime, Sour orange	Turkey	Cimen and Yesiloglu 2016; Yesiloglu *et al.*, 2017
Guava	Allahabadi, Allahabad Safeda, Chittidar, Chinese Guava, Lucknow-49, Lalguda, Seedless, KG-1	Nepal	Gotame *et al.*, 2020
Bael	5 accessions	Sri Lanka	Pathirana *et al.*, 2020
	Narendra Bael-5 NB-5, Narendra Bael-7 NB-7, Narendra Bael-9 NB-9, Narendra Bael-17 NB-16	Nepal	Gautam *et al.*, 2020
Ber	Kaithal, Umran, Gola	Mali	Kalinganire *et al.*, 2012
	Banarasi, Gola, Umran, Jingzhao, Chuizhao	Nepal	Gotame *et al.*, 2014; ARS-Hort 2014
Phalsa	Faizabadi	Nepal	Gotame *et al.*, 2014; ARS-Hort 2014
Apricot	Blenheim Indian, Shakarpara, Tilton Indian,	Nepal	Gotame *et al.*, 2014; ARS-Hort 2014
Aonla	Chakaiya, Kanchan NA-4, NA-6, NA-7, NA-10	Nepal	Gotame *et al.*, 2014; ARS-Hort 2014
Jackfruit	Khajawa, Rasdar	Nepal	Gotame *et al.*, 2014; ARS-Hort 2014

Table 5: Mango varieties released in other countries utilizing Indian germplasm

Variety	Indian germplasm utilized	Institute	Reference
Embrapa Roxa 141	Amrapali × Tommy Atkins	CARC, Brazil	Pinto *et al.* 2000
Embrapa Alfa 142	Mallika × Van Dyke	CARC, Brazil	Pinto *et al.* 2000
BRS Beta	Amrapali × Winter	CARC, Brazil	Pinto *et al.* 2000
Embrapa Lita	Amrapali × Tommy Atkins	CARC, Brazil	Pinto *et al.* 2000
BRS Omega	Amrapali × Tommy Atkins	CARC, Brazil	Pinto *et al.* 2000
Bailey's Marvel	Haden × Bombay	Florida, USA	Schnell *et al.* 2006
Anderson	Sandersha × Haden	Florida, USA	Schnell *et al.* 2006
Haden-JIRCAS	Mulgoba × Turpentine	Florida, USA	Campbell 1992; Knight *et al.* 2009; Schnell *et al.* 2006
Haden-OPARC	Mulgoba × Turpentine	Florida, USA	Campbell 1992; Knight *et al.* 2009; Schnell *et al.* 2006
Jacquelin-OPARC	Haden × Bombay	Florida, USA	Schnell *et al.* 2006
Jacquelin-JIRCAS	Haden × Bombay	Florida, USA	Schnell *et al.* 2006

Challenges in Exchanges and Utilization of Fruit Germplasm

Sharing of germplasm beyond boundaries and their proper utilization often face certain challenges. Major challenges faced are discussed hereunder.

Handling of Planting Material in Transportation

In majority of fruit crops except papaya, citrus rootstock and macadamia nut which are propagated through seeds, the planting material are bud woods, cuttings, tissue cultured plantlets which are perishable and need extra care for handling and maintenance. The cuttings/budwoods coming from long distance undergo transportation shock or due to poor handling during transportation process loses viability or the condition deteriorates to an extent that they hardly survive at the indenter's site. Therefore, the percentage of successful sprouting and establishment is very poor. That is why in many cases, a single variety needs to be reintroduced several times.

Maladaptation and Challenges in Utilization

Many a times the new germplasm does not establish well or perform well in new environment due to maladaptation. These materials may not be of present value but certainly these can be of future values. Hence, these need to be maintained for future use. Moreover, when a new germplasm/crop is introduced in the region away from its centre of origin, it is expected to be often subjected to biotic and abiotic stresses, which needs extra care. In order to tackle these problems, the breeders must show concern for a balance between potential risks and rewards while selecting a germplasm for introduction from other countries/places. Moreover, in annual crops introduction of novel germplasm is more amenable to rapid evaluation and introgression as compared to fruit crops with long generation times which requires a lapse of time that can last several decades in developing/ breeding a genotype utilising any new material. This hinders efficient utilization of germplasm.

Monoculture Restricting Genetic Diversity

Though, an introduced variety exhibits its success through adaptation and adoption by farmers, sometimes it leads to monoculture and thus restricts the diversity. We have several examples in temperate fruits where the Indian fruit production is dominated by few exotic cultivars such as Starking Delicious, Red Delicious, Rich-a-Red, Golden Delicious, Red Chief and Oregan Spur in apple; Max Red, Bartlett, Manning Elizabeth, China Pear in pear; New Castle, Royal and Nugget in apricot; Elberta, July Elberta, Alton, J.H. Hale in peach; Santa Rosa, Satsuma, Methley in plum; Non Pareil, Drake, Merced, Ne Plus Ultra in almond; Tioga Torrey, Fern, Chandler in strawberry and Black Heart, White Heart, Early Rivers, Pink Early in cherry which are accepted as commercial cultivars, leading to monocultures. Santa Rosa is the predominant plum cultivar grown over 90 percent of the total area under plum cultivation in Himachal Pradesh and this is leading to monoculture like situation and creating gluts in the market (Sharma *et al.*, 2018).

Introduction of Invasive Pests and Diseases

In the last decade, India has steadily replaced licensing and discretionary controls over imports with deregulation and simpler import procedures. Most of import items fall within the scope of India's EXIM Policy regulation of Open General License under which they are freely importable without restrictions and without a license, except to the extent that they are regulated by the provisions of the policy or any other law. Many a time, introduction of new material is associated with the risk of introducing exotic pests and diseases. A number of new diseases and pests have been recorded over the years which created havoc (Table 6). For small samples and research purposes, the quarantine examination makes sure that there is no escape of pests or diseases; however commercial imports of fruit crops in large quantities pose a threat for such escape. Moreover, the online sites selling planting materials of horticultural crops pave a route for such a challenge. Plant biosecurity and Policy regulations shall also keep in view these aspects to prevent /minimise and control the introduction and spread of these pests in the course of international trade. We have the example of spread of *Phylloxera*, when a North American aphid accidentally introduced into continental Europe around 1865 and devastated much of Europe's grape-growing regions. This led to the establishment of first international plant protection convention namely *Phylloxera* Convention signed by five countries in 1881.

Table 6: Diseases/ pests of fruit crops introduced in India from other countries

Disease/ pest	Scientific name	Host	Date of first record	Introduced from
Wooly apple aphid	*Eriosoma lanigerum*	Apple, pear	1909	England
San Jose scale	*Quadraspidiotus perniciosus*	Apple	1900	Italy
Downy mildew of grape	*Plasmopora viticola*	Grape	1910	Europe
Fluted scale	*Iceryu purchuse*	Citrus	1928	Sri Lanka
Crown gall	*Agrobacterium tumefaciens*	Apple, pear	1940	England
Hairy root of apple	*Agrobacterium rhizogenes*	Apple	1940	England
Bunchy top	*Bunchy top of banana viroid*	Banana	1940	Sri Lanka
Fire blight of pear	*Erwinia amylovora*	Pear	1940	England
Apple canker	*Sphaeropsis malorum*	Apple	1943	Australia
Banana mosaic	*Banana mosaic virus*	Banana	1961	-
Apple scab	*Venturia inequalis*	Apple	1978	UK
Codling moth	*Cydia pomonella*	Apple	1989	Pakistan and Afghanistan

Disease/ pest	Scientific name	Host	Date of first record	Introduced from
Banana bract and streak virus	*Banana bract and streak virus*	Banana	1995	Sri Lanka
Coconut eriophid mite	*Aceria gurreronis Keifer*	Coconut	1997	South America
Papaya Mealy bug	*Paracoccus marginatus*	Papaya	2008	Central America
Banana mealy bug	*Pseudococcus jackbeardsleyi* Gimpel and Miller	Banana, papaya, custard apple	2012	Neotropical
Woolly whitefly	*Aleurothrixus floccosus*	Guava, *Citrus* species	2019	Neotropical
Neotropical white fly	*Aleurotrachelus atratus*	*Cocos nucifera* and *Dypsis lutescens*	2019	Neotropical

Source: Paroda *et al.*, 1987; Khetrapal and Gupta 2007; Muniappan *et al.*, 2008; Vijay Laxmi *et al.*, 2014; Gupta *et al.*, 2019; Shivakumara *et al.*, 2020; Latha and Sathyanarayana 2013; Khan *et al.*, 2017; Singh *et al.*, 2020; Ranjan *et al.*, 2022

Challenges of Regulatory Mechanism

Though, the complexities of inter-ministerial and regulatory mechanism formulated to regulate import and export sometimes slow down the process of exchange of germplasm for research, with new agreements and norms the exchange process has become more vibrant. However, actual cases of benefit sharing as per the mechanism provided in the BDA 2002 is still to reach milestones. The complex web of MTAs, licenses, contracts, patents, and phytosanitary restrictions that follow plant germplasm today may inhibit exchange of material among plant breeders and farmers around the world (Luby *et al.* 2015). Most elite material is now proprietary which restricts the plant breeders' ability to access diverse cultivars and the traits they encompass, thus threatening the exchange of germplasm that underpins the development of new cultivars.

Conclusion

Enhanced utilization and smooth flow of germplasm around the world is the need for conservation and sustainable utilization of these resources for food and nutritional security. The present germplasm flows are not restricted to their centers of diversity, but find their sources in international and national genebanks and breeding programs based in other parts of the world. India has been and continues to be highly interdependent for germplasm share with a number of countries. In last 25 years, India developed work plans for exchange of germplasm with many institutions/ countries which has benefitted both the countries and continued so today. India is a party to CBD, Treaty, Nagoya Protocol on Access and Benefit Sharing and these instruments provide a pathway to enhance collaboration on mutual exchange benefitting both the parties. Addition of more crops to the list of Annex I crops may facilitate better exchange of germplasm for research use. India is looking forward for enriching its fruit basket with more introduction of new crops/germplasm like blueberries, pawpaw, cassabanana etc. which may add diversity and help to achieve food and nutritional security.

References

Amin, M., and Hanif M. (2002). Cultivation of mango in Dera Ismail Khan. Agricultural Research Institute, Ratta, D.I. Khan, Pakistan. pp.1-18.

ARS-Hort. (2014). Annual Report-2070/2071 (2013/2014). Agriculture Research Station (Horticulture), Rajikot, Jumla

Blatter, E. (1926). The palms of British India and Ceylon. Oxford University Press, London.

Bose, T.K. (2001). Fruit: Tropical and Subtropical, 1 (pp.721), Naya Yug, India.

Chadha K. L., and Pareek, O.P. (1993). Fruit research in India-History, Infrastructure and achievements. In: (Eds.) Advances In Horticulture, Vol.1- Fruit Crops, pp 1-30, Malhotra Publishing House, New Delhi, India.

Chadha, K.L. (1995). In: Proc. of Expert Consultation on Tropical Fruit Specially Asia, 17-19 May, MARDI Serdang Kuala Lumpur

Chaudhary, J.N., Pandey, Y.R., Gautam, I.P., & Basnet, R.B. (2012). Identification of suitable papaya genotypes for commercial cultivation in Nepal. Proceeding of the Fourth SAS-N Convention, 4-6 April 2012. Society of Agricultural Scientist- Nepal, (SASN), Khumaltar, Lalitpur.

Cimen B. and Yesiloglu, T. (2016). Rootstock Breeding for Abiotic Stress Tolerance in Citrus. In: A K Shanker & S.Chitra (Eds), Abiotic and Biotic Stress in Plants - Recent Advances and Future Perspectives pp.521-563, Tech publisher.

Cooke, T. (1908). The flora of the presidency of Bombay. Taylor & Francis, London.

Gautam, I. and Gotame, T. (2020). Diversity of Native and Exotic Fruit Genetic Resources in Nepal. Journal of Nepal Agricultural Research Council 10:3126/jnarc.v6i0.28114

Ghosh, S. P. (2012). Deciduous fruit production in India. http://fao.org/3/ab985e/ab985e07.htm. Accessed 5 July 2020.

Gotame, T.P., Gautam, I. P., Shrestha, S. L., Shrestha, J. and Joshi, B. K. (2020). Advances in fruit breeding in Nepal. Journal of Agriculture and Natural Resources, https://doi.org/10.3126/janr.v3i1.27183.

Gupta, N., Verma, S. C., Sharma, P. L., Thakur, M., Sharma. P. and Devi, D. (2019). Status of invasive insect pests of India and their natural enemies. Journal of Entomology and Zoology Studies, 7(1): 482-489.

Kaini, B. R. (1994). Status of fruit plant genetic resources in Nepal. In M. P. Upadhya, H. K. Saiju, B. K. Baniya and M. S. Bista (Eds), In: Plant Genetic Resources, Nepalese Perspectives. Proceeding of the national workshop on plant genetic resource conservation, use and management organized by NARC at Kathmandu 28 Nov-1 Dec, 1994. NARC and IPGRI. pp. 103

Kalinganire A, Weber, J., and Coulibaly, S. (2012). Improved Ziziphus mauritiana germplasm for Sahelian smallholder farmers: First steps toward a domestication programme. Forests Trees and Livelihoods 21: 128-37. 10.1080/14728028.2012.715474.

Kalinganire, A., Weber, J. C., Uwamariya, A., & Kone, B. (2008). Improving rural livelihoods through domestication of indigenous fruit trees in parklands of the Sahel. In F.K. Akinnifesi, R.B.B. Leakey, C. Oluyede, O.C. Ajayi, G. Sileshi, Z. Tchoundjeu, P. Matakala, F.R. Kwesiga (Eds). Indigenous fruit trees in the tropics: Domestication, utilization and commercialization pp186– 203. Wallingford (Oxfordshire, UK): CAB International Publishing.

Khan, H.H., Verma, S.S., Saleem, M., Verma, S.K., Usmani, T.A., Naz, H. and Naz, A. 2017. Role of history of Plant Quarantine in India- A review. International Journal of Chemical Studies. 5(6): 2034-37.

Khetarpal, R., & Gupta, K. (2007). Plant biosecurity in India - Status and strategy. Asian Biotechnology and Development Review 9: 83-107.

Khokher, A. K. (2015). Conventional breeding approaches for postharvest quality improvement of fruits. In M. W. Siddiqui Postharvest Biology and Technology of Horticultural crops: principles and practices for quality maintenance, pp. 89-140. Apple Acadaemic Press, Inc, USA.

Knight, R.J., Campbell, R.J., & Maguire, I. (2009). Important mango cultivars and their descriptors. The Mango, 2nd Edition: Botany, Production and Uses. 42-66. 10.1079/9781845934897.0042.

Kone, B., Kalinganire, A., Doumbia, M. (2009). La culture du jujubier: un manuel pour l'horticulteur sahe´lien [Growing ber: a manual for the Sahelian horticulturist]. Nairobi (Kenya): World Agroforestry Centre, Manuel Technique No. 10.

Latha, S., and Sathyanarayana N. (2013). "An overview of the status and the potential impact of the exotic pathogens on Indian horticulture." Pest Management in Horticultural Ecosystems, 18: 88-93.

Luby, C.H., Kloppenburg, J., Michaels, T. M. and Goldman, L. L. (2015). Enhancing freedom to operate for plant breeders and farmers through open source plant breeding. Crop Science 55(3): 2481-88.

Mehrotra, R.C., Dilcher, D. L., and Awasthi, N. A. (1998). Paleocene Mangifera - Like leaf fossil from India, Phytomorphology, 48: 91-100.

Mehta, I. (2017). History of mango-King of fruits. International Journal of Engineering Science, 6(7): 20-24

Mitra, S.K. and Dinesh, M.R. (2019). Papaya breeding in India-achievements and future thrust. ActaHorticulturae ,https://doi.org/10.17660/ActaHortic.2019.1250.4

Muniappan, R., Shepard, B. M., Watson, G. W., Carner, G. R., Sartiami, D., Rauf, A., and Hammig, M. D. (2008). First report of the papaya mealybug, Paracoccus marginatus (Hemiptera: Pseudococcidae), in Indonesia and India. Journal of Agricultural and Urban Entomology 25: 37-40.

Nafees, M., Ahmad, S., Anwar, R., Ahmad, I., Maryyam & Hussnain, R. R. (2013). Improved horticultural practices against leaf wilting, root rot and nutrient uptake in mango (Mangifera indica L). Pakistan Journal of Agricultural Sciences 50: 393-398.

NCRP (2018). Annual Report 2074/75 (2017/18). National Citrus Research Programme, Paripatle, Dhankuta.

Olmo, H. P. (1976). Grapes. In T. K. Bose (ed.). Fruits of India: tropical and subtropical, Naya Prokash, Calcutta, India.

Paroda, R.S., Mathur, V.K. and Chandel, K.P.S. (1987). Use of tissue culture in plant quarantine for exchange of germplasm and planting materials in India. FAO Expert Consultation Use of Tissue Culture in Plant Quarantine for Exchange of Planting Materials, New Delhi, 26 February-2 March 1987.

Pathirana, C. K., Ranaweera, L.T., Madhujith, T., Ketipearachchi, K. W., Gamlath, K. L., Eeswara, J.P. et al. (2020). Assessment of the elite accessions of bael [Aegle marmelos (L.) Corr.] In Sri Lanka based on morphometric, organoleptic, and elemental properties of the fruits and phylogenetic relationships. PLoS ONE, 15(5), e0233609. https://doi.org/10.1371/journal.pone.0233609

Pinto, A. C. Q., Andrade, S. R. M. and Venturoli, S. (2004). Fruit set success of three mango cultivars using reciprocal crosses. Acta Horticulturae 645: 299-301.

Pinto, A.C. D., Vargas Ramos, V. H. and Junquueira, N.T.V. (2000). New varieties and hybrid selections from mango hybridization program in central region of Brazil. Acta Horticulturae, https://doi.org/10.17660/ActaHortic.2000.509.20

Rajan, S., and Hudedamani, U. (2019). In P. E. Rajasekharan & V. R. Rao (Eds.), Conservation and Utilization of Horticultural Genetic Resources. Springer Nature Singapore Pte Ltd. 2019. http://doi.org/10.1007/978-981-13-3669-0-1

Ranjan, P., Brahmi, P., Tyagi V., Ranjan J. K., Srivastava V., Yadav S.K., Singh S.P., Singh S., Binda P.C., Singh S.K. and Singh K. (2022) Global interdependence for fruit genetic resources: status and challenges in India. Food Security- Springer https://doi.org/10.1007/s12571-021-01249-6

Reynolds, P. K. (1951). Earliest evidence of banana culture. Journal of American Ornamental Society 71 (Supplement)

Riaz, R., Khan, A. S., Ziaf, K., & Cheema, H. (2018). Genetic diversity of wild and cultivated mango genotypes of Pakistan using SSR markers. Pakistan Journal of Agricultural Sciences, 10.21162/PAKJAS/18.6360.

Schnell, R.J., Brown, J.S., Olano, C.T., Meerow, A., Campbell, R.J., & Kuhn, D.N. (2006). Mango Genetic Diversity Analysis and Pedigree Inferences for Florida Cultivars Using Microsatellite Markers. Journal of the American Society for Horticultural Science, 131: 10.21273/JASHS.131.2.214.

Shah, J. (2014). Date palm cultivation in India: an overview of activities. Emirates Journal of Food and Agriculture, doi:https://doi.org/10.9755/ejfa.v26i11.18986.

Shahzadi, K., Naz, S. and Ilyas, S. (2016). Genetic diversity of citrus germplasm in Pakistan based on Random Amplified Polymorphic DNA (RAPD) markers. The Journal of Animal & Plant Sciences 26:(4), 1094-1100.

Sharma, D. D., Kumaar, M., Singh, N., & Shylla, B. (2018). Plant growth and fruiting behavior of newly introduced plum (Prunus salicina Lindl.) cultivars under mid-hills conditions of Himachal Pradesh. The Pharma Innovation Journal 7(4): 408-13.

Sherman, A., Rubinstein, M., Eshed, R. et al. (2015). Mango (Mangifera indica L.) germplasm diversity based on single nucleotide polymorphisms derived from the transcriptome. BMC Plant Biology, https://doi.org/10.1186/s12870-015-0663-6

Shivakumara, Kt, Joshi, S., Venkatesan, T., Pradeeksha, N., Polaiah, A. and Manivel, P. (2020). A Report on the occurrence of invasive papaya mealybug, Paracoccus marginatus Williams & Granara de Willink (Hemiptera: Pseudococcidae) on a medicinal herb, Gymnema sylvestre (R.Br) in Gujarat, India. shodhganga.inflibnet.ac.in.

Simmonds, N.W. (1993). Introgression and incorporation. Strategies for the use of crop genetic resources. Biological Review 68:539-62.

Singh, A. K., Srinivasan, K., Saxena, S. and Dhillon, B. S. (2006). Hundred years of plant genetic resources management in India. National Bureau of Plant Genetic Resources, Pusa Campus, New Delhi, India.

Singh, B. P., & Rana, R.S. (1993). Promising fruit introductions. In: Advances in Horticulture, Vol. 1, Fruit crops Part 1, pp. 43-66, Malhotra Publishing House, New Delhi.

Singh, H. P., Parthasarathy, V. A. and Prasath, D. (2009). Horticultural Crops-Varietal wealth (563). Stadium Press (India) Pvt Ltd.

Singh, I. P. and Sharma, C. K. (1996). HPSC-3: a high yielding papaya hybrid for Tripura. Journal of Hill Research, 90(1), 73-75.

Singh, R. (1969). Fruits, pp. 152-70. National Book Trust, India,

Singh, S., Sharma, J. H., Udikeri, A. and Ansari, H. (May 5th 2020). Invasive Insects in India, Invasive Species - Introduction Pathways, Economic Impact, and Possible Management Options, Hamadttu El-Shafie, IntechOpen, DOI: 10.5772/intechopen.91986.

Vavilov, N.I. (1926). Centres of Origin of Cultivated Plants, Bulletin of Applied Botany of Genetics and Plant-Breeding, 16,1-248.

Vijay Laxmi, R., Geetanjaly and Sharma, P. (2014). Plant Quarantine: An Effective Strategy of Pest Management in India. Research Journal of Agriculture and Forestry Sciences 2(1): 11-16.

Walter T. (1904). On Yuan Chwang's Travel in India. Royal Asiatic Society, London.

Warschefsky Emily, J. and Wettberg Eric, J. B. von. (2019). Population genomic analysis of mango (Mangifera indica) suggests a complex history of domestication. New Phytologist 222: 2023–2037

Yamanaka, S., Hosaka, F., Matsumura, M., Onoue-Makishi, Y., Nashima, K., Urasaki, N., Ogata, T., Shoda, M. and Yamamoto, T. (2019). Genetic diversity and relatedness of mango cultivars assessed by SSR markers. Breeding Science, 69. 10.1270/jsbbs.18204.

Yeşiloğlu, T., Yılmaz, B., İncesu, M. and Çimen B. (2017). The Turkish citrus industry. Chronica Horticulturae 57(4): 17-22.

3

Fruit Crop Wild Relatives and Their Importance in Breeding

Vartika Srivastava[1*], Pragya Ranjan[1], Monika Singh[1] and Chavlesh Kumar[2]

[1]ICAR- National Bureau of Plant Genetic Resources (NBPGR), New Delhi
[2]ICAR- Division of Fruits and Horticultural Technology, IARI, New Delhi

Indian subcontinent is a mega-center of crop plant diversity, as well as wild relatives and other useful taxa (Zeven and de Wet 1982). Diverse agroclimatic conditions, agricultural practices, and ethnic diversification have resulted in the development of a rich diversity of crop plants among native and introduced taxa (Arora 1991). Fruit crops contribute significantly to nutrition and health because they are high in dietary energy and contain a significant amount of vitamins and minerals (Pareek *et al.*, 1998). Most of these species serve multiple functions, such as food, shelter, timber, fuel, medicine etc. As a result of their numerous applications, tropical fruits significantly contribute to food and nutritional security, income generation, ecosystem and environmental sustainability.

Crop wild relatives (CWR) refer to the wild taxa which are closely related to crop plants, this includes wild progenitors and/or wild forms of crops. The CWRs function as a source of important genes for improving agricultural production and sustaining agro-ecosystems. Climate change and increased ecosystem instability are likely to make CWRs a critical resource in ensuring food security. In the early twentieth century, Russian botanist Nikolai Vavilov recognized the importance of crop wild relatives. For thousands of years, humans have used genetic material from wild relatives of various crops to improve crop quality and yield. Many useful genes have been contributed to crop plants by CWRs, and modern varieties of most major crops now contain genes from their wild relatives. The CWRs conservation and long-term use is critical for increasing agricultural production, improving food security, and protecting the environment. Wild fruit trees not only provide habitat for pollinators such as insects, birds, bats, and other animals, they also serve as an important ecosystem function. During times of food scarcity, certain wild fruit trees provide essential food resources for animals, potentially leading to the extinction of species that rely on them.

Diversity of fruit crops in India

With over 300 fruit species reported, the Indian subcontinent is characterized by a rich diversity of fruit crops, with mango, banana, citrus, guava, grape, pineapple, papaya, sapota, litchi, and apple accounting for roughly 75% of total area under fruits (Mitra *et al.*, 2010). India is home to a number of globally important fruit crop species, including mango, banana, and citrus. The country is endowed with a tremendous diversity of banana, citrus species, as well as a number of wild/ minor species. Several fruits, including *bael*, custard apple, jackfruit, *karonda*, *kokum*, *lasoda* and *phalsa* can be found in the wild or in local markets as they are becoming more popular in major cities and are now available in supermarkets. A total of 135 species are identified as CWRs belonging to 36 fruit crops (John and Pradheep, 2019) including mango, citrus, banana, jackfruit, walnut, apricot, pomegranate, ber, jamun, aonla etc. These CWRs may be a source of traits of breeders' interest (Table 1).

Rich diversity of *Musa* germplasm is reported in India particularly in the north-eastern region (NER), Western Ghats and Andaman and Nicobar Islands. *Musa acuminata* and *Musa balbisiana,* the wild ancestors of the majority of modern-day bananas, have been found to be widely distributed in India. Some of the important species of *Musa* known to occur in the NER are *Musa aurantiaca*; *Musa itinerans*; *Musa nagensium*; *Musa sikkimensis*; *Musa cheesmanii*; *Musa thomsonii* (Uma, 2006). *Musa nagalandiana* of Nagaland is a very rare species (Dey *et al.*, 2014); *Musa markkui* from Lohit valley of Arunachal Pradesh (Gogoi and Borah, 2013). *Musa kamengensis* (named after Kameng, the locality where it was first observed), grows at altitudes as high as 1,500 m indicating its cold tolerant nature (Gogoi and Häkkinen, 2013). *Musa argentii* found growing in Lohit District, Arunachal Pradesh. *Musa chunii* a critically endangered species found only from one location in Arunachal Pradesh and no populations were located in from other Northeastern states (Mamiyil *et al.*, 2013). Apart from these, around eleven wild species of *Musa* were found to occur in Assam, viz. *M. aurantiaca* subsp. burmanica, *M. cheesmanii* (found in Bhalukpung and in Assam, Arunachal Pradesh bordering areas of Likabali and Jonai), *Musa flaviflora* (common in Jorabat area of Kamrup district, Kaziranga, Bokakhat and areas bordering to Nagaland in Golaghat and Johrat District of Assam), *Musa itinerans* (widely found in the forest areas near Bordumsa, Jagun in Tinsukia District of Assam), *Musa laterita* (indigenous to NE India), *M. aurantiaca* (first described from a specimen collected in the state of Assam), *M. mannii* (first described from a specimen collected in the northeastern state of Assam), this variety, namdangensis, is named after Namdang of Changlang district of Arunachal Pradesh (Gogoi and Borah 2014) the only locality where it has been seen, *Musa nagensium* (found wild near Singpho villages from Jagun to Bordumsa in Tinsukia District), *Musa sanguinea* (reported from bank of river Buree Dihing, Tinsukia district), *Musa velutina* (found in Lakhimpur district, Dibrugarh district, Tinsukia district and border areas of Nagaland)

Citrus is distributed from North-eastern India and Southern China to Northern Australia and New Caledonia. The cultivated species are native to Southeast Asia's tropical and subtropical regions. Citrus, *Fortunella*, and *Poncirus* are the three genera that are commonly grown in India and which are closely related, have intergeneric fertility, and easily hybridise, resulting in the development of a variety of unusual plant forms. Citrus species and cultivars differ greatly due to frequent bud mutation, interspecific and intergeneric hybridization, and have a long history of cultivation (Shahsavar *et al.* 2007). The existence of intergeneric hybrids is common among these three genera, viz. Tangor (mandarin x sweet orange, Tangelo (mandarin x grapefruit), Lemonime (lemon x lime), Citrange (sweet orange x *Poncirus trifoliata*), Citromelo (grapefruit x *Poncirus trifoliata*), Limquat (lime x *Fortunella* spp.), Citrangequat (*Fortunella margarita* x Rusk citrange), and Calamonsi/ Calmondin (mandarin x *Fortunella* spp.), etc. There are five citrus groups that are commercially important and these include sweet orange, mandarin (including Satsuma), grapefruit, lemon and lime and numerous varieties and cultivars exist. Kumquat (*Fortunella* spp.) is grown to a limited extent for fresh fruit and processing. Pummelos are of economic importance in many areas within Southeast Asia and China.

In addition to *Mangifera indica*, other *Mangifera* species reported from India include *M. andmanica*, *M. khasiana*, *M. sylvatica*, and *M. camptosperma* (Mukherji, 1985). This diversity is mostly attributed to the seedling races derived from mono-embryonic mango stones. Almost all commercial mango cultivars are the result of seedling selection. India has around 1000 distinct varieties, with approximately 30 of them being commercially grown. Apart from the major fruit crops, potential under-utilized fruit crops includes, jackfruit (*Artocarpus heterophyllus*), bael (*Aegle marmelos*), jamun (*Syzygium cuminii*), carambola (*Averhoa carambola*), aonla (*Emblica officinalis*), kokum (*Garcinia indica*), karonda (*Carissa carandas*), phalsa (*Grewia asiatica*), ker (*Capparis decidua*), chironji (*Buchanania lanzan*), lasoda (*Cordia myxa*), pilu (*Salvadora oleoides*), wood apple (*feronia limonia*) and kaphal (*Myrica nagi*) etc., could also contribute to the nutritional security and economy of the country.

Table 1: Important CWRs for fruits and nuts of Indian region

Genus	Wild species	Area of diversity
Mangifera	*M. sylvatica, M. khasiana, M. anadamanica, M. comptosperma*	North Eastern States, Andaman and Nicobar Islands; Western Ghats (Sirsi Forests, Karnataka)
Musa	*Musa arunachalensis, Musa nagalandiana, Musa kamengensis, Musa puspanjalia, Musa aurantiaca, Musa manni, Musa flavicarpa, Musa acuminata, Musa balbisiana, Musa cheesmani and M. velutina*	North-Eastern states, Western Ghats, Eastern Ghats and the Andaman and Nicobar Islands, West Kameng District, Arunachal Pradesh, Nagaland
Citrus	*"Soh Nairiang"* a wild sweet orange, wild Indian mandarin (*C. indica*) , *C. assamensis, C. ichangensis, C. latipes* and *C. macroptera*	Naga hills, Garo hills of Meghalaya and Kaziranga forests of Assam
Malus	*M. baccata* (L.) Borkh., *M. baccata* var. *himalaica (*Maxim.) Schneid., *M. baccata* var. *dirangensis, M. sikkimensis*(Wenz.) Koehne ex Schneider	North-Western and Eastern Himalayan regions
Pyrus	*P. jacquemontiana, P. khasiana, P. pashia, Pyrus kumaonii, P. polycarpa, P. serotina, P. thomsoni*	North-Western and Eastern Himalayan regions
Sorbus	*S. aucuparia* L., *S. cuspidata* (Spach.) Hedlund., *S. foliolosa* (Wallich.) Spach., *S. granulosa* (Bertol.) Rehd., *S. insignis* (Hook. f.) Hedlund, *S. lanata* (D. Don) Schner., *S. microphylla* Wenzi, *S. rhamnoides* (Decne.) Rehder	North-Western and Eastern Himalayan regions
Cotoneaster	*C. acuminata, C. acuminatus, C. affinis, C. bacillaris, C. buxifolia, C. falconeri, C. frigida, C. microphylla, C. multiflora, C. nummularia, C. roseus, C. rotundifolia, C. vulgaris*	North-Western and Eastern Himalayan regions
Docynia	*D. hookeriana, D. indica*	North-Western and Eastern Himalayan regions
Garcinia	*G. indica, G. morella, G. xanthochymus*	Western Ghats, Assam, Eastern Himalayas
Pyracantha	*P. crenulata*	North-Western and Eastern Himalayan regions
Prunus	*P. cerasoides, P. armeniaca, P. persica, P. cornuta, P. salicina, P. nepalensis, P. wallichi, P. jacquemontii, P. mira, P. undulata and P. cerasoides*	North-Western and Eastern Himalayan regions, Shillong plateau of Khasi hills in Meghalaya
Ribes	*R. alpestre, R. glaciate, R. griffithii, R. nigrum, R. rubrum*	North-Western and Eastern Himalayan regions

Genus	Wild species	Area of diversity
Rubus	*R. ellipticus, R. foliosus, R. fruiticosus, R. lasiocarpus, R. nepalensis, R. niveus, R. paniculatus, R. purpureusi R. hexagynus, R. ferox, R. calycinus, R. pungens, R. rosaefolius, R. saxatilis, R. macilentus* and *R. acuminatus*	North-Western and Eastern Himalayan regions
Vitis	*V. himalayana, V. lanata, V. latifolia, V. parviflora, V. capreolata, V. divaricata, V. semicordata* and *V. trifolia*	North-Western and Eastern Himalayan regions; and other parts of India
Fragaria	*F. nubicola* and *F. indica*	North-Western and Eastern Himalayan regions
Hippophae	*H. rhamnoides, H. salicifolia* and *H. tibetana*	North-Western and Eastern Himalayan regions
Viburnum	*V. corylifolium, V. cotinifolium, V. grandiflorum, V. jacquemontii* and *V. mullaha*	North-Western and Eastern Himalayan regions
Carissa	*C. congesta, C. grandiflora* and *C. spinarum*	North-Western and Eastern Himalayan regions
Elaeagnus	*E. angustifolia, E. latifolia* and *E. umbellata*	North-Western and Eastern Himalayan regions
Diospyros	*D. kaki* and *D. lotus*	North-Western and Eastern Himalayan regions
Corylas	*C. colurna* and *C. ferox.*	North-Western and Eastern Himalayan regions

Source: *Ranjan et al. (2022)*

Threats to Natural Diversity of Fruit Crops

The diversity of wild and cultivated fruit species are under threat from rapid genetic erosion caused by natural habitat destruction and other economic and cultural pressures. Threats include: a) habitat destruction, b) agricultural expansion, c) wetlands filling, d) conversion of biodiversity-rich sites for human settlement and industrial development, and e) uncontrolled commercial exploitation. Deforestation and land use conversion have put external pressure on wild fruit tree resources in Asia over the last century, resulting in high levels of deforestation and loss of target fruit wild relative.

Important Breeding Objectives

Fruit breeding goals are determined by fruit crops, location, and consumer needs. The main goals of fruit breeding are to achieve maximum quality production per unit area at the lowest possible cost, as well as tolerance to biotic and abiotic stresses. The goals of breeding for rootstocks and scions are distinct and variable as mentioned below:

Rootstock

- Easy to propagate using vegetative propagation
- Wide adaptability for different geographical regions and edaphic conditions
- Wider compatibility with most of the scion cultivars by having a strong scion stock union leading to longevity
- Resistance to biotic and abiotic stresses prevailing in the particular crop
- Imparting dwarfness to fit in to the high density planting system
- Strong root system for better anchorage and absorption of water and nutrients

Scion

- Dwarfing stature
- Regular, precocious and prolific bearing
- High productivity with good fruit quality traits
- Resistance to biotic and abiotic stresses prevalent in the crop
- Attractive fruit colour and pleasant aroma
- Suitable for processing and export
- Good keeping and transport quality

Challenges in Fruit Breeding

- Since most fruit crops are woody and have a long juvenile period of 2-10 years depending on species, screening the population for desired traits is time consuming.
- The majority of fruit species are highly heterozygous.
- Polyembryonic nature of some fruit species (citrus, mango, etc.) cause emergence of two types of seedlings, viz. zygotic and nucellar
- In certain species presence of parthenocarpy (stenospermocarpy in grape) and seedlessness (banana, pineapple) lead to the development of rudimentary seeds that are non-viable, and hence, to obtain hybrid, special techniques of embryo rescue essentially required.
- Presence of sexual incompatibility e.g. mango, apple, pear, loquat etc.
- More number of chromosome hinders genetic analyses e.g. ber, mulberry.
- Excessive fruit drop (mango, citrus) leading to retention of very less hybrids for further evaluation.
- Presence of single seed in most of the cases warrants number of crosses e.g. mango, litchi, mahua etc.

Use of CWRs in Development of Commercial Varieties

The CWRs have played a significant role in the evolution of the cultivated form of the fruit crop and the diversification of its varieties. It was well deciphered through historical, morphological and molecular data that the wild species, *Malus sieversii* has played a seminal role in the evolution of cultivated apples (Cornille *et al.*, 2012). Several wild apples have significantly contributed to the genetic improvement of the apple to counter several biotic and abiotic stresses. For instance, the Japanese crabapple *Malus floribunda* 821 has been widely utilized for identifying the gene(s) for scab resistance and their genetic improvement of the apple cultivars for scab resistance (Brown and Maloney, 2013). It was well elucidated in fruit crop like grapevine. *Vitis* species were explored for the development of the rootstock varieties and genetic improvement of the scion genotypes. The development of commercial rootstock genotypes, Dogridge and Salt creek from *Vitis champini* and other *Vitis* species used as rootstock indicated the importance of wild species for sustaining the grape production (Satisha and Prakash, 2006; Satisha *et al.*, 2007). Similarly, a rootstock in the apple was selected from *Malus baccata* bio-type from Shillong (Meghalaya, northeast India) for commercial cultivation of apples in the Indian Himalayan region (Pramanick *et al.*, 2012). Likewise, the different CWRs are utilizing many fruit crops to improve the varieties, including the scion and rootstock. Therefore, the conservation, characterization and utilization of CWRs in fruit crops are essential.

Modern tools of fruit breeding

Genetic Modification in Fruit Crops for Desirable Traits

In past improved varieties for better yield, fruit quality, aroma, antioxidants and nutritional traits have been developed through conventional breeding methods; however, the process is time consuming due to lengthy breeding process which is also resource-intensive. Moreover, perennial and woody nature of most of the fruit crops, long juvenile period, high heterozygosity and reproductive barriers limit the implementation of classical breeding practices (Kramer and Redenbaugh, 1994). New breeding techniques and biotechnological approaches such as transgenic cisgenesis and genome-editing technologies have the potential to accelerate the process of introducing beneficial traits into fruit crops, such as yield, quality, biotic and abiotic stress resistance and nutritional aspects. Apart from these, many physiological aspects related to fruit maturity, texture and ripening, which are directly affecting shelf-life, significantly contribute to its market value, consumer acceptability, and quality can be specifically and precisely addressed using modern breeding/ biotechnological tools. Genetic modification in plants refers to the alteration in the genetic makeup of plants for desired trait(s) using new plant breeding tools. The historic timeline for genetic modification in fruit crops can be divided into different phases depending upon the standardization of protocols and application of technologies (Fig. 1).

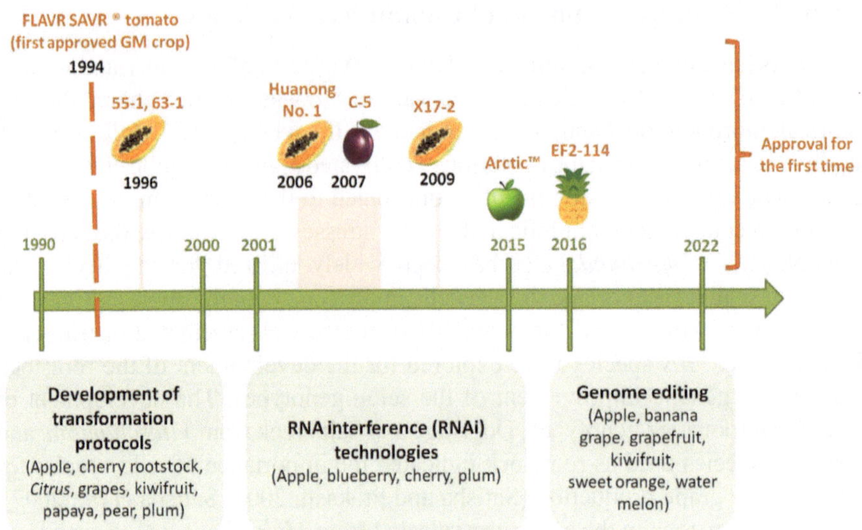

Fig. 1: Historic timeline for genetic modification in fruit crops
Source: *Song et al., 2019; Lobato-Gómez et al., 2021*

1. Transgenic Approach in Fruit Crops

Transgenic technology allows modification of an existing trait or introduction of a new trait for desired characters to develop transgenic or biotech or genetically modified (GM) crops using genetic engineering or recombinant DNA tools. *Agrobacterium tumefaciens* mediated genetic transformation is widely employed based on the efficient tissue and cell culture conditions, somatic embryogenesis and plant regeneration (Litz and Padilla, 2012; Penna *et al.,* 2023).

Since the commercialization of first transgenic crop FLAVR SAVR® tomato in 1994 (Kramer and Redenbaugh 1994), the global area under adoption of GM crops has reached to more than 190 million hectares (ISAAA, 2019). More than 30 GM crops have been approved for commercial planting or for use as food or feed in different countries but approval of GM fruit crops is limited to GM apple *(Malus domestica*), papaya (*Carica papaya*), pineapple (*Ananas comosus*), plum (*Prunus domestica*) (Fig. 2, https://www.isaaa.org/gmapprovaldatabase/cropslist/default.asp). The approved GM fruit crops have been developed for resistance to disease, delayed ripening, reduced browning or anti-browning, colour or texture of fruits.

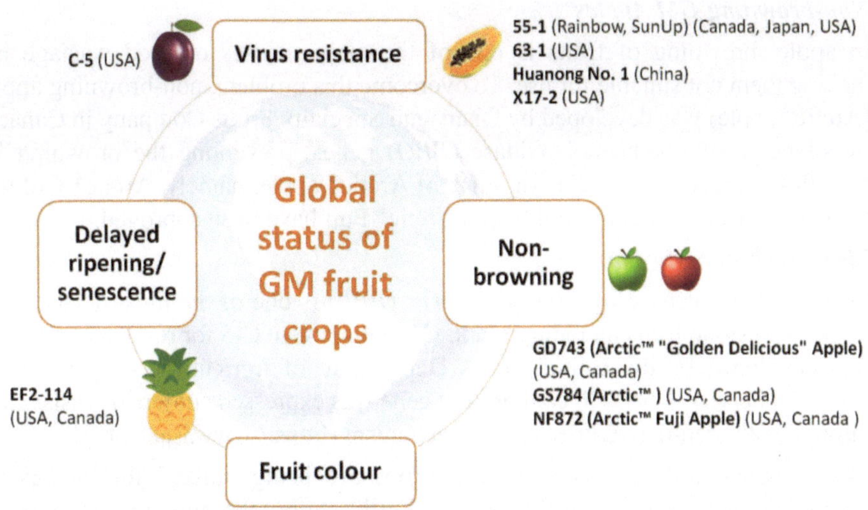

Fig. 2: Global adoption and trait diversification of GM fruit crops

*(**Source:** https://www.isaaa.org)*

GM papaya (SunUp and Rainbow): a Success Story on GM fruit

Virus-resistant GM papaya was commercialized 25 years ago (Lobato-Gómez *et al.* 2021). GM papaya line 55-1 expressing the *Papaya Ringspot Virus* (*PRSV*) coat protein gene under the control of *Cauliflower Mosaic Virus* derived promoter and terminator regions was developed by Cornell University in collaboration with University of Hawaii, USA (https://bch.cbd.int/en/database/40294). Ferreira *et al.* (2002) reported that two GM papaya varieties derived from 55-1, namely, SunUp and Rainbow, expressing *PRSV* coat protein gene, has provided a good solution to the problem of *PRSV* in Hawaii. Nearly all the area under papaya cultivation in Puna district of Hawaii was destroyed due to attack by *PRSV* since 1992. In 1995, field trials of SunUp and Rainbow were conducted where *PRSV* infection was not observed in these GM varieties. After adoption of GM papaya, the problem of *PSRV* was reduced, contributing to the increase in papaya production. Therefore, GM papaya has become a successful example of the use of modern biotechnology for economic gains by the large companies (Gonsalves *et al.* 2004).

GM Pineapple (Pinkglow™) with Pink Fruit Colour

The GM pineapple event, EF2-114 (Pinkglow™) was developed by Del Monte Fresh Produce Company, Florida, USA, in which the fruit color was modified from yellow to pink by the insertion of *Citrus unshiu* phytoene synthase (*psy*) gene for enhanced expression of lycopene, and an RNAi (RNA interference) construct for silencing of gene involved in the ethylene biosynthesis was also introduced for control of flowering (https://bch.cbd.int/en/database/record?documentID=111940).

Non-browning GM Apples

In apple, browning of fruits is one of the major causes of food wastage by making them not suitable for use. To overcome this problem, non-browning apple (Arctic® apple) was developed by Okanagan Specialty Fruits Company in Canada by silencing of polyphenol oxidase (*PPO*) genes, preventing the browning in fruit flesh. Three commercial varieties of Arctic® apple, namely, Arctic® Golden Delicious, Arctic® Granny Smith, and Arctic® Fuji have been approved.

Virus Resistant GM Plum

Sharka disease caused by *Plum Pox Virus* (*PPV*) is one of the most devastating diseases of stone fruits including plum. GM plum event C-5 for resistance to *PPV* was developed by the United States Department of Agriculture - Agricultural Research Service. The *PPV* coat protein gene was expressed conferring resistance through pathogen-derived resistance mechanism (https://www.isaaa.org).

The commercial GM events of fruit crops are being utilized for increasing production and product quality traits. Furthermore, the research in genetic modification in these crops as well as in other fruit crops such as banana, *Citrus* sp. has been reported for beneficial traits (Table 2). The updated progress of GM fruit crops has also been reviewed by Lobato-Gómez *et al.* (2021).

Genome editing in fruit crops

Genome-edited plants are developed through use of specialized enzymes to mutate, insert, replace or remove a part of DNA from the targeted site with high level of specificity employing genetic engineering/ advanced molecular biology tools. The specialized enzymes used for gene editing are considered as gene editors, which are categorized into zinc finger nucleases (ZFNs), transcription activator-like effector nucleases (TALENs), meganucleases (homing endonucleases) and the clustered regularly interspaced short palindromic repeats/ CRISPR-associated 9 (CRISPR-Cas9) (Boch *et al.,* 2009; Boissel *et al.,* 2014; Ricroch, 2019; Urnov *et al.,* 2010; Zhang *et al.,* 2018).

In the present era of genome editing, CRISPR-Cas9 technology has been employed for targeted modification due to its higher efficiency, ease-of-use, comparatively low cost and ability of introducing edits at multiple sites in the targeted genomes in a single procedure (Ricroch 2019). CRISPR is an integral part of a bacterial defense system, which forms the basis of the CRISPR-Cas9 system. Genome editing using CRISPR-Cas9 is being employed in crop improvement programmes in fruit crops for tolerance to biotic stress, nutritional enhancement and yield improvement (Table 2).

Table 2: Selected reports of use of transgenic technology and genome editing for desirable traits in fruit crops

Crop	Gene/target	Trait	Technology	References
Apple	*DIPM-1, DIPM-2, DIPM-4*	Resistance to fire blight disease	GE	Malnoy *et al.*, 2016
	MdDIPM 4 (a susceptibility gene)	Resistance to fire blight disease	GE	Pompili *et al.*, 2019
	MdPGT1 (phloretin-specific UDP-2′-O-glucosyltransferase)	Reduced foliar phloridzin	GE	Miranda *et al.*, 2023
Banana	Overexpression of *MusaPIP1*, a native plasma membrane aquaporin	Tolerance to different abiotic stresses	GM	Sreedharan *et al.*, 2013
	RAS-PDS1 and *RAS-PDS2* (phytoene desaturase genes)	Functional disruption of *RAS-PDS* genes	GE	Kaur *et al.*, 2018
	eBSOLV	Resistance to *Banana Streak Virus*	GE	Tripathi *et al.*, 2019
	LCYε (fifth exon of the lycopene epsilon-cyclase gene)	Enhanced β-carotene content	GE	Kaur *et al.*, 2020
Blueberry	Overexpression of the VcDDF1, blueberry dwarf and delayed flowering gene	Increased freezing tolerance	GM	Song and Walworth, 2018
Citrus sp. (Sweet orange)	*NPR1* (non-expressor of Pathogenesis Related genes 1) gene from *A. thaliana*	Enhanced resistance to Huanglongbing	GM	Dutt *et al.*, 2015
Citrus sinensis	*cecropin B* gene	Decreased susceptibility to Huanglongbing	GM	Zou *et al.*, 2016
Citrus	*CsLOB 1* (a susceptibility gene promoter)	Resistance to citrus canker in *Citrus sinensis* Osbeck – Orange	GE	Peng *et al.*, 2017
	DMR6 orthologue (downy mildew resistance 6, a susceptibility gene)	Resistance to Banana *Xanthomonas* wilt	GE	Tripathi *et al.*, 2021
Grape	Overexpression of *VviNAC17*, a grapevine transcription factor	Induces the synthesis of berry quality-related secondary metabolites	GM	Badim *et al.*, 2022
Grape	*MLO-7* (a susceptible gene)	Resistance to powdery mildew	GE	Malnoy *et al.*2016
Plum	RNA interference or silencing of *PPV* coat protein gene	Resistance to Sharka disease caused by *PPV*	GM	Polák *et al.*, 2017; Sidorova *et al.* 2019
Sweet cherry	Overexpression of *PcMPK3-HA* gene	Root growth	GM	Zong *et al.* 2022
Watermelon	*ClBG1*(β-glucosidase gene)	Reduced seed size and promoted seed germination	GE	Wang *et al.*, 2021

GM: genetically modified, GE: genome edited

GM and Genome-edited Fruit Crops in India

In India, genetically modified organisms (GMOs) and GMO-derived products are regulated as per the Rules for the Manufacture, Use/ Import/ Export and Storage of Hazardous Micro Organisms/ Genetically Engineered Organisms or Cells, 1989 notified under Environment Protection Act, 1986. Till date, no GM fruit crop has been approved in the country. However, GM banana lines with *OsNAS1* or *OsNAS2* genes and *Apsy2a* or *NEN-DXS2* genes for nutritional enhancement, developed by National Agri-Food Biotechnology Institute (NABI), Mohali has been recently approved for event selection trials (147th GEAC meeting dated 18.10.2022) (http://geacindia.gov.in/decisions-of-GEAC-meetings.aspx). Studies have also been reported for development of genome-edited banana by NABI, Mohali (Kaur *et al.* 2018, Kaur *et al.* 2020). In accordance with the guidelines for safety assessment of genome edited plants, 2022 (Department of Biotechnology, Ministry of Science & Technology, Government of India), genome-edited plants (SDN-1 and SDN-2) are exempted under Rule 20 of Rules 1989. Based on the genome editing technologies using site-directed nucleases (SDN), genome-edited plants can be classified as SDN-1, SDN-2 and SDN-3 (Friedrichs *et al.* 2019, Ricroch 2019). SDN-1 and SDN-2 mediated alterations generally represent the possible outcomes from the cellular mechanisms used to repair DNA breaks that occur naturally, whereas in SDN3, insertion of foreign gene may lead to formation of new proteins.

Global adoption of GM fruit crops is very limited, but genetic modification and other biotechnological/ new breeding techniques could be used for production of modified fruits crops with desirable traits to cater the increasing population and for nutritional security.

Conclusion

The CWRs are a critical source of novel traits and genes for future fruit crop breeding programmes. The importance of CWRs is well realized in the fruit crops that the wild relatives are used to improve both scion and rootstock in many fruit crops. Therefore, collection, conservation and characterization of these CWRs of fruit crops are essential to prevent their extinction and substantial utilization in fruit crop breeding under climate change situations. Although, fruit crop breeding is tedious because it has a long juvenile phase and heterozygous genetic background. Besides difficulties, CWRs are well utilized in fruit crop breeding, and successes have been achieved in the development of rootstock and scion cultivars using traditional breeding approaches. The new biotechnological tools could be used for the identification of the candidate genes from the CWRs and subsequent their utilization in fruit crop breeding. Furthermore, genome editing offers to fruit breeders to precisely design novel fruit crop cultivars to address rising demand, climate change, and market challenges in these perennial fruit tree species.

References

Arora RK (1991) Plant diversity in Indian gene Centre. In: Plant Genetic Resources Conservation and Management, Paroda, R.S. and Aorora, R.K. (Eds.), pp. 25–54. New Delhi: IPGRI.

Badim H, Vale M, Coelho M, Granell A, Gerós H and Conde A (2022) Constitutive expression of VviNAC17 transcription factor significantly induces the synthesis of flavonoids and other phenolics in transgenic grape berry cells. Frontiers in Plant Science 13: 964621.

Boch J, Scholze H, Schornack S, Landgraf A, Hahn S, Kay S, Lahaye T, Nickstadt A and Bonaset U (2009) Breaking the code of DNA binding specificity of TAL-type III effectors. Science 326: 1509-12.

Boissel S, Jarjour J, Astrakhan A, Adey A, Gouble A, Duchateau P, Shendure J, Stoddard BL, Certo MT, Baker D and Scharenbergn AM (2014) megaTALs: a rare-cleaving nuclease architecture for therapeutic genome engineering. Nucleic Acid Research 42(4): 2591-01.

Brown SK and Maloney KE (2013) An update on apple cultivars, brands and club-marketing. New York Fruit Quarterly, 21(1): 3-10.

Cornille A, Gladieux P, Smulders MJ, Roldán-Ruiz I, Laurens F et al., (2012) New insight into the history of domesticated apple: secondary contribution of the European wild apple to the genome of cultivated varieties. PLoS genetics 8(5): e1002703.

Dey S, NS Jamir, R Gogoi, SK Chaturvedi, HY Jakha, ZP Kikon (2014) Musa nagalandiana sp. nov. (Musaceae) from Nagaland, northeast India. Nordic Journal of Botany 32: 584–588. https://doi.org/10.1111/njb.00516

Dutt M, Barthe G, Irey M, Grosser J (2015) Transgenic citrus expressing an Arabidopsis NPR1 gene exhibit enhanced resistance against Huanglongbing (HLB; Citrus Greening). PLoS ONE, 10(9): e0137134.

Ferreira SA, Pitz KY, Manshardt R, Zee F, Fitch M, Gonsalves D (2002) Virus coat protein transgenic papaya provides practical control of Papaya Ringspot Virus in Hawaii. Plant Disease 86(2):101-105.

Friedrichs S, Takasu Y, Kearns P, Dagallier B, Oshima R, Schofield J and Moreddu C (2019) Meeting report of the OECD conference on "Genome Editing: Applications in Agriculture—Implications for Health, Environment and Regulation". Transgenic Research 28: 419-63.

Gogoi R and S Borah (2014) Musa mannii var. namdangensis (Musaceae) from Arunachal Pradesh, India. Taiwania 59(2): 93-97

Gogoi R and Borah S (2013) Musa markkui R. Gogoi & S. Borah, a new species from Arunachal Pradesh, India. – Gard. Bull. Singap. 58: 19-26.

Gogoi R and Häkkinen M (2013) Musa kamengensis (Musaceae), a new species from Arunachal Pradesh, India. Acta Phytotax. Geobot 64: 127 – 131.

Gonsalves D, Gonsalves C, Ferreira S, Pitz K, Fitch M, Manshardt R. and Slightom, J. (2004). Transgenic virus-resistant papaya: from hope to reality in controlling papaya ringspot virus in Hawaii. APSnet Features. Online. doi:10.1094/APSnetFeature-2004-0704

ISAAA (2019) Global status of commercialized biotech/GM crops in 2019: Biotech crops drive socio economic development and sustainable environment in the new frontier. ISAAA Brief No. 55. ISAAA: Ithaca, NY.

John JK and K Pradheep (2019) Wild relatives of horticultural crops: PGR Management in Indian context. In: P. E. Rajasekharan, V. R. Rao (eds.), Conservation and Utilization of Horticultural Genetic Resources, pp. 451- 482. https://doi.org/10.1007/978-981-13-3669-0_15

Kaur N, Alok A, Shivani, Kumar P, Kaur N, Awasthi P, Chaturvedi S, Pandey P, Pandey A, Pandey AK and Tiwari S (2020) CRISPR/Cas9 directed editing of lycopene epsilon-cyclase modulates metabolic flux for β-carotene biosynthesis in banana fruit. Metabolic Engineering 59:76-86.

Kaur N, Alok A, Shivani, Pandey P, Awasthi P and Tiwari S (2018) CRISPR/Cas9-mediated efficient editing in phytoene desaturase (PDS) demonstrates precise manipulation in banana cv. Rasthali genome. Functional and Integrative Genomics 18(1): 89-99.

Kramer MG and Redenbaugh K (1994) Commercialization of a tomato with an antisense polygalacturonase gene: The FLAVR SAVR™ tomato story. Euphytica 79: 293-97.

Litz RE and Padilla G (2012) Genetic transformation of fruit trees. In Genomics of Tree Crops; Springer International Publishing: New York City, NY, USA, pp. 117-54.

Lobato-Gómez M, Hewitt S, Capell T, Christou P, Dhingra A, Girón-Calva PS (2021) Transgenic and genome-edited fruits: background, constraints, benefits, and commercial opportunities. Horticulture Research 8(1):166.

Malnoy M, Viola R, Jung MH, Koo OJ, Kim S, Kim JS, Velasco R and Nagamangala Kanchiswamy C (2016) DNA-free genetically edited grapevine and apple protoplast using CRISPR/Cas9 ribonucleoproteins. Frontiers in Plant Science 20(7):1904.

Mamiyil S, Joe, A and Sreejith PE (2013). Musa chunii Häkkinen (Musaceae): An addition to the wild banana flora of India and notes on conservation of a Critically Endangered species. Annals of Plant Sciences 2(5): 160-62.

Miranda S, Piazza S, Nuzzo F, Li M, Lagrèze J, Mithöfer A, Cestaro A, Tarkowska D, Espley R, Dare A and Malnoy M (2023) CRISPR/Cas9 genome-editing applied to MdPGT1 in apple results in reduced foliar phloridzin without impacting plant growth. The Plant Journal 113(1): 92-105.

Mitra SK, Pathak PK, and Chakraborty I (2010) Potential underutilized tropical fruits of India. Acta Horticulturae 864, 6.

Mukherjee SK (1985) Systematic and Biogeographic Studies of Crop Genepools. Vol.1, Mangifera L. IBPGR, Rome.

Pareek OP, Suneel Sharma and RK Arora (1998) Underutilized Edible Fruits and Nuts: An Inventory of Genetic Resources in Their Regions of Diversity. IPGRI Office for South Asia, New Delhi, India. 235 p

Peng A, Chen S, Lei T, Xu L, He Y, Wu L, Yao L and Zou X (2017) Engineering canker-resistant plants through CRISPR/Cas9-targeted editing of the susceptibility gene Cs LOB 1 promoter in citrus. Plant Biotechnology Journal 15(12):1509-1519.

Penna S and Jain SM (2023) Fruit crop improvement with genome editing, *in vitro* and transgenic approaches. Horticulturae 9: 58.

Polák J, Kundu JK, Krška B, Beoni E, Komínek P, Pívalova J and Jarošová J (2017) Transgenic plum Prunus domestica L., clone C5 (cv. HoneySweet) for protection against sharka disease. Journal of Integrative Agriculture 16(3): 516-522.

Pompili V, Dalla Costa L, Piazza S, Pindo M and Malnoy M (2020) Reduced fire blight susceptibility in apple cultivars using a high-efficiency CRISPR/Cas9-FLP/FRT-based gene editing system. Plant Biotechnology Journal 18(3): 845-858.

Pramanick KK, Kishore DK, Singh R and Kumar J (2012) Performance of apple (Malus x domestica Borkh) cv. Red Spur on a new apple rootstock in high density planting. Scientia Horticulturae 133: 37-39.

Ranjan P, Brahmi P, Tyagi V, Ranjan JK, Srivastava V, Yadav SK, Singh SP, Singh S, Binda PC, Singh SK, Singh K (2022) Global interdependence for fruit genetic resources: status and challenges in India. Food Security 14(3):591-619.

Ricroch A (2019) Global developments of genome editing in agriculture. Transgenic Research 28: 45-52.

Satisha J and Prakash GS (2006) The influence of water and gas exchange parameters on grafted grapevines under conditions of moisture stress. South African Journal of Enology and Viticulture 27 (1): 40-45.

Satisha J, Ramteke SD and Karibasappa GS (2007). Physiological and biochemical characterization of grape rootstocks.

Shahsavar AR, K Izadpanah, E Tafazoli, BE Sayed Tabatabaei (2007) Characterization of citrus germplasm including unknown variants by inter-simple sequence repeat (ISSR) markers. Scientia Horticulturae 112: 310–314.

Sidorova T, Mikhailov R, Pushin A, Miroshnichenko D and Dolgov S (2019) Agrobacterium-mediated transformation of Russian Commercial Plum cv. "Startovaya" (*Prunus domestica* L.) with virus-derived hairpin RNA construct confers durable resistance to PPV infection in mature plants. Frontiers in Plant Science 10:286.

Song GQ and Walworth A (2018) An invaluable transgenic blueberry for studying chilling-induced flowering in woody plants. BMC Plant Biology 18(1): 265.

Song GQ, Prieto H and Orbovic V (2019) Agrobacterium-mediated transformation of tree fruit crops: methods, progress, and challenges. Frontiers in Plant Science 10:226.

Sreedharan S, Shekhawat UK and Ganapathi TR (2013) Transgenic banana plants overexpressing a native plasma membrane aquaporin MusaPIP1;2 display high tolerance levels to different abiotic stresses. Plant Biotechnology Journal 11(8): 942-52.

Tripathi JN, Ntui VO, Ron M, Muiruri SK, Britt A and Tripathi L (2019) CRISPR/Cas9 editing of endogenous banana streak virus in the B genome of *Musa* spp. overcomes a major challenge in banana breeding. Communications Biology 2(1):46.

Tripathi JN, Ntui VO, Shah T and Tripathi L (2021) CRISPR/Cas9-mediated editing of DMR6 orthologue in banana (Musa spp.) confers enhanced resistance to bacterial disease. Plant Biotechnology Journal 19(7):1291-93.

Uma S (2006) Farmers' Knowledge of Wild Musa in India. A0327/E, Plant Production and Protection Division, FAO, Rome, 55 p.

Urnov FD, Rebar EJ, Holmes MC, Zhang HS and Gregory PD (2010) Genome editing with engineered zinc finger nucleases. Nature Reviews Genetics 11(9): 636-646.

Wang Y, Wang J, Guo S, Tian S, Zhang J, Ren Y, Li M, Gong G, Zhang H and Xu Y (2021) CRISPR/Cas9-mediated mutagenesis of ClBG1 decreased seed size and promoted seed germination in watermelon. Horticulture Research 8: Article No. 70.

Zeven AC and De Wet JMJ (1982) Dictionary of Cultivated Plants and Their Regions of Diversity: Excluding Most Ornamentals, Forest Trees and Lower Plants. Wageningen: CAPD.

Zhang Y, Massel K, Godwin ID and Gao C (2018) Applications and potential of genome editing in crop improvement. Genome Biology 19: 210.

Zong X, Xu L, Tan Y and Wei H (2022) Development of genetically modified sweet cherry rootstock 'Gisela 6' with overexpression of PcMPK3-HA gene by Agrobacterium-mediated genetic transformation. Plant Cell Tissue and Organ Culture, 151: 375-84.

Zou X, Jiang X, Xu L, Lei T, Peng A, He Y, Yao L and Chen S (2017) Transgenic citrus expressing synthesized cecropin B genes in the phloem exhibits decreased susceptibility to Huanglongbing. Plant Molecular Biology, 93: 341-353.

4

Role of Rootstock in Production of Citrus Fruits

Awtar Singh and Kanhaiya Singh

Division of Fruits and Horticultural Technology, ICAR-Indian Agricultural Research Institute, New Delhi – 110 012

Perennial fruit crops are largely cultivated under borderline conditions leading to various physiological disarrays. An environment-friendly technique for escaping or reducing losses in production due to biotic and abiotic stresses is their propagation onto rootstocks, which can reduce the adverse effects of external stresses. The scion and rootstock grow together, although genetically different but working infusion and grow as a single entity. The rootstock provides the root system, thus anchoring the tree and acts as an absorbing organ for water and mineral nutrients. The rootstock also influences the tree size, precocity in bearing, fruit quality and resistance to biotic and abiotic stresses. More than twenty-two horticultural characteristics are influenced by citrus rootstocks, including vigour and size, depth of rooting, low temperature tolerance, adaptation to hostile soil conditions, disease resistance and fruit quality (Castle, 1987). In certain cases, rootstocks are solitary determining factor that allows fruit trees to be grown in environments that may else not be suitable for fruit growing.

Importance of Rootstock

Rootstock is the most important factor in defining the vigour and ultimate size of the tree. Although rootstocks have many applications in the enhancement of fruit quality, imparting adaptability to climatic and edaphic conditions, etc., but major emphasis for rootstock selection and breeding have been given to managing vigour and securing regular and high yield of quality fruits. The rootstock condenses youthfulness period and the plants propagated with a rootstock tied with a pathogen-free scion bring perfection in the homogeneity and steadiness of an orchard. For achieving all these effects and extended life of the trees, a successful rootstock should have compatibility with the scion cultivar propagated on it. Fruit maturation, fruit holding on tree and post-harvest preservation are also affected by rootstock. The rootstocks directly affect the uptake ability of plants for water and nutrients.

The quest for new rootstock in fruit production is required to sustain the production under the unavoidable biotic and abiotic stresses in many different ecological situations of the fruit-growing areas of the world (Castle, 2010; Habran *et al.*, 2016). Also, new diseases, the spread of known diseases, and different environmental conditions affected by climatic alteration, force the demand for developing new rootstocks. Therefore, there must be continuous development of new citrus rootstocks to meet the growing demand of the citrus industry and to sustain the fruit production. The use of rootstocks in citrus decreases the long juvenility period and allows the cultivation of citrus under several abiotic and biotic stress conditions. The physiology of the whole tree is affected by rootstock, including traits of economic relevance such as fruit yield, fruit size, juice quality, tree vigour, and resistance against biotic and abiotic stresses (Castle *et al.*, 1993). Fruit maturation, fruit holding on tree, and postharvest preservation are also affected by rootstock (Hodgson, 1967).

Owing to variety of the climatic and ecological conditions in citriculture, a variety designated for its proper qualities should have a transnational appeal while a rootstock would be more adapted to precise soil conditions and soil pathogens present in each area. Furthermore, rootstocks should favourably alter the variety features for many characters. An outstanding new rootstock having resistance/tolerance to citrus tristeza virus, *Phytophthora* spp., citrus nematodes along with displaying good adaptability to high salinity, calcareous soils and drought is still needed. By using the rootstocks, various plant growth and fruit characteristics of the scion variety can be modified that are not possible through simple breeding approaches as the two components of rootstock and scion cultivars can be improved separately and then combined in a single tree.

Rootstocks play an important role in speedy expansion of citrus in the world as well as breeding new cultivars. The use of citrus rootstocks provides many choices to growers for improving fruit quality and yield, obtaining early fruiting and uniform cropping, avoiding juvenility, controlling the tree size, having the opportunity for high-density planting, *etc*. All these factors are responsible for many economic important advantages to the growers and, as a result, the citrus fruits are the most produced fresh fruits in the world for several decades (Tuzcu *et al.*, 2005). Selecting a right rootstock is an important decision, and local climatic and soil conditions play important role in their selection. Although any citrus variety can be used as a rootstock, some of them are better suited to specific conditions than the others (Davies and Albrigo, 1994; Lawrence and Bridges, 1974). Some of the important characteristics required in a desirable citrus rootstock should be as a good adaptation to all kinds of soils, tolerance to salinity, iron chlorosis, flooding, drought, high affinity with commercial species/cultivars, high yields of good fruit quality, reduced tree size, resistance to citrus tristeza virus (CTV), resistance to fungal diseases (*Phytophthora* spp., *Armillaria mellea,* etc.), resistance/tolerance to bacterial diseases and resistance to nematodes.

Genetically improved rootstocks with favourable horticultural characteristics can be a long-term approach for sustainable citrus industry. Hence, screening studies for biotic and abiotic stresses should be carried out properly and regularly. In addition, continuous improvement of rootstocks is necessary to sustain cultivating citrus trees under stressed environments (Haddou *et al.* 2000; Romero *et al.* 2006; Garcia-Sanchez *et al.*, 2007; Rodriguez-Gamir *et al.*, 2010). The use of rootstocks in fruit production includes not only a stronger resistance against pathogens but also a higher tolerance to abiotic stress conditions (Rouphael *et al.*, 2012). Photosynthesis is the basic determinant of physiological processes responsible for plant growth and productivity (Lawlor, 1995).

The citrus rootstocks have disparity in supplying shoot tissues with water and carbon, improving the resistance to biotic and abiotic stresses and affecting plant water status and photosynthesis (Pedrosoa *et al.*, 2014). The rootstocks also alter the physiological performance of scion varieties under water stress conditions (Magalha es Filho *et al.*, 2008; Rodriguez-Gamir *et al.*, 2011). Polyploidy in citrus leads to better adaptation to adverse environmental conditions (Allario *et al.*, 2013). The drought tolerance in autotetraploid Rangpur lime (*Citrus limonia*) was better than the diploid Rangpur lime when tested with a common scion variety.

Types of Rootstocks

If rootstocks are obtained from seeds other than the apomictic seeds, they are called as seedling rootstocks and if obtained through vegetative means, micropropagation or through apomictic seeds, they are called clonal rootstocks. The seedling rootstocks have better anchorage, but they may not be uniform, whereas, the clonal rootstocks are more uniform, but they may have poor anchorage. The seedling rootstocks can be mass-produced simply and economically. They tend to be disease free because viruses are usually not transmitted through seed production. They tend to be more deep-rooted and firmly anchored than rootstocks grown from cuttings. These rootstocks have genetic variability due to sexual reproduction, which can be limited using elite seed source trees; uniformity can be controlled through managing production (grading, culling, *etc.*). Further, the use of apomictic seeds can result in clonal, seedling rootstocks- particularly important in the production of citrus. In citrus, majority of the rootstocks have the phenomenon of nucellar polyembryony and thus, their propagation from seed is very easy and reliable as the seedlings obtained are uniform and ultimately the grafted plants exhibit uniformity in the orchards. There is a need to identify the mother tree sources which produce majority of the nucellar seeds and this practice is specifically important for the mother plant identification in citrus and particularly in Carrizo and Troyer citranges, as the extent of nucellar embryony may vary even within the plants of the same cultivar. Establishment of seed source trees and seed extraction are relatively specialized cultural practices. Seed source trees of the major rootstocks are generally grown from seed, although some may be budded on rootstocks. Fruits for seed extraction are collected after seeds have matured and seeds extracted immediately or fruits are stored under refrigerated conditions, else the germination will reduce.

Drawbacks of Own Rooted Trees

The own rooted plants are vigorous and not suitable for modern production systems and are generally responsible for delayed start of fruit bearing. They lack in tolerance to soil-borne pests and diseases and the soil-borne fungal diseases caused by *Phytophthora* spp. are responsible to the introduction of rootstocks in citrus cultivation. They display non-uniformity in the growth, tree size, effective canopy, precocity, yield efficiency, adaptability to numerous biotic and abiotic stresses. Therefore, there is a necessity to find/breed rootstocks suitable for varying conditions. Rootstock choice offers a powerful tool for the sustainable strengthening of fruit production because while the scion genotype can be used to select fruit characteristics, the adaptation to water deficit and high salinity, tolerance to alkaline soils and susceptibility to pest and pathogens (foot rot/ collar rot in citrus) can be influenced by the rootstock selection. Thus, the rootstocks are important mainly for two main reasons: 1 many of the citrus cultivars are difficult to root by cuttings or by layering; 2 rootstocks are employed for fighting dangerous soil pathogens and to help plants in adaptation to the precise areas and management situations.

Bearing in mind these problems, the reasons making the use of citrus rootstocks imperative can be summarized as: propagation of the fruiting scion cultivar onto a root system, gaining uniformity in fruiting portion or the scion growth in comparison to seedlings, controlling tree size or vigour of the scion cultivar, adapting scion to adverse soil conditions, *etc.* providing tolerance against various biotic stresses prompted by soil pests.

Selection of Right Rootstock

There exists a wide variation with respect to climate, soil, topography and altitude for fruit producing areas and as such the requirements for rootstocks for different fruits vary from region to region depending upon the existing agroclimatic conditions. There is a general lack of rootstocks for different citrus fruits suitable for specific agro-climatic conditions and a need is felt to find/ develop rootstocks to suitingsuch situations. Seedling rootstocks exhibit a general lack of uniformity due to genetic variability, but is citrus, this problem is generally overcome by the phenomenon of polyembryony, which is responsible for uniformity in the seedling population of rootstocks. But, in the new rootstock programmes, for early release of the rootstocks and for combining superior characteristics from many genotypes, the requirement for the nucellar polyembryony is not essential as the new rootstocks can be easily and safely propagated vegetatively (Bowman *et al.*, 2016). With the increasing costs of land, labour and other production inputs, the fruit growers all over the world are seeking methods of increasing their productivity per unit area by adopting more intensive systems of orchard management such as higher tree densities, new training and pruning systems and greater use of irrigation and fertilizers *etc*. Such changes in production practices require rootstocks that are better adaptable to more intensive production systems.

The dwarf, semi-dwarf rootstocks and certain wild species, which have the potential for use as dwarfing rootstocks for citrus, can be tried for high density plantings. Some of the genera of the true citrus fruits like trifoliate orange and polyploid citrus genotypes can be tried as dwarfing rootstocks. Keeping the prevailing biotic and abiotic conditions in consideration, responses of grafted plants to external stimuli and to deficient or excessive concentrations of nutrients and heavy metals will depend on the scion as well as the rootstock genotype, and the rootstock-scion physiological interaction (Pareek, 2011). The new strategies in the era of spreading huanglongbing (HLB), the rootstocks which can be used for higher density plantings and are precocious in bearing are preferred so that good fruit harvests can be done before the decline of the plants due to the disease (Bowman *et al*., 2016). The choice of rootstock should be based on the most important limiting factors to production in a particular region, local climate and soil conditions, cultivars and intended use of the crop. Moreover, rootstock found suitable at one time may fail in the future, hence the research continues for the development of new rootstock.

Genetic Advantage of a Rootstock

It is very difficult to combine all the desirable traits in the scion cultivar of citrus because of the breeding problems connected with these fruits. The rootstock is a short-cut to the prolonged breeding process as many of the scion traits can be influenced by the rootstock (Prabhu, 2011). Thus, some of the required traits can be merged in the scion cultivar and some in the rootstock and when the plants are propagated, they accomplish as a solitary entity. A rootstock enables the mass propagation of fruiting plants and trees sharing common evolutionary ancestry and anatomy, keeping intact the heritable heterogeneity, a unique feature compared to other breeding mechanisms. It is the best method in those fruit species like citrus,which are particularly disease and pest susceptible when grown on their own roots. *e.g*., the Rangpur lime and Cleopatra mandarin improve the salt tolerance limit of the scion cultivars, whereas trifoliate orange and its hybrids improve the tolerance/resistance against many biotic stresses. Rootstock provides vigorous root system to the scion, which can develop the feeder roots and reach farther from stem and thus help in drought tolerance.The roots of rootstock increase the nutrient uptake efficiency.They also modify the shoot vigour and improve the fruit quality and yield.

Selection Criteria for Rootstocks

The rootstock must be compatible with the soil in the target orchard. It should be able to establish itself well and transfer nutrition to a scion to improve its genetic potential of fruit production and fruiting characteristics. Rootstocks need to be categorized with a comprehensive guide to their perfect soil and climate. It should be suitable for adaptation in a range of pH, mineral content, nematode population, salinity, water availability, pathogen load and soil type. The genetic testing through robust techniques should be available to differentiate variability of generic nature in any recognized genetic stock for rootstock purposes. A rootstock's capability for excluding nutrients is more important than their accumulation. Rangpur lime, sour orange and Cleopatra are sodium and chloride excluders, owing to which they are tolerant of high salinity (Kirkpatrick and Bitters, 1968). Similarly, *Severinia buxifolia* is Cl and B excluder and trifoliate orange is a chloride accumulator. Trifoliate orange (*Poncirus trifoliata*) has been reported to be a dwarfing rootstock for Pongan (Xian *et al.,* 1994). Flying Dragon rootstock was found successful as a dwarfing rootstock for encouraging high density planting (Rabe *et al.,* 1994). *C. macrophylla* was found dwarfing rootstock for Temple mandarin (Levy *et al.,*1993). Alemow (*C. macrophylla*) has also been recommended for Nagpur mandarin and acid lime in India. Many species of citrus and its relatives have the potential for their utilization as new rootstocks or their use in rootstock breeding programmes for the improvement of specific drawbacks of a particular rootstock.

Ideal Rootstocks for Citrus

Propagation of citrus plants on the suitable rootstocks has several advantages over the propagation of its own roots as cutting, stool or layer. The rootstock development is mainly concentrated for three characters, namely dwarfing, compatibility with scion cultivars and easy and inexpensive to propagate. The main rootstocks recommended for mitigating various types of stresses are mentioned in Table 1.

Future Thrusts

In most of the rootstocks, the genetic diversity and inheritability of rootstock characters are not known enough and the gene-pool utilized for rootstock breeding is narrowed to genotypes. There are still tasks to discover the genetic diversity of species used as rootstocks in their original gene-centre.Rapid progress is needed in the marker-assisted breeding of rootstocks. The development of orchard systems may modify the targets of rootstock breeders; thus, a permanent communication is essential between researchers of orchard systems and rootstocks. The rootstock selection for the sustainable fruit growing is an important opportunity to improve the orchard resistance to pests and diseases and tolerance for abiotic stress conditions. The climate changes also bring challenges for the fruit growing. Requirements for an ideal rootstock are many and no single rootstock possess all the desired attributes, several superior rootstocks are sought, each suited to a specific set of conditions. The future requirements of rootstocks in citrus would focus on

- Introduction and large-scale multi-location testing of new hybrid/cybrid rootstocks having multiple tolerance against several biotic and abiotic stresses.
- Development of complex hybrids through inter-, intra-specific and intergeneric hybridization to develop more versatile rootstocks to increase their usefulness and adaptation with respect to compatibility, size control, precocity, productivity and resistance to biotic and abiotic stresses.
- Developing hybrids with at least one of the parents as scion variety to overcome the problem of incompatibility between stock and scion varieties.
- Survey, collection, selection and evaluation of many indigenous *Citrus* species to explore their possibility to use as rootstocks under different agro-climatic conditions.
- Development of virus-free material for commercial rootstocks to reduce virus-related incompatibility problems and to maintain sustainability in productivity and fruit quality.
- Development of rootstocks which are fully dwarf like Flying Dragon trifoliate orange or semi-dwarf like majority of the tetraploid rootstocks developed in many countries (Bowman *et al.*, 2016).
- Establishment of certified mother blocks of rootstocks from the plants having proven records of performance (*e.g.*,Carrizo and Troyer citranges).
- Use of modern biotechnological approaches and particularly marker-assisted selection and gene editing for the development of rootstocks that have tolerance to different biotic and abiotic stresses. The development of tetraploids and tetrazyg citrus rootstocks, which have tolerance/resistance to various biotic and abiotic stresses is possible through the protoplast fusion and manipulation, and such type of rootstocks are urgently needed in the modern citriculture that is being challenged by the diseases like huanglongbing (Grosser *et al.*, 2000; Grosser and Gmitter, 2011)

Table 1: Characteristics of commercial citrus rootstocks for various stress conditions

Rootstock	Parentage	Important characteristics	References
Rough lemon	*Citrus jambhiri* (Lush.)	Tolerant to virus diseases, tristeza, xylopsorosis and exocortis but susceptible to gummosis, deep rooted	Lee *et al.*, 2009
Cleopatra mandarin	*Citrus reticulata* (C. reshni)	Salt tolerant (Cl- excluder), tolerant to tristeza, citrus nematode, salt and calcareous soils, cold hardy and deep rooted	Legua *et al.*, 2011b
Alemow	*Citrus macrophylla* Wester	Vigorous, precocious but poor juice quality. Suitable for cool and dry climate, drought tolerant, *Phytophthora* resistant, can tolerate high soil boron, Cl and Ca, slow growng, excellent juice quality. Susceptible to CTV.	Legua *et al.*, 2011a
Carrizo citrange	Sweet orange [(*C. sinensis* (L.) Osbeck] X *P. trifoliata* (L.) Raf.	Tolerant to gummosis and tristeza, good productivity and quality of fruits. Susceptible to salinity and lime-induced chlorosis.	Lee *et al.*, 2009; Legua *et al.*, 2011b; Bowman and Joubert, 2020
Flying Dragon	*Poncirus trifoliata* (L.) Raf.	Most dwarfing, highly resistant to *Phytophthora* and citrus nematode	Lee *et al.*, 2009
Rangpur lime	*Citrus limonia* Osbeck	Tolerant to tristeza, suitable for heavy, deep and high pH soils, susceptible to foot rot, exocortis and burrowing nematode.	Lee *et al.*, 2009
Sour orange	*Citrus aurantium* (L.)	Suitable for light soils, cold hardy, susceptible to tristeza, *Phytophthora*	Lee *et al.*, 2009
Trifoliate orange	*Poncirus trifoliata* (L.) Raf.	Very cold hardy, resistant to heavy non-calcareous soils, foot rot, tristeza and nematodes. Likes slightly acidic soils, moderately dwarfing, precocious and productive but susceptible to CEV	Chen and Wan, 1993
Troyer citrange	Sweet orange [*C. sinensis* (L.) Osbeck] *X P. trifoliata* (L.) Raf.	Tolerant to gummosis and tristeza, good productivity and quality of fruits. Susceptible to salinity and lime-induced chlorosis.	Castle *et al.*, 1993; Lee *et al.*, 2009

Rootstock	Parentage	Important characteristics	References
Swingle ctirumelo	Duncan grapefruit (*Citrus paradisi*) X *Poncirus trifoliate.*	Cold hardy, tolerant to CTV, blight, root rot, resistant to citrus nematode. Also has exocortis and xyloporosis tolerance, tolerant to water-logged conditions and has moderate drought tolerance. Susceptible to iron-induced chlorosis. Good productivity	Castle *et al*., 1993; Lee *et al*., 2009
C-35 citrange	Ruby Blood orange X Webber-Fawcett trifoliate orange	Tolerant to *Phytophthora* and CTV and resistant to citrus nematode. Frost tolerance. Sensitive to calcareous soils and high salinity, suitable for replant sites, susceptible to Zn and Mn deficiencies	Cameron and Soost, 1986; Castle *et al*., 1993b
F& A5	Cleopatra mandarin (*Citrus reticulata*) X Rubidoux trifoliate orange (*Poncirus trifoliata* (L.) Raf.	Tolerant/Resistant to CTV, nematodes, salinity, lime-induced chlorosisand flooding.	Forner *et al*., 2003
F& A13	Cleopatra mandarin (*Citrus reticulata*) X Rubidoux trifoliate orange (*Poncirus trifoliata* (L.) Raf.	Tolerant/Resistant to CTV, nematodes, salinity, lime-induced chlorosisand flooding.	Forner *et al*., 2003
F& A418	Troyer citrange X common mandarin	Dwarfing, resistant to CTV, nematodes, salinity and flooding. Susceptible to calcareous soils, *Phytophthora* spp. (root rot). High productivity with excellent fruit quality	Castle *et al*., 1993
F& A517	King mandarin X *P. trifoliata*	Dwarfing, Resistant to CTV, nematodes, calcareous soils, salinity and flooding. Excellent productivity and fruit quality	Castle *et al*., 1993
US-1279	Changsha mandarin (*C. reticulata* (L.) Blanco X Gotha Road #6 trifoliate orange (*P. trifoliata* (L.) Raf.	Superior production of good quality fruits after trees are infected with greening. Produce medium-sized trees, adapted to flatwood soils, less prone to decline under HLB condition	Bowman and McCollum, 2014a

Rootstock	Parentage	Important characteristics	References
US-1281	Cleopatra mandarin (*C. reticulata* (L.) Blanco X Gotha Road #6 trifoliate orange (*P. trifoliata* (L.) Raf.	Superior production of good quality fruits after trees are infected with greening. Produce medium-sized trees, adapted to flatwood soils, less prone to decline under HLB condition	Bowman and McCollum, 2014b
US-1282	Cleopatra mandarin (*C. reticulata* (L.) Blanco X Gotha Road #6 trifoliate orange (*P. trifoliata* (L.) Raf.	Superior production of good quality fruits after trees are infected with greening. Produce medium-sized trees, adapted to flatwood soils, less prone to decline under HLB condition	Bowman and McCollum, 2014c
US-1283	Ninkat mandarin (*C. reticulata*) X Gotha Road #6 trifoliate orange (*P. trifoliata* (L.) Raf.	Superior production of good quality fruits after trees are infected with greening. Produce medium-sized trees, adapted to flatwood soils, less prone to decline under HLB condition	Bowman and McCollum, 2014d
US-1284	Ninkat mandarin (*C. reticulata*) X Gotha Road #6 trifoliate orange (*P. trifoliata* (L.) Raf.	Superior production of good quality fruits after trees are infected with greening. Produce medium-sized trees, adapted to flatwood soils, less prone to decline under HLB condition	Bowman and McCollum, 2014e
US-802	Siamese pummelo (*C. grandis* Osbeck) X Gotha Road trifoliate orange (*P. trifoliata*)	Supports strong vigour and development of large tree. Best yield in HLB infected trees. Tolerant to endemic diseases and pest problems. Resistant/tolerant to CTV, citrus blight, *Phytophthora palmivora* and *Diarepes* root weevil and produces good quality fruits	Bowman, 2007a
US-812	Sunki mandarin (*C. reticulata*) X Benecke trifoliate orange (*P. trifoliata*)	Provide moderate vigour to the scion, produces good yields under HLB infection. Highly productive, moderate-sized tree and tolerant/resistant to CTV and citrus blight	Bowman and Wutscher 2001; Bowman and Rouse, 2006
US-942	Sunki mandarin X Flying Dragon trifoliate orange	Provide moderate vigour, best yields under HLB infection	Bowman, 2010; Bowman and McCollum, 2010

Rootstock	Parentage	Important characteristics	References
US-897	Cleopatra mandarin X Flying Dragon trifoliate orange	Induce relatively dwarf trees, good under HLB free condition. Good productivity per unit canopy volume, good fruit quality, resistance/tolerance to CTV, *P. palmivora* and *Diarepes* root weevil, dwarfing effect on scion	Bowman, 2007b
US-852	Changasha mandarin X English Large Flowered trifoliate orange	Resistant to *Phytophthora nicotianae*, citrus tristeza virus (CTV) and citrus nematode. Budded trees are semi-dwarf, high yielding, but cannot be propagated efficiently by seeds	Bowman, 1999
Volkamer lemon	*C. volkameriana*	Produces vigorous trees, precocious bearing, but susceptible to *Phytophthora* spp. Tolerant to CTV and exocortis, susceptible to xyloporosis and woody gall, tolerant to iron-induced chlorosis and flooding	Castle *et al.*, 1993; Legua *et al.*, 2011a
Gou Tou Chen	Probably sour orange (*C. aurantium* (L.)) hybrid	Not very productive and fruit quality low	Castle *et al.*, 1993; Legua *et al.*, 2011a
Bitters (C 22)	Sunki mandarin X Swingle trifoliate orange	Small tree, high yield, tolerance to freezing, CTV, moderately tolerant to *Phytophthora*, but not nematodes, tolerant to calcareous soils	Federici*et al.*, 2009
Carpenter (C 54)	Sunki mandarin X Swingle trifoliate orange	Replacement rootstock, suitable for calcareous soils	Federici *et al.*, 2009
Furr (C 57)	Sunki mandarin X Swingle trifoliate orange	Medium to large tree, good yield, freezing tolerant, tolerant to CTV, very tolerant to *P. parasitica* and nematodes and moderately tolerant to calcareous soils	Federici *et al.*, 2009
X-639	Cleopatra mandarin X Trifoliate orange	Foot rot and tristeza virus tolerance, citrus blight tolerance	Castle *et al.*, 1993b; Lee *et al.*, 2009
Benton citrange	Ruby Blood orange X Trifoliate orange	Suitable for replant situations, moderately cold tolerant and higher yielding	Lee *et al.*, 2009
C-32 citrange	Ruby sweet orange X Trifoliate orange	Tolerant to *Phytophthora*, citrus nematode and tristeza. Its low seed content makes propagation difficult	Castle *et al.*, 1993b; Hodgson, 1967

Rootstock	Parentage	Important characteristics	References
Rangpur X Troyer	Rangpur lime X Troyer citrange	Suitable for close planting	Castle *et al*., 1993b
Smooth Flat Seville (SFS)	Probably sour orange hybrid	Apparently citrus blight tolerant, less susceptible to tristeza than sour orange	Castle *et al*., 1993b
US-1516	African pummelo X Flying Dragon trifoliate orange	Induction of superior tree health and fruit productivity and good fruit quality in trees infected with HLB and exhibits improved HLB tolerance	Anonymous, 2015
US Super Sour 1	Hirado Buntan seedling pummelo X Cleopatra mandarin	Induction of superior tree health and fruit productivity and good fruit quality in trees infected with HLB and exhibits improved HLB tolerance	Anonymous, 2018a
US Super Sour 2	Benecke trifoliate orange X [Chinotto sour orange X *C. ichangensis*]	Induction of superior fruit productivity, good tree health, good fruit quality in trees infected with HLB, good HLB tolerance	Anonymous, 2018b
US Super Sour 3	Sunki mandarin X US-802	Induction of superior fruit productivity and tree health, good fruit quality in trees infected with HLB, improved HLB tolerance	Anonymous, 2018c
Flhorag 1	Intergeneric somatic hybrid between Willow leaf mandarin and Pomeroy trifoliate orange	Dwarfed scion size, highly tolerant to iron deficiency	Dambier *et al*., 2011
Rough lemon 8166	*Citrus jambhiri* Lush.	Resistant to foot rot	Castle *et al*., 1993; Arora, 2000

Summary

The rootstocks are an essential constituent in modern fruit production because of their capability of adjusting a particular cultivar to diverse environmental conditions and cultural practices. Rootstocks can provide numerous traits to the scion that are absent in the scion, such as soil's pest and disease resistance, better anchorage, improved nutrient uptake, better tolerance to soils with high salinity content or drought, as well as other limiting soil conditions. On the other hand, they can modify the performance of scion, like for example, by reducing tree vigour, and modify canopy structure that would allow the establishment of high-density orchards. Rootstocks can also reduce or extend the fruit maturation period; improve yield and fruit quality increasing profit returns. Therefore, each rootstock/scion combination can generate a plant with characteristic that neither component exhibits if grown separately. Many rootstocks have been selected and developed for different types of citrus fruits following conventional breeding and biotechnological approaches. A good progress has been made in the development of rootstocks for citrus, but still a lot needs to be attempted. There is an urgent need for the integration of conventional breeding approaches to the biotechnological approaches for speeding up the process of citrus rootstock improvement/ development.

References

Allario T, Brumos J, Colmenero-Flores J, Iglesias DJ, Pina JA, Navarro, L, Talon M, Ollitrault, P and Morillon, R. 2013. Tetraploid Rangpur lime rootstock increases drought tolerance via enhanced constitutive root abscisic acid production. Plant, Cell & Environment 36: 856–868.

Arora, R.K. 2000. A conclusion from a thirty-year-old rootstock trial in grapefruit with respect to growth and mortality. International Society of Citriculture Congress, Orlando Florida, USA.

Bowman, K.D. 1999. Notice to fruit growers and nurserymen relative to the naming and release of the US-852 citrus rootstock. US, Department of Agriculture, ARS, Washington, DC.

Bowman, K.D. 2007a. Notice to fruit growers and nurserymen relative to the naming and release of the US-802 citrus rootstock. US, Department of Agriculture, ARS, Washington, DC.

Bowman, K.D. 2007b. Notice to fruit growers and nurserymen relative to the naming and release of the US-897 citrus rootstock. US, Department of Agriculture, ARS, Washington, DC.

Bowman, K.D. and Joubert, J. 2020. Citrus rootstocks. In: The Genus Citrus; Talon, M., Caruso, M., Gmitter, F.G., (Eds.) Woodhead Publishing, Cambridge, Mass.

Bowman, K.D. and McCollum, G.M. 2010. Notice to fruit growers and nurserymen relative to the naming and release of the US-942 citrus rootstock. U.S. Department of Agriculture, ARS, Washington, D.C.

Bowman, K.D. and McCollum, G.M. 2014a. Release of the US-1279, citrus rootstock. U.S. Dept. Agric, ARS, Washington, D.C.

Bowman, K.D. and McCollum, G.M. 2014b. Release of the US-1281, citrus rootstock. U.S. Dept. Agric., ARS, Washington, D.C.

Bowman, K.D. and McCollum, G.M. 2014c. Release of the US-1282, citrus rootstock. U.S. Dept. Agr., ARS, Washington, D.C.

Bowman, K.D. and McCollum, G.M. 2014d. Release of the US-1283, citrus rootstock. U.S. Dept. Agr., ARS, Washington, D.C.

Bowman, K.D. and McCollum, G.M. 2014e. Release of the US-1284, citrus rootstock. U.S. Dept. Agr., ARS, Washington, D.C.

Bowman, K.D. and Rouse, R.E. 2006. US-812 citrus rootstock. HortScience 41:832–36.

Bowman, K.D. and Wutscher, H.K. 2001. Notice to fruit growers and nurserymen relative to the naming and release of the US-812 citrus rootstock. U.S. Dept. Agric., ARS, Washington, D.C.

Bowman, K.D., G. McCollum, G.M. and Albrecht, U. 2016. Performance of 'Valencia' orange (Citrus sinensis [L.] Osbeck) on 17 rootstocks in a trial severely affected by huanglongbing. Sci. Hort. 201:355–61.

Cameron, J.W. and Soost, R.K. 1986. C35 and C32: Citrange rootstocks for citrus. HortScience 21 (1): 157-58.

Castle, W.S. 1987. Citrus Rootstocks. John Wiley and Sons, New York, pp. 361-99.

Castle, W.S., Tucker, D.P.H., Krezdorn, A.H. and Youtsey, C.O. 1993. Rootstocks for Florida citrus. University of Florida, Gainesville.

Castle,W.S. 2010. A career perspective on citrus rootstocks, their development and commercialization. HortScience 45:11–15.

Chen, Z.S. and Wan, L.Z. 1993. Rootstocks. Chen ZS and Wan LZ, In: The Atlas of Major Citrus Cultivars in China. Science and Technology. Press of Sichuan Province, Chengdu, 94–106.

Dambier, D., Benyahia, H., Pensabene-Bellavia, G., Kacar, Y.A., Froelicher, Y., Belfalah, Z., Lhou, B., Handaji, N., Printz, B., Morillon, R., Yesiloglu, T., Navarro, L. and Ollitrault, pp. 2011. Somatic hybridization for citrus rootstock breeding: an effective tool to solve some important issues of the Mediterranean citrus industry. Plant Cell Rep. 30 (5): 883-900.

Davies, F.S. and Albrigo, L.G. 1994. Citrus Crop Protection Science in Horticulture, No: 2, CAB International, Wallington, p. 254.

Federici, C.T., Kupper, R.S., Roose, M.L. 2009. "Bitters", "Carpenter" and "Furr" Trifoliate Hybrids: Three New Citrus Rootstocks' Online at https://plantbiology.ucr.edu/faculty/new%20citru s%20rootstocks%202009.pdf.

Forner, J.B., Forner-Girnar, M.A., and Alcaide, A. 2003. Forner-Alcaide 5 and Forner-Alcaide 13: Two New Citrus Rootstocks Released in Spain. HortScience 38 (4): 629-30.

Garcia-Sanchez, F., Syvertsen, J.P., Gimeno, V., Botia, P. and Perez-Perez, J.G. 2007. Responses to flooding and drought stress by two citrus rootstocks seedlings with different water-use efficiency. Physiol. Plantarum 130: 532–42.

Grosser, J.W. and Gmitter, F.G. Jr. 2011. Protoplast fusion for production of tetraploids and triploids: applications for scion and rootstock breeding in citrus. Plant Cell Tissue and Organ Culture 104: 343–57.

Grosser, J.W.; Ollitrault, P., Olivares-Fuster, O. 2000. Invited review: somatic hybridization in Citrus: an effective tool to facilitate variety improvement. In Vitro Cellular and Developmental Biology – Plant 36: 434-49.

Habran, A., Commisso, M., Helwi, P., Hilbert, G., Negri, S., Ollat, N., et al. 2016. Rootstocks/scion/nitrogen interactions affect secondary metabolism in the grape berry. Frontiers in Plant Science 7:1134.

Haddou, M. Ait, Rochdi, A., Bousrhal, A., Ben yahia, A and Benazzouz, A. 2000. Study of tolerance of three citrus rootstocks to chloride and sodium ions. Paper presented in ISC Congress. Abstract 132: 230.

Hodgson, R.W. 1967. Horticultural varieties of citrus. In: Reuther W, Webber HJ, Batchelor LD, editors. The Citrus Industry vol. 1. Berkeley: University of California Press, pp. 431–591.

Kirkpatrick, J.D. and Bitters, W.P. 1968. Physiological and morphological response of various citrus rootstocks to salinity. Proceedings of the International Society of Citriculture 1: 391-400.

Lawlor, D.W. 1995. The effect of water deficit on photosynthesis. In: Smirnoff, N. (editor). Environment and Plant Metabolism, Flexibility and Acclimation London: BIOS Scientific Publisher, pp. 129–160.

Lawrence, F.P. and Bridges, D. 1974. Rootstocks for Citrus in Florida. Florida Cooperative Extension Service, Institute of Food and agricultural Sciences, University of Florida, Gainesville. p. 13.

Lee, A.T.C., Joubert, J. and van Vuuren, S.P. 2009. Chapter 6: Rootstock Choice. Integrated Production Guidelines for Export Citrusvol. 1, Citrus Research international.

Legua, P., Bellver, R., Forner, J.B. and Forner-Giner, M.A. 2011a. Plant growth, yield and fruit quality of 'Lane Late' navel orange on four citrus rootstocks. Spanish J. Agril. Res.9 (1): 271-79.

Legua, P., Bellver, R., Forner, J.B. and Forner-Giner, M.A. 2011b. Trifoliata hybrids rootstocks for 'Lane Late' navel orange in Spain. Sci. Agric. (Piracicaba, Braz.) 68 (5): 548-53.

Levy, Y., Lifshitz, J. and Bavli, H. (1993). Alemow (Citrus macrophylla Wester) - a dwarfing rootstock for old line temple mandarin. Scientia Hort. 53(4): 289-300.

Magalhaes Filho, J.R., Amaral, L.R., Machado, D.F.S.P., Medina, C.L. and Machado, E.C.M. 2008. Water deficit, gas exchange and root growth in 'Valencia' orange tree budded on two rootstocks. Bragantia 67: 75–82.

Moya, J.L., Tadeo, F.R., Legaz, F. Primo - Milo, E. and Talon, M. 2000. Salt tolerance in citrus is influenced by morphological, physiological and phenological factors. Differences between Carrizo citrange and Cleopatra mandarin. International Society of Citriculture congress Abstract 162: 396.

Pareek, O.P. 2011. Rootstock resources of arid zone fruits to overcome biotic and abiotic stresses. In: Singh,A.K., Awasthi, O.P., Dubey, A.K. and Nagaraja, A. (Eds.), Advances in Rootstocks for Overcoming Biotic and Abiotic Stresses in Fruit Crops, Div. Fruits & Hort. Tech., IARI, New Delhi, pp. 29-38.

Pedrosoa, F., Prudentea, D.A., Buenoa, A.C.R., Machadoa, E.C. and Ribeiroa, R.V. 2014. Drought tolerance in citrus trees is enhanced by rootstock-dependent changes in root growth and carbohydrate availability. Environ. Exp. Bot. 101: 26–35.

Prabhu, K.V. 2011. Breeding approaches for rootstock improvement for biotic and abiotic stress tolerance. Singh, A.K., Awasthi, O.P., Dubey, A.K. and Nagaraja, A. (Eds.), In: Advances in Rootstocks for Overcoming Biotic and Abiotic Stresses in Fruit Crops. Div. Fruits & Hort. Tech., IARI, New Delhi, pp. 29-38.

Rabe, E., Cook, N., Jacob, G. and Vander Walt, H.P. 1994. Current status of research on citrus tree size control in Southern Africa. Proceeding of International Society of Citriculture 2: 714-720.

Rodriguez-Gamir, J., Ancillo, G., Aparicio, F., Bordas, M., Primo-Millo, E. and Forner-Giner, M.A. 2011. Water-deficittolerance in citrus is mediated by the down-regulation of PIP gene expression in the roots. Plant Soil 347: 91–104.

Rodriguez-Gamir, J., Primo-Millo, E., Forner, J.B. and Forner-Giner, M.A. 2010. Citrus rootstock responses to water stress. Sci. Hortic. 126: 95–102.

Romero, P., Navarro, J.M., Pérez-Pérez, J., García-Sánchez, F., Gómez-Gómez, A., Porras, I., Martinez, V. and Botía, P. 2006. Deficit irrigation and rootstock: their effects on water relations, vegetative development, yield, fruit quality and mineral nutrition of Clemenules mandarin. Tree Physiol. 26: 1537–48.

Rouphael, Y., Cardarelli, M., Rea, E. and Colla, G. 2012. Improving melon and cucumber photosynthetic activity, mineral composition, and growth performance under salinity stress by grafting on to Cucurbita hybrid rootstocks Photosynthetica 50: 180–188.

Tuzcu,O., Yesiloglu, T.and Yildirim, B. 2005. Citrus Rootstocks in Mediterranean and Some Suggestions for Their Future. Book of Abstracts of the 7th International Congress of Citrus Nurserymen. 10th International Citrus Congress, 17-21 September 2005; Cairo, Egypt, pp. 82–84.

USDA, Washington 2015. Release of US-1515, Citrus rootstock. USDA, Agricultural Research Service Washington, D.C.

USDA, Washington 2018a. Release of US Super Sour 1, Citrus rootstock with improved tolerance to Huanglongbing. USDA, Agricultural Research Service Washington, D.C.

USDA, Washington 2018b. Release of Super Sour 2, Citrus rootstock. USDA, Agricultural Research Service Washington, D.C.

USDA, Washington 2018c. Release of Super Sour 3, Citrus rootstock. USDA, Agricultural Research Service Washington, D.C.

Xian, X.C., Huang, S.R., Ma, P.Q., Wu, W., Liang, W.Y. and Li, S.U. 1994. Techniques for producing higher yields of good quality cv. Pongan fruits. China Citrus 23 (1): 19-20.

5

Impact of Climate Change on Mango Productivity in Indian Subtropics

P.L. Saroj, Dinesh Kumar and Ashish Yadav

ICAR-Central Institute for Subtropical Horticulture, Lucknow - 226 101 (UP), India

Prevailing climate and weather conditions play crucial role in production and quality of various fruits including mango (*Mangifera indica* L.). In India, mango has 2.3 million ha area with annual production of 20.5 million tonnes and average productivity of 8.9 t/ha. However, erratic precipitation and high temperature during flowering and fruit setting induced mainly due to climatic disturbance in the recent past had adversely impacted the production potential of mango in subtropics. Analysis of twenty years (2001-02 to 2021-22) weather data of ICAR-CISH, Lucknow, indicated that average annual rainfall of the region is 956.41 mm spread over 25-30 number of rainydays. The mean maximum temperature during May was 39.15°C and minimum temperature 6.31°C in January. In general, physiology of flowering in different varieties of mango is influenced by atmospheric temperature and precipitation but no single climatic factor can decide physiological performance of a plant. For example, photosynthesis depends primarily on radiation but it is also influenced by temperature, carbon dioxide concentration, water availability and mineral elements.

The mango starts flowering during second fortnight of February under subtropical region and fruit setting takes place in second fortnight of March. After setting, fruit starts developing and continued up to end of May. The harvesting takes place after attaining physiological maturity June onward depending upon varieties and climatic conditions. The fruit yield varied from 10 to 20 t/ha depending upon varieties, planting density, management and prevailing climatic conditions. The observations on flowering pattern in recent past showed significant fluctuations in flower initiation and fruit setting due to climatic aberrations. The delay in flowering of early varieties was observed due to prolonged winter. Similarly, sudden rise in temperature disturbed fruit setting in spite of normal flowering; thereby poor crop load of mango. Therefore, long-term strategy may be framed to address these problems to develop suitable management techniques as well as evolving new varieties which can withstand under changing climatic scenario.

Impact of Climate Change

Although climate change impacts are witnessed all over the world but South Asia is categorized as most vulnerable. The vulnerability of India to climate change can be very well understood as the huge population depend on agriculture and excessive pressure on natural resources. Within the country, the eco-regions like; hill and mountain, arid and coastal ecosystems are more prone to climate change. The ill impact of climate change is aggravating year after year with high perturbations of weather parameters. The emission of greenhouse gasses like Carbon dioxide (64%), methane (19%), chlorofluorocarbon (11%) and nitrous oxide (6%) are major cause of global warming. Among anthropogenic activities, burning of fossil fuels is major contributor (49%), of global warming, followed by industrial processes (24%) and deforestation (14%) while agriculture (13%) though significantly but contributing minimum to climate change. Rise in temperature is not only the problem but also melting of glacier is posing a serious threat. The Himalayan glaciers have lost about 40% of their area in the last several hundred years, or an estimated 390 - 586 cubic kilometers of ice-enough to raise global sea levels 0.92 - 1.38 millimeters. The carbon dioxide is also acidifying sea water.

As per IPCC Report (2021) "India will suffer more frequent and intense heat waves, extreme rainfall vents and erratic monsoons, as well as more cyclonic activity among other weather-related calamities, in the coming decades". Again on 18/4/ 2022 IPCC warning came that "India will likely face irreversible impact of climate change, with increasing heat waves, droughts and erratic rainfall events in the coming years if no mitigation measures are put in place". The major indicators of climate change impact are:

- Increase in greenhouse gasses
- Increase in temperature
- Melting of glacier
- Change in sea level
- Loss of arable land
- Drought/ Flood
- Vapor pressure deficit (VPD)
- Soil and water salinity
- Frequency of extreme events etc.

Phenological Response to Weather

Understanding about periodic events in biological life cycles (phenophases) and how these are influenced by seasonal and climatic changes in their environment is very essential before going for any management strategy. Such studies have only recently been considered as an area of climate change impacts research in several crops. The growth of root, shoot, flower and fruit as well as their development is highly influenced directly or indirectly by various environmental conditions. This is important in mango too, as production in recent years is getting adversely impacted by weather variables as both high as well as low temperatures, unseasonal rainfall, floods in some areas, sunshine hours, relative humidity, wind speed etc., These variables potentially affected different phenophases, *viz*. vegetative growth dynamics, fruit bud differentiation, flowering, pests and diseases dynamics, fruit setting and development, maturity, harvesting and eventually the markets and overall economics. Under the influence of climate change, early and delayed flowering is a common feature in mango.

Vegetative Growth Flushes

The pattern of vegetative growth flushes in mango is highly variable based on age, variety, growing region, management practices and prevailing environmental conditions. The vegetative growth requires comparatively higher temperature regimes while in reproductive phase inflorescence emergence starts just after coldest period of the winter in the subtropical region. In general, young plants give continue growth flushes irrespective of their regular or alternate bearing habits while older trees have differential growth flushing response. The bearing trees remain quiescent for several weeks at a time in the same environment. In northern subtropical region, there are three growth flushes *i.e.,* (i) March-April, (ii) June-July and (iii) September-October due to distinct winter and summer periods. Whereas, in southern tropical and eastern region, 2-5 growth flushes are observed due to mild temperature prevails for a longer period of time. On the other hand in western region, first vegetative flush comes in February-March, second in March-April and third in October and November. Under optimum temperature regimes with non-limiting nutrient and water availability, the growth flushes occurs at regular intervals. As far as growth cycle is concerned, tree grown at 20/15 ^{0}C required 20 weeks to complete growth and dormancy cycle while at 30/25 ^{0}C the same cycle is completed in 6 weeks with more leaves per flush (Whiley *et al.*, 1989).

Mango trees are considerably hardy plant species that can tolerate temperature as high as 48 ^{0}C and as low as 0^{0}C, which is not normally occurs. Occurrence of frost is a limiting factor for mango cultivation. The trees are adversely affected by frost and long cold spell during winter months owing to drying of upper leaves and branches. Besides air temperature, soil temperature also play vital role in vegetative and reproductive growth of mango. The number and size of leaves during growth flushes are also influenced by temperature. Occurrence of frost (-0.2^{0}C to -1.2^{0}C on 9th and 10th January, 2013) had adversely affected shoot apices including shoot portion of about 5-6 inches. Young mango plantations suffer more due to low temperature. Biennial bearing varieties, Langra, Chausa, Fazali etc put forth very little growth during fruiting period and also after harvest.

Intermittent rainfall can also increase random flushing. Drought and flood both have negative effect on the vegetative development of mango trees as they reduce tree growth. The unseasonal rains especially at the time of fruit bud differentiation could transform the already differentiated shoot meristems in favor of reproductive ones to vegetative ones in the tropics while, low temperature in the subtropics; occurrence of frost (-0.2^{0}C to -1.2^{0}C on 9th and 10th January, 2013) as it happened at Lucknow (Uttar Pradesh) had killed a portion of the shoots and triggered the emergence of new vegetative flushes from the lateral buds subtending the dead portion of shoots, thus contributing to transgression of tree physiology from an expected reproductive one predominantly. The occurrence of hailstorm on 18th and 19th January 2013 had further compounded the problem. The expected change of climatic variables and their impact on vegetative growth of mango are given in Table 1.

Table 1: Expected change of climatic variables and their impact on vegetative growth of mango

Climatic variables	Expected change	Impact
Temperature	Increase	Rapid and continue growth rhythm
Carbon dioxide	Increase	More dry matter
Drought	Increase	Poor growth
Flood	Increase/ decrease	Poor growth and death
Cold	Increase	Leaf and shoot drying

Flowering and Fruit Setting

Phenological pattern of vegetative and reproductive phases of mango are likely to be influenced by climatic aberrations leading to adverse impact on productivity and quality of mango. Mango is a day neutral plant whose flowering is unaffected by photoperiod but floral induction requires exposure of mature leaves to light and higher level of light intensity could have positive effect on mango flowering. The flowering in mango both in regular and biennial bearing varieties is the result of hormonal balances within the plant system under the strong influence of environmental factors especially the temperature. Though, other environmental factors could also impact mango phenophases, which needs variety and region-specific database. Among other environmental factors, viz light, water, sunshine hours, wind, salinity though exert considerable influence on productivity have not been systematically investigated across agro-ecologies and varieties about their specified roles at threshold level. Plant water stress has been presumed to provide the stimulus for flowering (Singh, 1960). Cessation of shoot growth favours the flower bud formation in most of the mango varieties. Occurrence of mild drought has positive effect on floral induction by promoting early growth cessation and vegetative rest required for floral induction. In subtropics, stress and temperature provides the strong environmental stimulus for flower induction, though the threshold being cultivar-specific. Contrary to this, dry period preceding flowering appears necessary for satisfactory flowering in the tropics. Development of floral buds is strongly influenced by cool night temperature (15 ^{0}C) followed by <20 ^{0}C day temperature (Ou, 1982). Under controlled condition, Whiley *et al.* (1988, 1989, and 1991) described vegetative induction at 30 ^{0}C day and 25 ^{0}C night temperatures and floral induction at 15 ^{0}C day and 10 ^{0}C night temperatures in mono and polyembryonic cultivars. The high temperature during flowering may also lead to flower drying.

In India, flowering in mango commences first in early December in southern part of Kerala and Tamil Nadu and by middle of December in coastal region of Karnataka, Maharashtra, Tamil Nadu and southern Andhra Pradesh. Thereafter, flowering starts in south-central part of India by the first week of January in Andhra Pradesh, Maharashtra, all over odisha, southern Madhya Pradesh and coastal Gujarat. By the first week of February mango trees flower all over eastern India, Madhya Pradesh and Gujarat and by mid February onwards flowering occurs in western Uttar Pradesh, Haryana and Punjab. In subtropics also, flowering occurs in between 15th February to first week of March based on temperature regime. In hilly tracts of Assam, Meghalaya, Tripura, southern part of Himachal Pradesh and Jammu and Kashmir particularly in valley region, flowering begins in the month of March.

Fruit setting is the transformation of ovary to a rapidly growing young fruit which is initiated after successful pollination and fertilization or, in its absence, through parthenocarpy. The proper fruit setting is prerequisite for better productivity in any crop, if other conditions remain normal. The mango is very sensitive crop to weather conditions prevailing during flowering and fruit setting stage. The low fruit set in mango is not only because of self incompatibility but numerous factors encountered in pollination and fertilization. Unfavourable weather conditions, viz. rains, high humidity, high or low temperature regime, poor sunshine and cloudy weather, high wind velocity, excessive moisture stress, flood and water logging etc during flowering and subsequent fruit setting period may result in failure of pollination, poor pollen germination and pollen tube growth and ovule abortion. Drought and higher vapour pressure deficit have negative effect on fruit setting and retention. The influence of climatic variables on flowering and fruit set in mango is given in Table 2.

Table 2: Expected change of climatic variables and their impact on flowering and fruit set in mango.

Climatic variable	Expected change	Impact
Temperature	Increase	Negative effect on floral induction but mild increase has positive effect on pollen viability and fruit setting
Light	Increase	Positive
Vapor pressure deficit	Increase	Negative
Drought	Increase/ decrease	Positive effect on floral induction but negative on fruit setting and retention

Fruit Growth and Development

Besides other factors, viz. good plant health, optimum moisture regime, timely flowering and fruit setting etc., the fruit growth and development of mango is highly influenced by prevailing weather parameters. Rise in temperature and high and low precipitation along with uneven distribution is predicted parameters of climate change. Moisture deficit during growth of fruits has negative impact on fruit size and overall productivity of mango. Increased precipitation during fruit growth and development may not only delay the number of days taken to maturity but also poor fruit appearance. The effect of drought on fruit quality is positive and negative and well known in non-irrigated mango orchards. Drought reduces fruit size (Spreer *et al*., 2009) and increases fruit quality particularly dry matter content and sugar concentration (Lechaudel *et al*., 2005).

Mild increase in temperature would have positive effect on mango fruit growth but extreme high temperature and low atmospheric humidity coupled with high wind velocity leads to dropping of developing fruits. The heat unit required for mango fruit maturity is cultivar and region specific. For calculation of heat unit summation, different baseline temperatures have been given by different workers ranging from 10-17.9 °C. The base line temperature for mango given by Whiley *et al*. (1991) is 15 °C while Oppenheimer (1947) has given 17.9 °C. The higher temperature may be beneficial in terms of fruit quality, as stress induced by higher temperature helps in synthesis of secondary metabolites. Higher temperature induces physiological changes within the mango fruit and build up of high temperature inside fruits may lead to breakdown of tissues, causing spongy tissue problem as in Alphonso variety of mango. The late harvestting of mango fruits, mainly Dashehari variety at higher temperature may cause of jelly seed formation. Under changing climatic pattern of higher temperature regime, this problem may be seen in other varieties also.

The availability of proper light intensity is very important in fruit development particularly skin colour development of coloured varieties. Moreover, higher light intensity and carbon dioxide concentration enhances photosynthesis could have a positive effect on fruit quality by accumulating higher fruit dry matter (Urban *et al*. 2003). The expected impact of climatic variables on fruit growth and quality is given in table 3.

Table 3. Expected impact of climatic variables on fruit growth and quality of mango

Climatic variable	Expected change	Impact
Temperature	Increase	Positive effect on rapid fruit growth but positive/ negative effect on quality
Light	Increase	Positive effect on fruit size, colour and quality of fruits
Carbon dioxide	Increase	Positive
Drought	Increase	Positive effect on quality but negative effect on size

Impact on Pest and Diseases

Understanding pest dynamics in the scenario of climate change is a very complex phenomenon. A large number of pest and diseases are reported from different mango growing areas owing to serious loss of mango production. However, the expected impact of climate change on pest and diseases are:

- Changes in spectrum of insect pests, diseases, weeds and natural enemies.
- Increased risk of invasion by exotic and migrant pests and pathogens,
- Noxious abundances of several species also in higher altitudes.
- Altered development, morphology and reproduction.
- Change in interspecific interactions.
- Loss of resistance in cultivars containing temperature-sensitive genes.
- Emergence of new bio types due to climate change.

In mango, higher rainfall after fruit setting coupled with high temperature and relative humidity may be favourable for pest and diseases. Also, unseasonal rains encourage pest and disease problem which lower fruit yield. Rains, heavy dew or foggy weather encourages insect-pest and diseases and also hampers activity of pollinators. Under humid and high rainfall areas anthracnose may be serious problem. Rains at maturity of fruits affected more with anthracnose and these fruits are less attractive because of blackening of peel. In subtropics, besides problem of mango hopper, mealy bug, malformation, anthracnose, powdery mildew etc., the problem of mango thrips, shoot gall psylla, blossom midge, jelly seed formation, wilting of trees etc. becoming serious problem now these days. Some pest and disorders of mango are given in Fig. 1.

Blossom Midge

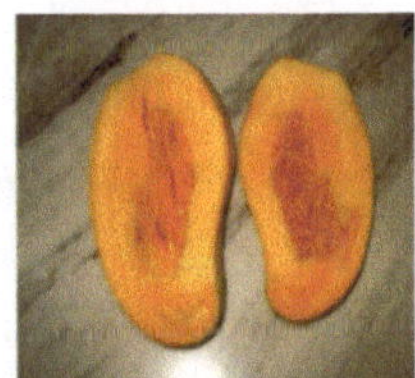

Jelly Seed in Amrapali

Shooting gall psylla

Spongy tissue in Mallika

Fig. 1: Some pest and physiological disorders of mango

Multiple Impact of Climate Disruption

The mango trees have wider edapho-climatic adoptability but various phenophases are very sensitive to climatic aberrations coupled with problem of pest and diseases which have negative impact on fruit yield and quality parameters. The impact of weather parameters on plant growth, flowering, fruit set, fruit development, yield and quality, pest and disease occurrence etc. can be very well visualized, as given in fig. 2. However, intensity of impact may vary based on growing region, variety, soil fertility status and moisture regime, management practices etc.

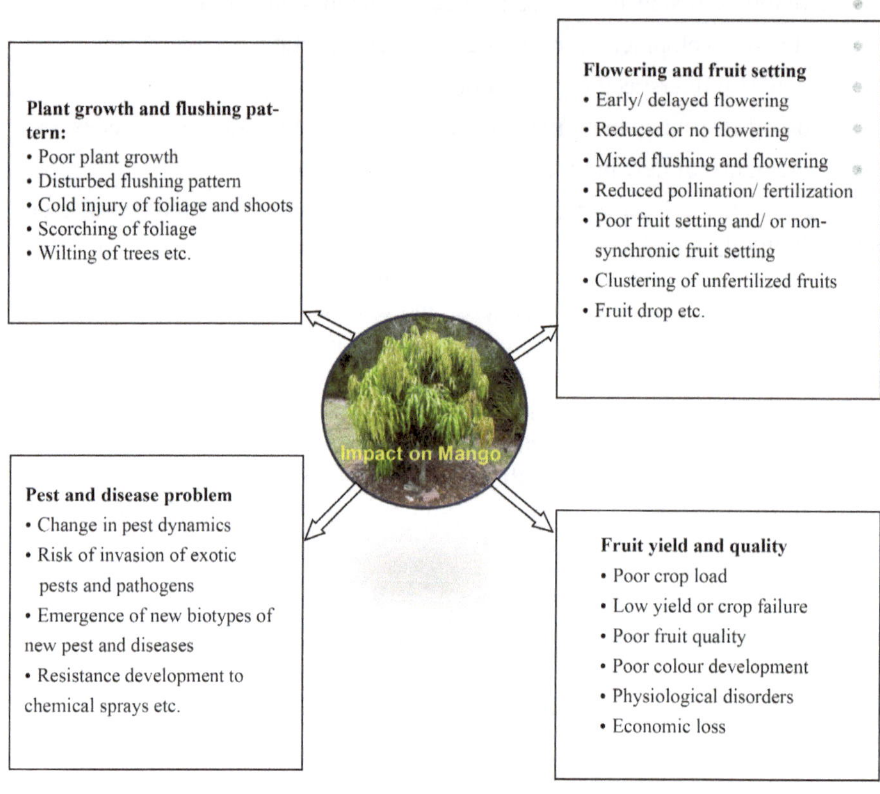

Fig. 2: Multiple impact of climate change on mango in Indian subtropics

Impact of Weather Aberrations a Case Study

As already stated that climate impact on different crops and commodity is a long range change in pattern of weather parameters, thereby before taking some observation on mango cultivar Dashehari; the annual precipitation and mean maximum and mean minimum temperature of Lucknow (Uttar Pradesh) located in Indian subtropics was analyzed over 21 years. The annual rainfall at Experimental Farm, ICAR-CISH, Lucknow (Uttar Pradesh) depicted in Fig. 3 shows wide range of variations from as high as 1843.84 mm in 2008-09 to as low as 553.70 mm in 2015-16. Such erratic pattern in annual rainfall has direct influence on soil moisture regime and relative humidity. Moreover, average rainfall over 21 years was 956.41 mm per annum which is good enough for mango growing areas. Also, in irrigated mango orchards, the productivity was not affected, even if rainfall was less than 700 mm per annum.

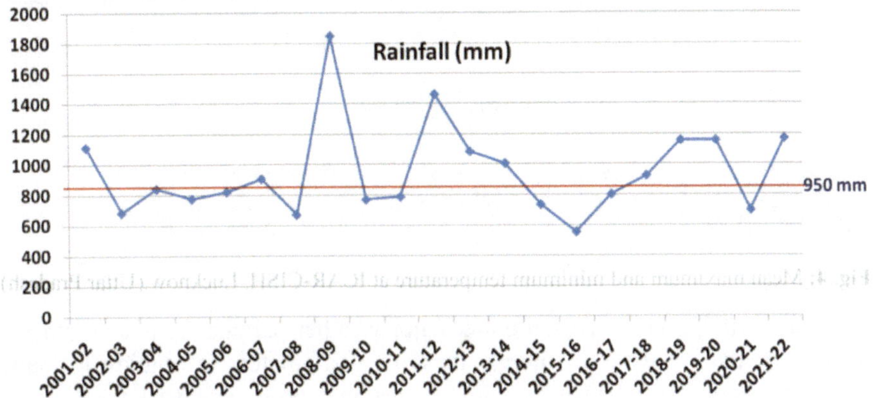

Fig. 3: Annual rainfall at ICAR-CISH, Lucknow (Uttar Pradesh)

Similarly, mean maximum and mean minimum temperature was also recorded during the same period and depicted in Fig. 4. There was a wide variation of about 15 °C between mean maximum and mean minimum temperature with average maximum temperature was 31.98 °C and minimum was 17.95 °C over 21 years (2001-02 to 2021-22). In mean maximum temperature, the range of variation was highest in 2005-06 (33.78 °C) and lowest in 2021-22 (30.88 °C) while in mean minimum temperature, highest temperature was in 2004-05 (19.49 °C) and lowest was in 2018-19 (16.25 °C) on monthly basis. The standard deviation was 0.86 °C for mean maximum temperature and 1.0 °C for mean minimum temperature. The analysis of variance over 21 years of meteorological data reveals that both mean maximum (-0.34%) and mean minimum (0.79%) temperatures were slightly decreased in Indian subtropics.

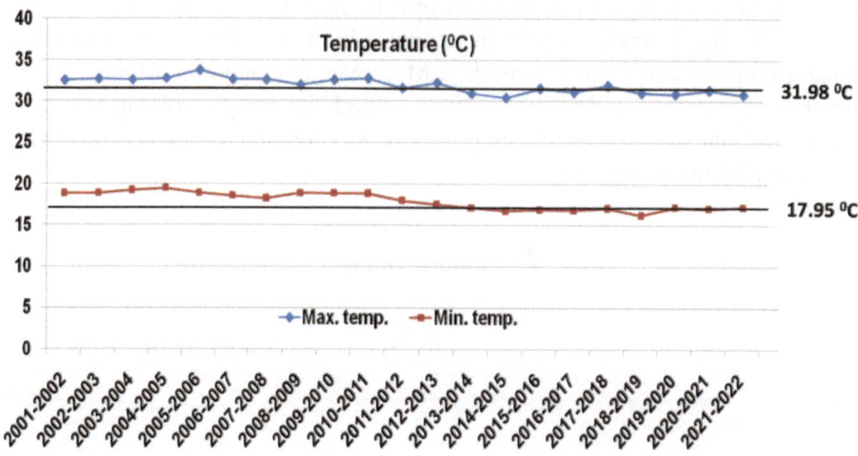

Fig. 4: Mean maximum and minimum temperature at ICAR-CISH, Lucknow (Uttar Pradesh)

Though, rainfall and temperature are major weather parameters but month wise long range data analysis of merely these two parameters in relation to mango production cannot be taken as indicator. Rather, meteorological week wise analysis of weather parameters at particular phenophase of a mango variety is more realistic to assess the flowering, fruit set, fruit growth and development as well as yield and yield attributing characters. The observations on mango flowering and fruit setting at Experimental Farm, ICAR-Central Institute for Subtropical Horticulture, Lucknow indicate that there was poor fruit set during 2022 and thereby poor fruit yield of mango cultivar Dashehari was recorded. This was mainly due to increase in temperature during 2nd fortnight of March, 2022. The weekly weather data on maximum and minimum temperature from December to March, 2021 and 2022 depicted in Fig. 5. During 2022, the day temperature was more than 30 ^{0}C from 2nd meteorological week and reached at peak of about 38 ^{0}C in fourth meteorological week while night was very cool and temperature gone below base line temperature of 15 ^{0}C required for mango flowering and fruit setting. Such an abrupt rise in temperature resulted flower drying, poor fruit setting, in spite of good flowering in mango cultivar Dashehari. Of course, there were varietal differences on flowering pattern in the same edaphoclimatic conditions (Fig. 6). Because of high temperature, the pollination and subsequent fertilization activities were also hampered, owing to poor fruit set. Also, fruit setting was observed in staggered manner, as there was normal fruit setting in case of early flowering panicles on the same tree but later failed. Therefore, there were marked differences among the size of fruits of mango cultivar Dashehari due to staggered fruit setting (Fig 7).

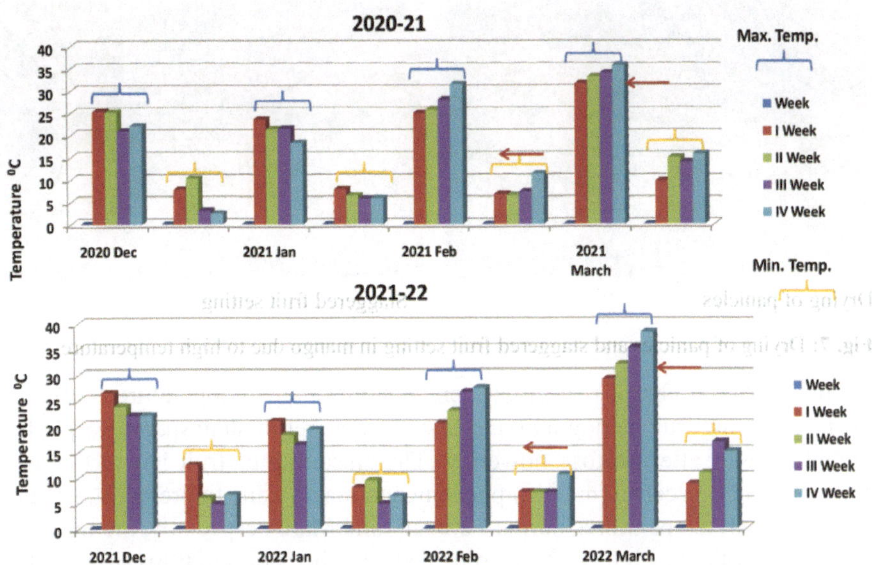

Fig. 5: Temperature regime from December to March (2020-21 and 2021-22)

Fig. 6: Varietal difference on flowering behavior of Dashehari (Left) and Langra (Right)

Drying of panicles Staggered fruit setting

Fig. 7: Drying of panicles and staggered fruit setting in mango due to high temperature

In subtropics, an abrupt rise in day temperature or low temperature (<10^{0}C) in the night during fruit setting may increase the number of stenospermocarpic fruit (nubbins) even after profuse flowering. The clustering of fruit lets at the distal end of the inflorescence (nubbin) phenomenon may be due to prevalence of high fluctuations in temperatures during February and March especially, the minimum temperature of cold winters which coincided with the period of flowering and fruit development probably impacted by hormonal interpolations. Such fruits invariably with aborted embryos prematurely drop off and are not carried to maturity (Fig. 8.)

Fig. 8: Clustering (Nubbins) of fruits in Dashehari variety of mango

Strategies

Since several thousand years, people and societies have adjusted themselves to climate change particularly draught and floods. The climate change (drought in particular) has been partly responsible for the rise and fall of civilizations in many countries. The main goal of adaptation is to reduce the risks from the harmful effects of climate change and to harness potential beneficial opportunities associated with climate change. Moreover, the mitigation strategy of reducing climate change impact involves reducing the flow of heat-trapping greenhouse gases into the atmosphere, either by reducing sources of these gases from burning of fossil fuels for electricity, heat, or transport or enhancing the "sinks" that accumulate and store these gases in oceans, forests, and soil. The human life is so adjusted to a stable climate that any deviation will adversely affect their performance. Similarly, for optimum output of any crop, threshold level of five primarily inputs i.e. space, water, energy, light and nutrients are essentially required. Stress of these inputs at any stage may adversely affect the crop performance. Therefore, reducing emissions of green house gases, effective use of renewable source of energy and improving soil cover by perennial trees and shrubs should be given major emphasis. *In-situ* rain water harvesting should be done to reduce run-off and soil loss. Use of different land configurations of rain water harvesting and provision of trenching are also useful measures to improve soil profile moisture in the orchard. *Ex-situ* water harvesting and storage should be done to recharge ground water table and use of storage water for irrigation under rainfed mango orchard. It was observed at Experimental Farm, ICAR-CISH, Locknow that mango fruit development stages were vulnerable under high pan evaporation regimes (>6 mm / day to 10 mm/day) particularly during the last week of March and onwards. High pan evaporation values indicated higher rate of evapo-transpiration resulting soil moisture stress, leading to abiotic stress complexes barring water and nutrient flow through xylem. Thus protective irrigations are required as an intervention strategy so as to maintain optimum soil moisture and leaf water potential in order to sustain fruit set, growth and development and minimizing pre-harvest fruit drop unless otherwise this period is intervened by unseasonal rains.

Globally, 80,000 plants are reported to be of possible use, about 30,000 plants are found edible in nature and approximately 7,000 plants have been cultivated by mankind at one time or another, of which 158 plants are cultivated by man at some point of time. Among these, 30 crops provide world's food and only 10 crops supply 75% of the world's food budget. Out of these, only three crops i.e. rice, wheat and maize provide 60% of the world's food requirement. This dangerously narrow level of food base prompts to widen the base of grains, pulses, oil seeds, vegetables, fruits, spices, tuber crops, plantation crops, mushrooms, medicinal and aromatic plants etc. In research agenda, the focus must be given on climate resilient crops. As far as mango is concerned, there is a lot of genetic diversity exists in nature and also *ex-situ* collections are maintained in different Research and Developmental organizations in the country are praiseworthy. In National Filed Repository of ICAR-Central Institute for Subtropical Horticulture, Lucknow; 775 germplasm including 143 of North, 187 of South, 78 of East and

37 of west Indian types; besides 18 exotics, 38 superior seedlings, 30 hybrids, 51 Dashehari clones, 25 farmers variety and 167 others are conserved and maintained. This huge genetic diversity needs evaluation from climate resilience point of view and promising types can be utilized directly for commercial cultivation if possess desired attributes or may be utilized as parent material for further breeding programme. Some of the exotic colored varieties viz., Tommy Atkins and Kent showing resilience to weather dynamics by such features as high percentage of hermaphrodite flowers, fruit set per panicle, fruit retention to final harvest, yield potential even under adverse weather dynamics as is being observed under the Konkan conditions. The in-depth studies are required to understand the association of genes responsible for color trait with the climate resilience trait(s). Also, mango varieties exhibit specific climatic requirements for proper vegetative growth, flowering and fruit development. The commercial varieties of a region behave differently when grown in other climatic zones of the country. North Indian variety Chausa does not flower and fruiting in Western Ghats. Alphanso, the commercial variety of Western Ghats, fails to perform under North Indian conditions. Moreover, varieties like Langra of North India and Banglora and Neelum of South India have much wider adaptability.

For adaptation and mitigation of adverse impact of weather aberrations, following strategies may be taken into consideration;

Identification of Varieties

- High temperature tolerant
- Drought tolerant
- Induction of flowers at temperature fluctuation regime

Use of Rootstocks

- Tolerant to moisture stress and salinity
- Impart dwarf stature and strong root system

Biotechnological Approaches

- Development of molecular markers
- Somatic hybridization
- Somaclonal variations
- Introgression of stress tolerance associated genes

Orchard Management

- Use of windbreaks
- *In-situ* grafting
- Maintenance of proper soil moisture regime

- Use of mulching
- Management of plant canopy architecture
- Proper nutrition & plant protection
- Induction of flowering by synchronizing shoots initiation through tip pruning coupled with use of KNO_3 and exposure to stress.
- Exogenous application of precursor sucrose (3%) for the synthesis of heat shock proteins, whenever the maximum temperature rises to 44-45 ^{0}C (Reddy and Singh, 2011).

Weather Forecasting

- Precise weather forecasting system
- Quick dissemination of weather based advisories.

Positive Impact of Climate Change

Often climate change is perceived as negative phenomenon but positive influence must be harnessed in right perspective. There are some areas of positive impact of climate change as mentioned below;

- There is a new opportunity of mango across the Globe along transitional temperate region.
- Rising temperature in Sub-mountain region offers new scope for mango cultivation in foothills/ valley areas of Himachal Pradesh, Uttarakhand and Jammu and Kashmir.
- Under projected climatic change scenario, off-season production and extended market availability may be possible.
- Increase in carbon dioxide, will influence better photosynthetic assimilation; needs to be harnessed.

Conclusion

Thus, following conclusions can be drawn for further course of action plan:

- Region specific long term climatic data base in relation to commercial mango cultivars are required to be analyzed in a comprehensive manner to draw valid conclusion on impact of climate change.
- Though, mango being woody perennial tree has wide range of edaphoclimatic adoptability but optimal threshold values of climatic variables for different cultivars are to be established.
- Interactions between climatic variables and mango phenological stages must be analyzed systematically.
- Besides climatic variables; cultivar, rootstock used, edaphic conditions of growing region and management largely influence growth, flowering,

fruit set, productivity and quality of mango. It may be also taken into consideration while assessing impact of climate change.

- The available genetic resources of mango must be utilized as a long term breeding strategy to develop cultivars that can withstand against adverse impact of climate change.
- Often, mango plants are more prone to adverse weather aberrations, where intensive cultural operations and over use of chemicals are practiced in mango cultivation.
- Also, positive impact of climate change must be harnessed in mango orcharding.

References

Kahlon, G.S. and Dhillon, W.S. 2011. Impact of climate change mitigation strategies for subtropical fruit crops. In: Impact of Climate Change on Fruit Crops (Dhillon, W.S. and Aulakh, P.S. Eds.), Narendra Publishing House, New Delhi, pp. 65-70.

Lechaudel, M., Joas, J., Caro, Y., Genard, M., and Jannoyer, M. (2005). Leaf: fruit ratio and irrigation supply affect seasonal changes in minerals, organic acids and sugars of mango fruit. Journal of the Science of Food and Agriculture, 85(2): 251-60.

Oppenheimer, C. 1947. The acclimatization of new tropical and sub-tropical fruits in Palestine. Bulletin No. 44, Jewis Agency for Palestine. Agricultural Research Station, Rehovot.

Ou, S.K. 1982. Temperature effect on differential shoot development of mango during flowering period. Journal of Agriculture Research in China, 29: 301-308

Reddy, Y.N. and Singh Omveer 2011. Role of heat shock proteins in the adaptability of plants to higher temperatures and alleviation of heat shock in mango. In: Impact of Climate Change on Fruit Crops (Dhillon W.S. and Aulakh P.S. (Eds.)), Pub. Narendra Publishing House, New Delhi, pp. 57-64.

Singh, L.B. 1960. The Mango: Botany, Cultivation and Utilization. Leonard Hill, London.

Urban, L., Le Roux, X., Sinoquet, H., Jaffuel, S. and Jannoyer, M. (2003). A biochemical model of photosynthesis for mango leaves: evidence for an effect of the fruit on photosynthetic capacity of nearby leaves. Tree Physiology 23:289-300.

Venkateswarlu, B. and Shanker, A.K. 2012. Dryland agriculture: bringing resilience to crop production under changing climate. In: Crop Stress and its Management: Perspective and Strategies, Springer, Netherlands, pp. 19-44.

Whiley A.W., Saranah, J.B., Rasmussen, T. S., Winston, E.C. and Wolstenholme, B.N. (1988). Effect of temperature on 10 mango cultivars with relevance to production in Australia. Proceeding of 4th Australian Conference on Tree and Nut Crops. ACOTANC, Lismore, pp. 176-85.

Whiley AW, Rasmussen TS, Saranah JB and Wolstenholme BN 1989. Effect of temperature on growth, dry matter production and starch accumulation in ten mango (Mangifera indicam L.) cultivars. Journal of Horticultural Sciences and Biotechnology, 64: 753-765.

Wihley AW, Rasmussen TS, Wolstenholme BN, Saranah JB and Cull B W 1991. Interpretation of growth responses of some mango cultivars grown under controlled temperature. Acta Horticultural, 291: 22-23.

6

Research and Development in Arid and Semi-arid Fruit Crops

A.K. Singh, D.S. Mishra, L.P. Yadav, Gangadhara K. and Jagdish Rane

Central Horticulture Experiment Station (ICAR-CIAH), Godhra, Gujarat

A large number of arid and semi-arid fruit crops are used by the local inhabitants. These fruits are better known as protective food as cultivation is restricted and are grown mainly as wild. Being tolerant to biotic and abiotic stresses, these fruit crops are suitable for growing in drought prone areas. Semi-arid fruits are the oldest fruit tree crops with wide distribution, reflecting their adaptation to a wide range of edapho-climatic conditions. Development of viable agro-techniques, propagation methods, canopy management, and biotic and abiotic stresses management are equally important for improving productivity and quality of arid and semi-arid fruit crops. Bael, mahua, chironji, jamun, tamarind, khirni, wood apple, custard apple, etc. are climate smart since they survive in harsh agroclimatic conditions, and can be established in degraded lands.

National Scenario

The vast land resources available in arid and semi-arid regions of country offer great deal of opportunities for cultivation of variety of these crops. In these areas, wherever assured irrigation facilities are available, fruits like date palm, pomegranate, fig, guava, acid lime, sweet oranges and Kinnow mandarin are grown. However, under rainfed conditions aonla, ber, Lasora, mulberry, ber,bael, jamun, khirni, aonla and karonda are grown without much difficulty (Chundawat, 1990; Singh, 2019; Singh and Singh 2019). Since last three decades, considerable area has come up under fruit crops like aonla, ber, pomegranate, custard apple, bael, fig, date palm, tamarind etc. in different tropical and sub-tropical fruit growing regions of country (Saroj, 2018 and Singh *et al.* 2021) (Table 1).

Table 1: Area and production of arid and semi-arid fruit crops in India

Crop	1993-94		2017-18		2030 (Estimated)	
	Area(ha)	**Production (tones)**	**Area(ha)**	**Production (tones)**	**Area(ha)**	**Production (tones)**
Aonla	26,000	2,86,000	93,000	10,75,000	1,00,000	13,00,000
Ber	41,256	3,30,048	50,000	5,13,000	76,500	8,03,250
Custard apple	-	-	46,000	4,01,000	60,000	6,00,000
Date palm	5,000	41,000	25,000	2,00,000	40,000	3,75,000
Pomegranate	4,500	45,000	2,34,000	28,45,000	3,50,000	50,00,000
Sapota	49,000	6,44,000	97,000	11,76,000	1,10,000	12,80,000
Tamarind	10,500	52,500	48,000	2,01,000	62,400	3,12,000
Others (annona, fig phalsa etc)	6,600	30,450	15,700	1,31,700	35,000	3,50,000
Total	1,42,856	14,28,998	6,08,700	65,42,700	8,33,900	100,20,250

Source: NHB, 2018

The area and production of date palm has increased swiftly due to introduction of new varieties, easy availability of tissue cultured plants and drip irrigation and fertigation. However, area and production in pomegranate has been increased manifold due to high export potential, development of new varieties and standardization advanced production technologies. Maharashtra, major pomegranate producing state in India, is considered as the "pomegranate basket of India". Maharashtra alone consists of more than 70% of the total area followed by Karnataka and Andhra Pradesh. By 2025, area, production and export of pomegranate are expected to reach 7.5 lakh ha, 114 lakh tonnes and 83,800 tonnes, respectively (Pal *et al*., 2014). Similarly, aonla is grown on 93,000 ha area with a total production of 10,75,000 tones. An increase in area and production has become possible due to research and developmental efforts made by the different organizations. Simultaneously, regular plantation of minor fruit crops like jamun, bael, chironji, karonda etc as commercial orchards is now coming up in several parts of country especially in Gujarat, Rajasthan, Uttar Pradesh, Haryana, Punjab and Maharashtra (Singh *et al*., 2020).

Research Advances

The journey of research and development in arid and semi-arid fruit crops of tropics and subtropics began with the initiation of *adhoc* scheme in which research on some selected crops began which was financed by AP Cess fund of ICAR in 1976. Later, this scheme was merged during 6th Five-Year Plan to form the Cell III of the All India Coordinated Fruit Improvement Project (AICFIP). However, during 7th Five-Year Plan, Cell III of AICFIP was again restructured to form All India Coordinated Research Project on Arid Zone Fruits. Currently, this project is having 18 centres running in eleven states of country, viz. Rajasthan (04), Maharashtra (04), Gujarat (02), Uttar Pradesh (02), Tamil Nadu (01), Karnataka (01), Andhra Pradesh (01), Punjab (01), Haryana (01) and Madhya Pradesh (01). The main objective of AICRP on Arid Zone Fruits is to develop suitable and sustainable technologies for growing fruit crops like ber, date palm, aonla, pomegranate, fig, custard apple, bael, tamarind and jamun. During 7th Five-Year Plan, Planning Commission of India approved the establishment of National Research Centre on Arid Horticulture (NRCAH) at Bikaner which came into existence on April 01, 1993 and then upgraded to Central Institute for Arid Horticulture (CIAH) on September 27, 2000. In the same year on October 01, 2000, Central Horticultural Experiment Station, Vejalpur, Panchmahal (Godhra), working on semi-arid fruits, was merged with CIAH as its Regional Station. However, this journey of research and development of arid and semi-arid fruit crops in India was further fuelled the book Arid Fruit Culture (Chundawat, 1990) which has paved the way of research in minor arid and semi-arid fruit crops.

Diversity and Conservation

India, the centre of origin for many tropical and subtropical fruit tree species. Most of them are not commercially cultivated but they are significant source of livelihood support for many local communities. The local inhabitants of Western Ghats, Maharashtra, Gujarat, Rajasthan and North Eastern States of India were traditionally reliant on local fruits like tamarind, bael, aonla, jamun, ber, chironji, custard apple etc. Due to unsustainable market pressures and rapid urbanization, majority of these species have come to near extinction. Therefore, conservation of genetic resources of arid fruit and semi-arid through holistic approach is required which includes both *in-situ* and *ex-situ* conservation strategies (Singh *et al.*, 2020). The diversity of some of underutilized fruits is well studied while for other underutilized fruits relatively less attention has been given so far. Therefore, crop-specific surveys in targeted diversity rich areas were undertaken in arid and semi-arid regions of Gujarat, Madhya Pradesh, Uttar Pradesh, Chhattishgarh, Haryana, Punjab, etc. A large number of germplasm of semi-arid fruits were collected for systematic evaluation, characterization and conservation of indigenous germplasm at CHES, Godhra (Singh *et al.*, 2020). The present status of underutilized fruit crops conservation in India is 1,717 at ICAR-NBPGR and its collaborative centres (10 regional stations) which is the main Indian statutory body responsible for plant genetic resource collection and conservation (Table 1), whereas the major research institutions like ICAR-CIAH, working on arid and semi-arid fruit crops are maintaining 1,127 accessions in their field gene banks (Meena *et al.*, 2020) (Table 2). Furthermore, 357 accessions of eleven underutilized fruit species are being cryo-preserved in gene banks (Malik *et al.*, 2010). Germplasm of various underutilized fruit crops were evaluated for development of varieties at CHES, Godhra, on the basis of desirable horticultural traits. Out of them, Goma Yashi (bael), Goma Priyanka (jamun), and tamarind (Goma Prateek) have been become popular at national level as evidenced by commercial scale plantation at the farmers' fields (Singh *et al.* 2018c; Singh *et al.* 2010a). A wide range of variability with regard to yield, qualitative and quantitative character in different underutilized fruit crops, viz. jamun (Mishra *et al.,* 2014, Singh *et al.*, 2010, 2011, 2018,2017,2020, 2022), bael (Singh *et al.* 2015, 2014, 2016b 2019b, 2019c and Parthasarathy *et al.*, 2021), aonla (Singh *et al.,*2019d), karonda (Singh *et al.* 2014),wood apple (Singh *et al.,* 2016c) khirni (Singh *et al.* 2016b), tamarind (Singh *et al.* 2006, Sharma *et al.* 2015, Singh *et al.* 2021), custard apple (Yadav *et al.* 2017 and 2018), mahua (Dhakar *et al.* 2015, Singh and Singh 2016), wild noni (Arya *et al.* 2014, Patel *et al.* 2014, Rathod *et al.* 2016, Singh *et al.* 2012, 2016a), chironji (Singh *et al.* 2019e, Malik *et al.,* 2010), phalsa (Malik *et al.,* 2010) and manila tamarind (Awasthi and Saroj 2006, Singh *et al.*, 2020) have been reported. At present, ICAR-CIAH (Table 2) and its regional Centre CHES, Godhra are maintaining a large number of diverse germplasm of underutilized semi-arid fruits in field repository (Table 2, 3 and 4).

Table 2: Germplasm conservation of semi-arid fruits at national field repository of CIAH, Bikaner

Crop	Scientific name	Number	Crop	Scientific name	Number
Bael	*Aegle marmelos*	21	Manila tamarind	*Pythocelobium dulcae*	03
Ber	*Ziziphus mauritiana*	318	Jharber	*Ziziphus rotundifolia*	22
Cactus pear	*Opuntia ficus-indica*	24	Jamun	*Syzygigium cuminii*	2
Phalsa	*Grewia subencequalis*	05	Lasora	*Cordya myxa*	15
Pomegranate	*Punica granatum*	154	Kair	*Capparis decidua*	06
Fig	*Ficus carica*	02	Karonda	*Carissa carandus*	05
Mulberry	*Morus* spp.	15	Wood apple	*Feronia limonia*	03

Table 3: Germplasm conservation of semi-arid fruits at national field repository of CHES, Godhra

Crop	Scientific name	No.	Crop	Scientific name	Number
Aonla (Indian gooseberry)	*Emblica officinalis G.*	26	Manila tamarind	*Pythocelubium dulcae*	25
Bael	*Aegle marmelos*	217	Jamun	*Syzygium cuminii*	72
Capegooseberry	*Physalis peruviana*	06	Palmyra palm	*Borassus flabellifer*	2
Phalsa	*Grewia subanaequalis*	25	Karonda	*Carissa carandus*	40
Badhal	*Artocarpus lacucha*	04	Fig	*Ficus carica*	07
Mulberry	*Morus* spp.	05	Chironji	*Buchanania lanzan*	30
Mahua	*Bassia latifolia*	30	Wood Apple	*Feronia limonia*	65
Tamarind	*Tamarindus indica*	25	Khirni	*Monilkara hexendra*	30
Custard apple	*Annona squamosa*	70	Lasoda	*Cordia myxa*	04
Mulberry	*Morus* spp.	03	Wood apple	*Feronia limonia*	58

Singh *et al.*, 2020

Table 4: Status of collection and conservation of underutilized fruits at ICAR-NBPGR and its collaborative centres and other institutions.

Crop	No. of accessions maintained			
	NBPGR, New Delhi	RS, HAU, Bawal	PAU, Abohar	CAZRI, Jodhpur
Ber	487	47	34	40
Aonla	159	6	-	-
Bael	57	10	-	-
Karonda	50	4	-	13
Timroo	24	-	-	-
Manila Tamarind	24	-	-	-
Mahua	153	-	-	-
Kair	-	22	-	20
Khirni	74	-	-	-
Phalsa	36	4	-	-
Pilu	207	-	-	-
Jamun	198	-	-	-
Tamarind	248	-	-	-
Total	1717	93	34	73

Meena *et al.*, 2022

Wide range of variability in leaf morphology, flower characters and phenology has been reported in different germplasm of semiarid fruits under rainfed semi-arid condition (Singh *et al.* 2013). Singh *et al.* (2015) observed inter-varietal morphological variability in terms of leaf base margin and apex in bael varieties under central Gujarat conditions. Morphological variations in terms of vivipary, metaxenia and cauliflory in bael germplasm have also been recorded (Singh *et al.*, 2018) under dryland conditions of western India.

Morphological, floral, phenological and pollination behaviour in different germplasm of hot semi-arid fruits have been studied in detail, viz. bael (Singh *et al.* 2011, 2014, 2017, 2018,2019b), jamun (Singh and Singh 2012b, Singh *et al.* 2007a, 2010a, 2011a and 2019d), khirni (Singh and Singh Singh *et al.* 2010, 2018), tamarind (Singh and Singh , 2017), chironji (Singh *et al.* 2006 and 2010), phalsa (Singh *et al.* 2019e), karonda (Singh *et al.* 2014, Singh *et al.,* 2019e), custard apple (Vikas *et al.* 2017), wood apple (Yadav *et al.* 2018), mahua (Singh *et al.* 2005 and 2008) and wild noni (Singh and Singh 2016a and 2014b) under rainfed hot environment of western India.

Varietal Wealth

The underutilized fruit crops grown in hot arid and semi-arid regions particularly western part of country play a critical role in improving nutritional and livelihood security in rural and tribal areas as the availability of fruits is either low or not accessible to them. During past several years, massive efforts have been made by various organizations in the development of improved varieties of underutilized fruit crops. Most of the resultant varieties are developed through clonal selections from locally adapted genotypes which have been proved worthy as they are potential source of stress related genes and have wider adaptability with desirable yield and quality traits under arid and semi-arid environments (Table 5).

Table 5: Improved varieties of semi-arid and arid fruits

Crop	Varieties
Bael	Goma Yashi, Thar Divya, Thar Neelkanth, Thar Srishti, Thar Prakriti, Thar Shivangi, NB-5, NB-7, NB-8 NB-9, NB-10, NB-16, NB-17, CISHB-1, CISHB-2, Pant Aparna, Pant Shivani, Pant Sujata and Pant Urvashi
Aonla	Goma Aishwarya, NA-5, NA-4, NA-6, NA-7, NA-10, Anand-1, Anand-2, Laxmi-52
Jamun	Goma Priyanka, Thar Kranti, Konkan Bahadoli, Jamwant, Paras, Rajamun, Rajendra Jamun-1, Jamwant
Poegranate	Ganesh, Bhagwa, Solapur Lal, Arakta, Mridula, Dholka, Jalore Seedless
Custard apple	Thar Amrit, Balanagar, Mammoth, Red Sitaphal, Yellow Sitaphal, Phule Janki and Sindhan
Mulberry	Thar Lohit, Thar Harit, Saharanpur Local-1, Saharanpur Chak Majra
Karonda	Pant Manohar, Pant Sudarshn, Pant Suverna, Konkan Bold, Thar Kamal
Tamarind	Goma Prateek, Prathisthan, PKM-1, T 263, Urigam, Ajanta, Yogeshwari, DTS 1 and DTS 2, Anant Rudhira
Lasoda	Thar Bold, Paras Gonda, Puskar Local, Maru Samridhi, Karan Lasoda
Khirni	Thar Rituraj
Phalsa	Thar Pragati
Chironj	Thar Priya
Mahua	Thar Madhu, NM-2, NM- 4, NM-7, NM- 9
Manila tamarind	PKM (MT) 1
Fig	Poona Fig, Dianna, Dinkar, Chalisgaon
Wood apple	Thar Gaurav, Thar Prabha

(Pal *et al.*, 2014; Singh *et al.*, 2020; Meena *et al.*, 2022)

Propagation

Looking into the importance of these fruits, the demand of their genuine planting material is increasing day-by-day. To meet the demand, vegetative propagation techniques have been standardized for commercial multiplication (Table 6). The variability has been observed in plants raised through seeds. Except a few plant species, vegetative methods of propagation are used for their multiplication. Propagation through vegetative methods, viz. stem cutting, layering, stooling and grafting have been standardized for many semi-arid fruits (Singh, 2018; Singh *et al.*, 2020). Patel *et al.* (2016) found seed priming treatments for improved seed germination and vigour of seedling in custard apple. Yadav *et al.* (2017) reported that GA_3 and cow urine can enhance seed germination and growth of custard apple seedlings under rainfed semi-arid conditions. Seeds treated with growth regulators (GA_3) enhanced per cent seed germination and growth of seedlings in *Pithecelobium dulce* under hot arid conditions (Singh *et al.,* 2011). In order to optimize the production of semi-arid fruit crops, propagation techniques of jamun, lasoda, khirni, wood apple, manila tamarind, custard apple, mahua, bael, chironji, etc. have been standardized for large scale multiplication of plants (Table 6). For better success and survival of semi-arid fruits, *in-situ* budding and grafting has been found better with vigorous growth of grafted plants under arid and semi-arid conditions (Singh *et al.* 2014e). However, the standardization of rootstocks needs attention to assess vegetative compatibility and vigour, fruiting, fruit quality and usefulness to wastelands (Table 6).

Table 6: Propagation techniques of arid and semi-arid fruit crops

Crop	Period of multiplication	Propagation methods
Aonla	May-June	Patch budding
Ber	May-June	Patch budding
Bael	May-June	Soft wood grafting and patch budding
Chironji	July-August	Soft wood grafting
Jamun	April -May	Soft wood grafting, patch budding
Lasora	June-July	Seed, Patch budding and micro-propagation
Tamarind	July-August	Soft wood grafting and patch budding
Mahua	March-April	Soft wood grafting
Custard apple	April- May	Soft wood grafting
Wood apple	May- June	Soft wood grafting, patch budding
Karonda	June -July	Seeds and cutting
Fig	October-November	Cutting
Morinda (Noni)	July-August	Air layering/seed
Manila tamarind	May-June	Seed and budding
Khirni	March-April	Softwood grafting
Timru	May-June	Seed and softwood grafting
Phalsa	July –August	Seed and cutting
Wild noni	July -August	Layering

Singh *et al.*, 2017,Awasthi *et al.*, 2005, Singh, 2014, 2018, Singh & Singh 2006, 2007, 2014a, 2014b, 2015, 2021b.

Orchard Establishment

Orchard establishment is a long-term investment and requires thorough planning starting like site selection, spacing, selection of species and varieties, quality of planting material, planting design and density. While site selection, important associated factors are local climate, and availability of quality water, availability of labours, proximity to the market, transport and processing units. The orchard should be protected from high wind velocity to create favourable microclimate. Wind breaks consist of planting of trees, shrubs or erection of artificial barriers on the windward side of an orchard for protection from wind. The fruit plants raised in nursery are generally used to establish new orchards. For success in dry lands, plants must have root architecture with a strong tendency to penetrate deep into the soil. Sometime plants raised in nursery beds or containers become soil or pot bound respectively and lost potential tap root during the process of plant lifting. Therefore, *in-situ* technique of orchard establishment is found suitable under arid and semi-arid conditions (Nath *et al.,* 2000). In this method, seeds of rootstock species are sown right there on the field at the recommended spacing during two months of July with the onset of monsoon.

The rootstock plants raised in this manner develop deep tap root system and hence they have more drought resistance. The seedlings are protected till next summer and when they attain pencil size thickness in monsoon, budding with suitable scion cultivars or elite genotype is done. This technique is found most suitable for the crops like ber, jamun, bael, gonad, wood apple, custard apple and aonla (Singh *et al.* 2009; Singh *et al.*, 2020). Among the fruit tree species, two methods of planting (i.e., auger hole and pit methods) were tested using 5 and 10 kg of gypsum in each auger hole and 10 and 20 kg of gypsum in each pit as soil amendments. After seven years *Ziziphus mauritiana, Syzygium cuminii, Emblica officinalis* and *Carissa caranandus* were the successful species for these soils showing good growth and also initiated fruit setting in semi-arid alkali-sodic Soils of Haryana, India (Dagar *et al.,* 2001). Innovative orchard establishment method for minimizing drought and edaphic stresses in fruit crops grown in shallow basaltic soils of semi-arid region has been standardized by the ICAR-National Institute for Abiotic Stress Management, Baramati, Maharashtra (Minhas *et al.,* 2015). In the medium rainfall region of eastern Uttar Pradesh, application of FYM, pond soil, gypsum, and pyrite in sodic soils resulted in better establishment and growth of aonla and bael plants (Sharma *et al.*, 2013).

High-Density Planting

The plant density mainly depends upon the plant type, soil fertility, varieties, growth habit of tree, rootstock used and management practices play important role in deciding optimum spacing in an orchard. Accordingly different systems of planting, *viz.* square, rectangular, traiangular (alternate), quincunx, hexagonal, contour and hedgerow system of planting may be chosen. Generally, adjacent planting is followed in arid regions and in poorly fertile soils. In this system, two plants are placed closely keeping double space between rows. In plains, planting, is generally done in square or rectangular system while on slopy lands, fruit trees are planted on contour terraces, half moon terraces, trenches and bunds, and micro-catchments. On marshy and wet areas mounding and ridge-ditch method of planting have been suggested. The trenches and bunds made across the slope are staggered (Saroj *et al.,* 1994). In a micro-catchment, which may be triangular or rectangular, trees are planted at the lowest point where run-off accumulates. In a micro-catchment, which may be triangular or rectangular, trees are planted at the lowest point where runoff accumulates (Shrma *et al.,* 2013). The planting distance 6m x 6 m or 8m x 8 m for ber cultivation is optimum. Date palm, bael and aonla are planted at 8 x 8 m or 10 x 10 m distance (Table 7).

Table 7: Recommended spacing for different fruit crops of arid and semi-arid regions

Crops	Spacing (m)	Number of plants/ha
Kinnow mandarin	5×5	400
Acid lime	5×5	400
Guava	5×5	400
Pomegranate	4×4	625
Date palm	5×5	400
Phalsa	2×2	2500
Ber	6×6	277
Karonda	4×4	625
Aonla	5x5	400
Fig	6×6	277
Mulberry	6×6	277
Bael	5x5	400
Gonda	6×6	277
Sweet orange	5×5	400

High-density planting studies in pomegranate revealed that the maximum plant height was recorded at 2m x 2 m spacing, whereas plant height, stem girth, average number of fruits, average weight of fruit and yield was obtained under 4.5m x 3.0 m spacing under Rahuri conditions (AICRP, 2013). High density planting is also beneficial in aonla (Singh *et al.,* 2010, Singh *et al.,* 2018) ber, bael, chironji (Singh *et al.,* 2011, 2016) and jamun (Singh *et al.,* 2018) fruit trees to achieve high yield under semi-arid conditions.

The high-density planting of Bhagwa pomegranate at spacing of 5m x 3m (666 plants ha) and 4m x 2.5 m (1000 plants/ha) under drip irrigation and fertigation has been adopted by the farmers of Rajasthan with a fruit yield of 15t/ha and net income of 1.5-2.0 lakh/ha. However, farmers are facing the problem of fruit cracking and nematode infestation in hot arid region of Rajasthan (Saroj, 2018). ICAR-CIAH, RS, CHES, Godhra, has released a semi dwarf bael variety Goma Yashi can be planted at 5m x 5m, accommodated 400 plants/ha which can yield 25t/ha at the age of 8th year of plantation with net return of 1.0 lakh/ha under rainfed and waste land conditions of Gujarat (Singh *et al.*, 2019). Under rainfed semi-arid conditions of Gujarat, aonla cv. NA-7 was planted in double hedge row system of planting by accommodating 260 plants/ha has given the fruit yield of 23t/ha with a net return of 2.43 lakh/ha at 11th year of planting (Saroj, 2018). High density planting systems has been successfully demonstrated for earliness, improved yield, smooth handling and cultural practices using double hedge row system of planting in aonla (Singh *et al.*, 2011). Moreover, by manipulating plant spacing using different planting systems like rectangular planting in hedge row, double hedge row, paired planting and cluster planting proved to be an important tool to achieve high quality produce.

Moisture Conservation and Mulching

Rainwater harvesting can help in supplying enough water to improve plant establishment and crop yield. Micro-catchment is one of the direct water harvesting system where small structures are constructed across land slopes which captures surface runoff and stores in plant root zone for subsequent use by the plant (Ali *et al.,* 2017). It has been observed that micro-catchment slopes greater than 5% did not significantly affect run off at Jodhpur. The highest ber yield was obtained when 0.5% and 5% slopes had 8.5 m and 7.0 m length of run, and 72 m^2 and 54 m^2 catchment areas per tree, respectively (Sharma *et al.*, 1986).

Effective moisture conservation practice also includes various types of mulching material. Mulches increase soil and water content under intermittent rain or irrigation. Generally, organic and inorganic mulches are commonly used to conserve soil moisture. Mulching helps in maintaining soil temperature and moisture, avoiding weed competition, improving soil structures and biological activities, resulting into increased crop yield. In addition, use of organic materials as mulch can enhance soil fertility, structure and other soil properties (Singh *et al.,* 2016c). Paddy straw mulch was found suitable for aonla crop to improve production under semi-arid conditions of Gujarat (Singh *et al.,* 2010). When compared to other mulches, plastic mulches are completely impermeable to water; it therefore prevents direct evaporation of moisture from the soil and thus limits the water losses. Thus, it plays a positive role in water conservation. Under arid condition, it is reported that the organic mulches like grasses, weeds and crop residues reduce the soil temperature considerably during summer months, while during winter months clear/ transparent plastic mulch increase soil temperature. Black polythene mulch as well as paddy straw mulch both is found very effective in initial plant establishment and improved plant growth of pomegranate, acid lime, aonla and guava ber orchards in western India.

The quality and efficiency of water management determine the yield and quality of fruits. Water demand is a function of weather conditions, crop species and variety, stage of growth, soil water retention capacity and texture and irrigation system management. If water is scarce and supplies are erratic, then irrigation at critical stage and soil moisture conservation are the most important agronomic interventions to maintain yields during stress. Among different systems, drip irrigation is an efficient tool of water application in most of fruit crops. In this system, water is delivered through drippers directly to the soil adjoining to the root system, which absorb the water immediately. Studies indicate that drip irrigation could save water from 33 to 55% with yield increase to a tune of 50% or more besides improving the quality of the produce in different fruit crops. Fertilizers and pesticides can also be delivered with drip irrigation system to the crops and thus minimizes their application losses. Most importantly, their split application can be made as per the requirement at different crop stages. The use of drip alone or in combination with mulching has been demonstrated as a successful technology for cultivation of pomegranate, kinnow, date palm etc. (Sharma *et al.,* 2013). Sharma *et al.* (2013) also emphasized the application of pitcher irrigation as water conservation approach for better establishment of fruit plants in dryland areas. It was attempted in cactus pear at CIAH, Bikaner, and the growth of cactus pear was better as compared to control. This technology can be used to establish other fruit crops too.

Water loss caused by transpiration can be reduced by use of radiation reflectants, stomata closing chemicals, and plastic films. Spraying of 4 to 6% kaolin, 0.5-1.0% liquid paraffin, and 1.5% power oil, after occasional rains in low rainfall areas, considerably reduces plant water losses (Pareek and Sharma, 1991). Chemicals such as phenyl mercuric acetate (PMA), decinyl succinic acid (DSA), abscisic acid (ABA), and cetylalcohol cause stomatal clousure thereby reducing transpiration. Shelterbelt and windbreaks can reduce evapo-transpiration by reducing the wind speed and stabilizing microclimate.

Nutrient Management

Soils of arid and semi-arid regions are poor in organic matter that affects nutrient use efficiency as well as soil moisture retention (Singh *et al.,* 2016). Fruit plants are nutrient exhaustive crops and deplete soil fertility extensively thus necessitates the judicious application of fertilizers. The balanced nutrition in fruit plants is required at appropriate time according to the age of plants. The application methods also play important role for availability of nutrients to the plants. In ber orchards, besides 10-15 kg organic manure, annual application of 100 g N, 50 g P_2O_5 and 50 g K_2O per tree is recommended. Fertilizer doses should be raised according to the age of plants and soil fertility of the region. Application of 15-20 kg FYM per tree has been found beneficial in aonla, custard apple, and tamarind. At MPKV, Rahuri, in addition to 50 kg FYM, 625 g N, 225 g P_2O_5 and 225 g K_2O has been recommended for application to 5-year-old pomegranate trees. In 6-7-year-old fig trees planted at 5 m x 5 m spacing, fertilization with 900 g N + 250 g K improved fruit production (Sharma *et al.*, 2013)

The organic manure application increases water as well as nutrient-use efficiency of soil due to increased organic matter content and other soil physical, chemical and biological properties which in turn favour better plant growth and development. Foliar application of nutrients has been also recommended as a means of improving plant growth especially in early stages of development as it provides required nutrient directly. Vermicompost gained popularity due to recent interest in organic farming and is mostly used in fruit crops. It is rich source of micro and macronutrients, vitamins, growth hormones and enzymes. Organic content is significantly higher with the application of FYM and vermicompost, while bulk density decreased. In recent past, use of different grade water-soluble fertilizer through drip has been increased due to its improved fertilizer-use efficiency and plant growth, yield and quality of produce. Micronutrients are often found deficient in semi-arid and arid soils, therefore playing significant role in production of high grades fruits. Foliar feeding of nutrients such as nitrogen (0.5-2.0% urea), zinc (0.05-1.0% zinc sulphate) and boron (0.05-1.0% borax) has give beneficial results (Pareek and Sharma, 1991).

Post-Harvest Management

Grading and packaging are important practices to fetch better price in the market. The packages protect the produce from damage loss as it maintains quality and retains freshness. Corrugated fibre board box (CFB), wood box with suitable cushioning materials are most suitable and economically-viable packing container for transportation of semi-arid horticultural produce. Cushioning material should be physiologically inactive. Moulded pulp tray, honeycomb, cell pack are better than the traditional material like straw and grasses (Singh *et al.,* 2020). A large quantity of fruits and vegetables produce goes waste due to unavailability of adequate storage facility in semi-arid dryland areas. Proper storage facility like cool storage, CA and ZECC storage can reduce the post-harvest loss to a greater extent and can improve the farm income (Singh *et al.*, 2018a, 2019a; Singh and Singh 2012).

The fruits grown in semi-arid regions are processed into various processed products, utilizing their acquired traditional knowledge like sun drying, pickling etc. However, with the application of modern techniques, the quality of products could be improved considerably. The pre-treatment of many fruits with hormone and harmless chemicals results in better quality end products (Meghwal, 2016). Solar drying and electric tray dehydration of fruits and vegetable help reduce dust load on the product and retain natural colour. Techniques for preparation of different products from underutilized fruits have also been standardized (Mishra, 2018; Reddy *et al.*, 2018).

Malnutrition in resource poor areas of semi-arid region is a major problem particularly in women and children. Tamarind, custard apple, bael, khirni, karonda, phalsa, mulberry, wild noni, wood apple etc. are a rich source of vitamins, minerals and dietary fibres(Singh *et al.,* 2021d). Bael fruits contain higher riboflavin than many fruits. Wood apple and custard apple are rich in carbohydrates and minerals which are vital for the maintenance of body and physiological function. These fruits are highly perishable in nature. The marketing is a major problem. Such fruits gets spoiled within 2–3 days of harvesting, if not consumed (Singh *et al.* 2007c, 2018a, 2019a). Due to glut in the market, the prices of these fruits drop down drastically making it uneconomical for farmers. There is a need to extend shelf-life of these fruits and to develop post-harvest value-addition technologies which are simple and adaptable at the farm level. This will not only result in developing small-scale industry but it will also provide employment to rural masses throughout the year resulting in increased income. Therefore, dried and dehydrated fruits, RTS, squash, fruit bars, candies, fruit concentrates, powders, wines, and condensed fruit juices through solar drying, are prepared for further commercialization. There is tremendous scope for preparing beverages from ripened fruit of chironji. Kernels are being used for the preparation of different kinds of sweets. Squash, RTS, and nectar may be prepared from pulp of there fruits. Value-added products of different hot semi-arid fruits are given in Table 8.

Table 8: Value-added products of semi-arid and arid fruits

Crop	Value-added products
Bael	Preserve, RTS, nectar, ice cream, slab, squash, cider, canned bael slices, pickles and powder
Ber	Dried ber, powder,jam candy, beverages,pickle, wine
Aonla	Murabba, candy, pickle, powder, laddu, burfy
Chironji	Dried kernels, fruit bar
Karonda	Pickle, candy, jelly, jam, preserve, wine, chutney
Wood apple	Squash, powder, pickle, chutney, jelly, fruit bar
Khirni	Dehydrated fruits, fruit bar, RTS, jam
Jamun	Juice, RTS, squash, syrup, carbonated drink and wine
Phalsa	Juice, squash, syrup
Lasoda	Pickle, culinary
Custard apple	Jam, beverages, ice cream
Tamarind	Tokku (chutney), panipuri masala, juice concentrate, pulp powder, jam, syrup, candy toffee, tamarind karnel powder
Timru	Bidi, dried fruit
Pilu	Squash, dried *peelu*, wines
Kair	Pickle, dried fruits
Mulberry	Juice, squash and syrup
Mahua	Biscuits, cakes dried powder, seed oil and wine
Manila tamarind	Biscuits, squash and syrup
Aloe	Candy, jelly, pickle, cold cream, crack cream, moisturizer, gel
Fig	Fig paste, concentrate, powder, nuggets, jam

Singh *et al.* 2016, Singh *et al.* 2010 Mishra, 2018

Way forward

- Genetic resources management
- Ethnobotanical and nutritional value also needs to be assessed with scientific validation.
- Genetic improvement and development of new varieties for specific traits.
- Standardization of propagation techniques

- Commercialization of processing, packaging and value addition in natural growing areas.
- Market linked production and shift in attitude from subsistence farming to commercial orcharding.
- Screening of genotypes for abiotic especially drought resistant and moisture stress.
- Developing suitable technologies for reducing post harvest losses and supporting cottage industries.
- To create awareness among the people regarding it nutritional, health and environmental security for its commercialization.
- It is need of hour to develop herbal products useful against various ailments with scientific validation.
- Emphasis should be given on post harvest technology to develop value added export oriented products. Small scale processing units should be established and promoted for commercialization of this fruit crop.

References

AICRP on Arid Zone Fruits, 2013. Annual Report, AICRP on Arid Zone Fruits, M.P.KV, Rahuri, Maharashtra

Altendorf S. 2018. Minor tropical fruits- mainstreaming a niche market. Food Outlook (July 2018), pp. 66-74.

Arya L, Narayanan K, Verma M, Singh A K and Gupta V. 2014. Genetic diversity and population structure analysis of Morinda tomentosa Heyne, with neutral and gene based markers. Genetic Resources and Crop Evolution, DOI10.1007/s10722-014-0168-4.

Awasthi O P, Saroj P L and Dhandar D G. 2005. Standardization of time of success of patch budding in tamarind (Tamarindus indica) under arid conditions. Progressive Horticulture, 37(2):294-97.

Chundawat B S. 1990. Arid Fruit Culture. Oxford & IBH Publ. Co. Pvt. Ltd., Delhi. p. 208.

Dhakar M K, Saloria D K, Kaushik R A, Kumavat K L, Singh Sanjay and Singh A K. 2015. Mahua. In: Breeding of Underutilized Fruit Crops Part 2, Ghosh S. N. (Ed.). Jaya Publishing House, New Delhi, pp. 305-23.

Diengngan S and Hasan M A. 2015. Genetic diversity of underutilized fruits in India for environmental sustainability. Advances in Plants & Agriculture Research 2(7):299–303.

Hiwale S. 2015. Problems of horticulture in semi-arid rainfed areas. In: Sustainable Horticulture in Semiarid Dry Lands. Springer New Delhi, pp. 5-13.

Krishna H, Saroj P L, Singh D and Singh R S.2018. Popularaizing underutilized arid fruits for nutrition. Indian Horticulture 63(5):30-32.

Malik, Choudhury, R., Dharial, O.P. and Bhandari, D.C. 2010. Genetic Resources of Tropical Underutilized Fruits of India. NBPGR, New Delhi. p. 168.

Meena,, V S, Gora J S, Singh A, Ram C, Meena N K, Pratibha A, Rouphael Y, Basile B. and Kumar P. 2022. Underutilized fruit crops of Indian arid and semi-arid regions: importance, conservation and utilization strategies. Horticulturae 8:171. https://doi.org/10.3390/horticulturae8020171.

Minhas P S, Singh Y, Nangare DD and Kumar PS. 2015. Establishing orchards in shallow soils. Indian Horticulture, Nov-Dec., 2015, pp. 5-8.

Mishra D S and Goswami, A. 2016.High density planting in fruit crops. Hort Flora Research Spectrum 5(3):261-64.

Mishra D S, Singh A, Kumar R, Singh S, Singh, A K. and Swamy GSK.2014. Jamun. In: Crop Improvement and Varietal Wealth, Part-2, Ghosh, S.N. (Ed.). Jaya Publishing House, New Delhi, pp.375-390.

Mishra DS. 2018. Enhancing income through value-addition. Indian Horticulture, 63(5):107-109.

Nath V, Saroj P L, Singh R S, Bhargava R. and Pareek, O P. 2000. In-situ establishment of ber orchards under hot arid eco-system in Rajasthan. Indian J. Hort, 57(1): 21-26

Pal R K, Babu K D, Singh NV, Maity A and Gaikwad N. 2014. Pomegranate research in India – status and future challenges. Progressive Horticulture, 46(2):184-01.

Pareek O P. and Sharma, S. 1991. Fruit trees for arid and semi-arid lands. Indian Farming, 41: 25-30.

Parthasarathy V A, Singh A K, Singh S, Mishra D S and Yadav V. 2021. Bael. In: Fruits: Tropical and Subtropica, Vol-3 (Parthasarathy et al. eds.) Daya Publishing House (Astral International Pvt. Ltd.), New Delhi, pp. 571-608.

Patel D D, Gaiwad S S and Patel K D.2016. Effect of seed priming treatments on germination and seedling vigour of custard apple. Current Horticulture, 4(2):21-24.

Patel M N, Parmar L D, Parihar A, Singh A K and Sheikh W A. 2014. A high throughput DNA Extraction Protocol and its utilization in molecular characterization of Noni (Morinda citrifolia L.) genotypes. Current Trends in Biotechnology and Pharmacy 8(2):166-74.

Rathod A H, Acharya S and Singh A K. 2016. Assessment of genetic variability and heritability in diverse accessions of wild noni (Morinda tomentosa Heyne ex Roth) in north Gujarat. Medicinal Plants 8(1):40-45.

Reddy V R, Meena R K. and Bhargava R. 2018. Doubling farmers income through value addition. Indian Horticulture 63(5):101-106.

Roy A. and Bauri, F K. 2019. Scope of minor fruit production in India. Acta Horticulturae 1241: 43-50.

Saroj P L. 2018. Exploiting potential of arid horticulture. Indian Horticulture, 63(5):3-16.

Singh A K, Pandey D, Singh Sanjay, Singh R S, Misra A K and Singh R K. 2019b.The bael in India. Indian Council of Agricultural Research (DKMA), New Delhi, pp.1-129

Singh A K, Singh S, Appa Rao VV, Bagle,B G. and More T A. 2011. Effect of high density planting systems on the productivity of NA-7 aonla under rainfed conditions. Indian J. Hort., 68(4): 461-465.

Singh A K, Singh S, Saroj P L, Mishra D S, Singh PP and Singh R K. 2019. Aonla (Emblica officinalis) in India: A review of its improvement, production and diversified uses. Indian Journal of Agricultural Sciences, 89(11): 1773-1781.

Singh A K, Singh S, Saroj P L, Mishra D S, Yadav V. and Kumar R. 2020. Cultivation of underutilized fruit crops in hot semi-arid regions: development and challenges- a review. Current Horticulture, 8(1): 12-23.

Singh A K, Singh S, Singh R S and Makwana P. 2016.Organoleptic scoring of RTS prepared from bael (Aegle marmelos) varieties, Indian Journal of Agricultural Sciences, 86(5): 611-4.

Singh A K, Singh S. and Makwana P. 2015. Intervarietal morphological variability in bael (Aegle marmelos) under rainfed semi-arid hot ecosystem of western India. Current Horticulture, 3(2):3-9.

Singh A K, Singh Sanjay and More T A. 2014. Preliminary evaluation of bael varieties under rainfed conditions of western India. Indian Journal of Horticulture 71(2): 264-68.

Singh A K, Singh Sanjay and Saroj P L. 2018.Exploring morphovariations in bael (Aegle marmelos). Current Horticulture, 6(2):52-57.

Singh A K, Singh Sanjay and Saroj P L. 2021c. Biodiversity in underutilized fruit crops of hot semi-arid region. In: Current Horticulture: Improvement, Production, Plant Health management and Value-Addition, Vol. 2. Brillion Publishing, New Delhi, pp. 175-192.

Singh A K, Singh Sanjay and Saroj P L.2017.Cultivating climate resilient bael for future. Indian Horticulture 62(4): Cover II-43-45.

Singh A K, Singh Sanjay, Joshi H K and Sisodia P S. 2012. Exploring diversity in Morinda for human welfare. Indian Horticulture, 57(6): cover II&III.

Singh A K, Singh Sanjay, Makwana P and Sharma S K. 2016a. Diversity in Morinda tomentosa Heyne ex Roth germplasm from Gujarat and Madhya Pradesh. International Journal of Noni Research 11(1&2):1-10.

Singh A K, Singh Sanjay, Makwana P and Sharma S K. 2016b.Evaluation of bael germplasm under rainfed semi-arid environment of western India. International Journal of Noni Research 11(1&2): 11-19.

Singh A K, Singh Sanjay, Mishra D S and Saroj P L. 2016c. More crop with minimal water. Indian Horticulture, 61(6): 86-91.

Singh A K, Singh Sanjay, Mishra D S and Saroj P L. 2021d. Nutritional security through underutilized semi-arid fruits. In: Current Horticulture: Improvement, Production, Plant Health management and Value-Addition, Vol. 1. Brillion Publishing, New Delhi, pp. 307-318.

Singh A K, Singh Sanjay, Saroj P L, Krishna H., Singh R S and Singh R K. 2019c. Research status of bael (Aegle marmelos) in India: A review. Indian Journal of Agricultural Sciences 84 (10):1563-71.

Singh A K, Singh Sanjay, Saroj P L.and Singh G. P. 2021d. Improvement and production technology of bael (Aegle marmelose) in India. Current Horticulture 9(1):3-14.

Singh A K, Singh Sanjay, Singh R S, Contractor K and Makwana P. 2014. In-situ patch budding for better establishment of bael in rainfed areas. Indian Horticulture 59(5):24-25.

Singh A K, Singh Sanjay, Singh R S, Joshi H K and Sharma S K. 2014. Characterization of bael varieties under rainfed hot semi-arid environment of western India. Indian Journal of Agricultural Sciences 84(10): 1236-42.

Singh A K, Singh Sanjay, Yadav V and Sharma B D. 2016d. Genetic variability in wood apple (Feronia limonia) from Gujarat. Indian Journal of Agricultural Sciences 86(11): 1504-8.

Singh A K, Singh, Sanjay and Joshi H K. 2013.Improving socio-economic through rainfed bael. Indian Horticulture 58(4): 14-17.

Singh A K. 2014. Noni. In: Propagation of Horticultural Plants of Arid and Semi-Arid Region. Singh R S and Bhargava R. (Eds.).New India Publishing Agency New Delhi, pp.379-388.

Singh A K. 2018.Propagating arid fruit commercially. Indian Horticulture 63 (5):82-88.

Singh R S, Singh A K, Maheshwari S K, Bhargava R and Saroj P L. 2021a. Underutilized Fruits of India. Brillion Publishing, New Delhi, pp. 1-475.

Singh R S, Singh A K, Singh Sanjay and Vikas Yadav. 2017. Underutilized fruits of hot arid Region. In: Biodiversity in Horticultural Crops, Vol. 6, Peter K V (Ed.). Daya Publishing House, New Delhi, pp.75-92.

Singh Sanjay and Singh A K. 2014.Chironji. In: Propagation of Horticultural Plants of Arid and Semi-Arid region, Singh R. S. and Bhargava R. (Eds.). New India Publishing Agency New Delhi, pp. 363-370.

Singh Sanjay and Singh A K. 2014. Standardization of softwood grafting in chironji under semi-arid environment of western India. Indian Journal Horticulture 71(1): 120-22.

Singh Sanjay and Singh A K. 2015.Standardization of time of softwood grafting in mahua and khirni under semi-arid environment of western India. Indian Journal of Agricultural Sciences, 85 (2): 166-70.

Singh Sanjay and Singh A K. 2016. Mahua (Bassia latifolia Roxb). In: Underutilized fruit crops: Importance and cultivation Part-2, Ghosh, S. N. Singh, Akath and Anirudh Thakur (Eds.). Jaya Publishing House, New Delhi, pp. 785-796.

Singh Sanjay and Singh A K.2007. Standardization of method and time of vegetative propagation in Tamarind (Tamarindus indica L) under semi- arid environment of western India. Indian Journal of Horticulture 64(1): 45-49.

Singh Sanjay and Singh, A. K. 2006. Standardization of method and time of propagation in jamun (Syzygium cuminii) under semi-arid environment of western India. Indian Journal of Agricultural Sciences 76(4):242-5.

Singh Sanjay Saroj P L, Mishra D S and Singh A K.2019e. Underutilized Fruit Crops: crop improvement and agro-techniques. Kaav Publication, New Delhi, pp1-306

Singh Sanjay, Singh A K, Bagle B G. and More T A. 2010. The Jamun. ICAR-DKMA, New Delhi. pp. 1-48.

Singh Sanjay, Singh A K, Mishra D S and Appa Rao V V. 2017. Effect of size of polythene bag on seedling growth and budding success of jamun. Indian Journal of Arid Horticulture 12(1&2): 58-61.

Singh Sanjay, Singh A K, Mishra D S and Saroj P L. 2021b.Quality planting materials for arid and semi-arid fruits. In: Current Horticulture: Improvement, Production, Plant Health management and Value-Addition, Vol. 1. Brillion Publishing, New Delhi, pp. 49-64.

Singh Sanjay, Singh A K, Mishra D S, Singh G P and Sharma B D. 2022. Advances in research in Jamun: a review. Current Horticulture, 10(1):8-13.

Singh Sanjay, Singh A K, Saroj P L, Mishra D S. 2019a. Research status for technological development of Jamun (Syzygium cuminii). Indian Journal of Agricultural Sciences 84 (11):1991-98

Singh Sanjay, Singh H P, Singh A K. and Sisodia P S. 2011.The jamun: Fruit for Future. Pub. Agro-tech Publishing Agency, Udaipur, Rajasthan, pp.1-100.

Singh Sanjay, Singh, A K, Mishra, D S, Saroj P L and Appa Rao V V. 2018. Jamun for health and wealth. Indian Horticulture 63 (5):33-36.

Singh, A. K. 2019. Bael. In: Handbook of Horticulture, Vol. 1, Verma A S, Dutt Som and Pradhan S (Eds.). ICAR-DKMA, New Delhi, pp.258-261.

Singh, A. K., Singh, Sanjay and Saroj, P. L. 2018. Bael (Production Technology). Technical Bulletin No.67, Pub. ICAR-CIAH, pp.1-53.

Singh, A. K., Singh, Sanjay, Singh, R. S. and Sisodia, P. S. 2011.The Bael: Potential Underutilized Fruit for Future. Agro-tech Publishing Agency, Udaipur, Rajasthan, pp.1-102.

Singh, A.K. and Singh, Sanjay.2019.Aonla. In: Handbook of Horticulture, Vol. 1, Verma A S, Dutt Som and Pradhan S (Eds). ICAR-DKMA, New Delhi, pp.239 242.

Yadav V, Singh A K, Appa Rao V V, Singh Sanjay and Saroj P L.2018. Wood apple variability: An Underutilized dryland fruit from Gujarat, India. International Journal of Current Microbilogy and Applied Sciences 7(6): 548-555.

Yadav V, Singh A K, Singh Sanjay and Appa Rao V V. 2017.Variability in custard apple (Annona Squamosa) genotypes for quality characters from Gujarat. Indian Journal of Agricultural Science 87 (12):1627-32.

7

Water and Energy-efficient Arid Fruit Production System

Akath Singh and Pradeep Kumar

ICAR-Central Arid Zone Research Institute, Jodhpur 342 003, India

Agriculture is biggest user of water, accounting 70% of all withdrawals. It is major water-consuming sector due to intensification of agriculture (CWC, 2021). With changing climatic condition, demand of water will increase. Hence, conservation and efficient management of limited water has become the need of the hour for sustainable fruit production. Since per unit returns from horticultural crops are higher, their production system improves the net return of farmers and would impart resilience to aberrant weather (Meghwal *et al.*, 2018). Therefore, immense opportunities for quality production of water-economizing horticultural crops. In arid and semi-arid regions, uncertain, erratic and scanty' rains coupled with meagre irrigation facilities leads to low and unstable yields. Inclusion of better crops and varieties, tissue-cultured seedlings, effective rain water harvesting from field in improved water harvesting structures, reducing evaporation losses by anti-evaporative materials, solar operated ferti-drip system and production under protected conditions are some of the tested tools and techniques for enhancing water productivity and higher monetization of available water.

Rainwater Harvesting and Its Management

Globally, water withdrawal for agriculture is about 70% and production of irrigated crops is a predominant water "consumer" given that ~70% (~3 trillion m^3) of totally abstracted fresh hydro-resources is exploited by agriculture sector (CWC, 2021). The scale, agricultural water losses are enormous (55%) of available irrigation water (FAO, 2017). They are caused by irrigation system, farm distribution, and field water application mismanagement. Only 45% of irrigation water is effectively used by crops to irrigate 25% of world's croplands (399 million ha) which supply 45% of global food (Koech and Langat, 2018). These issue become far graver in dryland regions, occupying about 41% of the Earth's terrestrial surface and are home to more than a third of the world's population, touching critical level in arid regions. Thus, imminent water problems of arid regions are aggravating now-days (Pravalie, 2016). Water is the most limiting factor for crop production in arid areas. Thus, the search of promising water management strategies involving irrigation water management, water conservation, and non-conventional water use for agriculture is foremost for achieving highly productive and sustainable agriculture. Arid region in western Rajasthan is characterized by scare rainfall, poor and limited availability of groundwater and frequent occurrence of drought. Annual rainfall in region varies from 100 to 500 mm. Under such water scarce situation, agricultural production system is highly unstable and often manifested through malnutrition and health hazards of dwelling people. The whole gamut of livelihood earning by the desert dwellers revolved around the success in harvesting and best utilization of rain water in traditional structures like *khadin, tanka, nadis/ farm pond* etc. since immemorial (Goyal *et al.*, 2007).

In arid and semi-arid parts, experiencing rainfall from 150-350 mm and 350-550 mm respectively, rainfall is not sufficient to ensure a good crop. Crop production is adversely affected due to low and erratic distribution. In dryland farming, the solution to soil moisture problem lies in the storage of rainfall by water harvesting methods (Oweis *et al.*, 2012). Therefore, water harvesting would be an alternative for increasing the available water supply for plant through microirrigation.

Enhancing Run-off from Orchard

Being light textured soil and low frequency rain, run off generation from cultivated field is low and uncertain. In fruit orchard, rows and inter rows may be mulched with polythene mulch during rainy season to enhance run-off. Row mulching the 100 micron black polythene and in inter rows, 200 micron black polythene now used. Inter row mulching removed after rainy season each year while row mulching in continued till harvesting and removed in April. Mulching significantly enhances run-off capacity of field and it is three time higher in row and inter row mulching compared to no mulching system.

Water R

Ex-situ rainwater harvesting is a promising technology for enhancing the availability of water in arid areas. *Ex-situ* water harvesting involves collecting run-off originating from rainfall over a surface away from the field and storing it in surface storage systems for later use in the field. This type of rainwater harvesting provides supplemental or protective irrigation during dry periods of the cropping season (Goyal and Sharma, 2000). Evaporation of water directly from soil surface is an important component of total water use of crops. Evaporation constitutes 19-46% of the total water on vertisols, 21% on alfisols and 12-18% on loamy sand soils. Water loses through evaporation is quite high in arid conditions (Pathak *et al.,* 2009). The evaporation loss is around 5-12 mm/day, average being 8 mm/day, volume of water loss = 672 litre/day (surface area of water is taken as 84 m^3). Anti-evapoprative material may be used to check such losses from farm pond or other open water structures. Thermacoal balls of two different sizes proved very effective and feasible for farm pond. Losses of water are checked by applying thermocol ball on surface, reducing 70% of the total loss.

Selection of Crop

The selection of crops involves integration of compatible technologies based on potential water availability, requirements of farmers, their resource inventory and available technologies. This involves activities related to sufficient runoff generation, its movement and storage in reservoirs through improved design and developing farm plan. Presently, stored water is mostly used as supplemental irrigation in *kharif* season (only if it is surplus over domestic use) and for growing post rainy season crops. Since perennial components are hardier and water economizing, niche horticultural crops like ber, pomegranate, gonda etc. needed to be integrated in system because their water requirement begins just after rainy season, i.e September onwards, hence maximum potential of harvested water would be utilized. The introduction of tissue cultured seedling of selected varieties of high returning crops like pomegranate (Bhagwa), Fig (Diana), dragon fruit under ferti-drip system may enhance net return of around Rs. 3.0-3.5lakhs/ha with given unit of water which is 4- 5-fold higher than traditional crops in arid subtropics.

Use of Mulches

Deficiency of water and high thermal regime of soils are two most important factors which adversely affect crop production in sandy soils. Surface mulches can be used to reduce evaporation loss, to prevent soil from blowing and washing away, to keep down weeds and ultimately to increase crop production (Kader *et al.*, 2019). Therefore mulches can be used for favourably manipulating soil environment for increasing crop production in the sandy plains of western Rajasthan.

Microirrigation Systems

Irrigation efficiencies under different methods of irrigation suggest reduction in water consumption due to drip method of irrigation over the surface method of irrigation varies from 30 to 70% for different crops (Postal *et al.,* 2001,). The productivity gain due to drip irrigation is estimated to be in the range of 20 - 90% for different crops (INCID, 1994). The main objective of the drip method of irrigation is to provide each plant with a continuous supply of soil moisture which is just sufficient to meet the evapo-transpiration demand. Thus, it is desirable to irrigate the field daily at the end of the day break. The unique feature in drip method of irrigation is that since the water loss of the day is duly compensated every evening, the root soil always remaining nearly at the field capacity and consequently the crop never suffers from moisture stresses. This is the main reason for higher yield as compared to other method of irrigations.

Deficit irrigation or regulated deficit irrigation approach seeks to avoid irrigation at less critical stages of crop growth and apply less water at the end so as to eliminate water stored at harvest. One has to well verse with the crop growth stages less sensitive to moisture stress and proper balancing between water given and ET demand of crop (Sam Geerts and Dirk Raes, 2009; Elias and Maria, 2007). In irrigation process, roots uptake nutrients (nitrogen, potassium) at a rate of up to 4-6 kg ha^{-1} day^{-1}. As in drip irrigation, the volume of root system is relatively small and the amount of nutrients available in soil is quite limited. Furthermore, water flowing through zone of root system washes away dissolved salts, including nutrients. Therefore, it is essential to resupply nutrients to the root system, particularly water soluble fertilizers, which is highly soluble in water and is rapidly depleted from the root zone, and will be washed downward at a slower rate, there is a need to replenish them, especially in sandy soil.

Fruit Bagging and Row Covering

In this technique, individual fruit or individual plant or complete row of plants are covered for a specific period to obtain desired results. Among various strategies, pre-harvest fruit bagging with different bagging materials and plant covering with non-woven fabric as a whole has been emerged as one of the best ecofriendly approaches in different parts of the world for enhancing marketability of quality fruit crops, e.g. apple, peach, pear, grapes etc (Sharma *et al.*, 2014). It is a physical protection that can improve the appearance and quality of fruit by avoiding physiological and pathological injuries and by modifying the micro-climate inside covered conditions which also reduce transpiration loss from plant canopy (Zhang *et al.*, 2015). This is evident by air temperature that is lower in row and individual covering while RH was higher than those in ambient condition. This resulted in significant reduction in transpiration losses from pomegranate canopy and ultimately enhances water productivity.

Water and Energy Harvesting Model

The returns from investment on rain water harvesting are less in prevailing conditions of arid region because it is difficult to maintain stored water after January onward while the popular annual crops of the region require irrigation up to April leading to water scarcity at pod grain filling stage and hence low productivity. However, availability of stored water in farm pond matches the irrigation period and water requirement of pomegranate (September-January) - a highly remunerative and popular crop of the region. An innovative energy and water harvesting model with solar PV based ferti-drip system has been developed for small and marginal resource poor farmers to enhance their economic return. The one acre model of Bhagwa variety was developed with a provision to harvest rain water of pomegranate field in low cost HDPE lined farm pond having storage capacity of 3.25 lakh litres. For enhancing run off, proper field levelling at around

1% slope followed by mulching of row and inter rows with plastic mulch. Anti evaporative material (thermocol balls of two different size) spread on pond water surface and reduced around 50% water loss from farm pond through evaporation. Harvested rain water is applied through ferti-drip system operated by 3HP solar PV pumping system with regulated deficit irrigation at ETc on daily basis. Sixty percent requirement of major nutrient is supplying through fertigation using water soluble fertizers. Individual plant or rows of plants are covered with non-woven polypropylene sheet (17gsm) to reduce transpiration loss from plant and also to create favorable microclimate. This system is able to harvest about 3.25 lakh liters of water sufficient to irrigate around 220 pomegranate plants with 30% higher water productivity without affecting crop productivity. Best quality pomegranate fruits with minimum agrochemical residue are being harvested which enhanced marketable yield. The system proved its potential in non traditional areas where ground water quantity and quality are not much suitable for cultivation of pomegranate. This model may also provide opportunities to resource poor farmers of water scarce areas in cultivating high value crops for enhancing their livelihood and net returns.

References

CWC, New Delhi. 2021. Water and Related Statistics. Ministry of Jal Shakti, Central Water Commission (CWC), Government of India, New Delhi.

Elias Fereres, María Auxiliadora Soriano. 2007. Deficit irrigation for reducing agricultural water use. Journal of Experimental Botany 58 (2): 147-59.

FAO, Rome 2017. Water for Sustainable Food and Agriculture. Report for the G20 Presidency of Germany. Pp.27 (www.fao.org/publications).

Goyal, R.K. and Sharma, A.K. 2000. Farm Pond: A well of wealth for the dryland dwellers. Intensive Agriculture XXXVIII(5-6): 12-14 & 26.

Goyal R K Bhati T K and Ojasvi P R 2007. Performance evaluation of some soil and water conservation measures in hot arid zone of India. Indian Journal of Soil Conservation 35(1): 58-63.

INCID. 1994. Drip Irrigation in India. Indian National Committee on Irrigation and Drainage, New Delhi.

Kader Mohammad Abdul, Singha Ashutus,Begum, Mili Amena, Arif Jewel, Khan Ferdous Hossain and Khan Nazrul Islam. 2019. Mulching as water-saving technique in dryland agriculture: review article. Bulletin of the National Research Centre 43(147): https://doi.org/10.1186/s42269-019-0186-7

Koech, Richard and Langat Philip. 2018. Improving Irrigation Water Use Efficiency: A Review of Advances, Challenges and Opportunities in the Australian Context. Water. 10(12), 1771; https://doi.org/10.3390/w10121771

Meghwal, P. R., Singh Akath, Kumar Pradeep, Khapte P. S. and Saxena A. 2018. Upliftment of arid zone economy through horticultural and protected cultivation. Indian Farming 68(09): 46–51.

Oweis T, Prinz D, Hachum A 2012. Rainwater Harvesting for Agriculture in the Dry Areas. CRC Press, Taylor and Francis Group, Leiden, 366 pp.

Pathak, P., Sahrawat, K.L., Wani, S.P., Sachan, R.C., Sudi, R. 2009. Opportunities for water harvesting and supplemental irrigation for improving rainfed agriculture in semi-arid areas. Suhas P Wani, Johan Rockstrom and Theib Oweis (Eds.), In: Rainfed Agriculture: Unlocking the Potential. CAB International, Wallingford, UK. pp.197-221.

Postal, S., Polak, P., Gonzales, F., and Keller, J. (2001). Drip Irrigation for Small Farmers: A New Initiative to Alleviate Hunger and Poverty. Water International. 26 (1): 8

Pravalie Remus. 2016. Drylands extent and environmental issues. A global approach. Earth Science Reviews. 161: 259-275.

Sam Geerts and Dirk Raes. 2009. Deficit irrigation as an on-farm strategy to maximize crop water productivity in dry areas. Agricultural Water Management. 96(9):1275-1284

Sharma R.R., .Pal, R.K, Sagar, V.R., Parmanick, K.K., Paul, V., Gupta, V.K. and Rana, M.R. 2014. Impact of pre-harvest fruit-bagging with different coloured bags on peel colour and the incidence of insect pests, disease and storage disorders in 'Royal Delicious' apple. The J Horti Sci and Biotech. 89(6): 613-618

Zhang B.B., J.Y. Guo, R.J. Ma, J.X. Cai, J. Yan and C.H. Zhang. 2015. Relationship between the bagging microenvironment and fruit quality in 'Guibao' peach [Prunus persica (L.) Batsch]. J. Hort Sci Biotech. 90(3): 303–10

8

Conservation and Utilization of Wild Fruit Species in India

Prakash Chandra Tripathi

Principal Scientist (Horticulture), Division of Fruit Crops, ICAR-Indian Institute of Horticultural Research, Hessaraghatta Lake Post Bangalore-560 089

India is floristically extremely rich with about 33 per cent of its botanical wealth (over 15,000 species of higher plants) being endemic. There are about 141 endemic genera distributed over 47 families (Nayar, 1980) the 4,900 endemic species, a large number are localised in the Himalayas (about 2,532 species), compared with the peninsular tract (1,788 species), and the Andaman and Nicobar Islands (185 species). Indian region is a major centre of domestication and diversity of crop plants (Zeven and de Wet, 1982; Arora, 1991). About 33 per cent of the cultivated plant species have theirorigin in this region (Damania, 2002). The Indian sub-continent is a centre of domestication and diversification of several economically useful wild plant species comprising about 3,000 plants of edible value, 4,000 species having known reputed medicinal value, 700 plants of traditional and social significance, 500 fibre-yielding species, 400 fodder plants, 40 plants having insectivorous uses, 300 gum and dye yielding plants and 100 aromatic and essential oil-yielding species (Arora, 1991). The distributional pattern in different phyto-geographical regions reveals eight zones viz. Western Himalaya, Eastern Himalaya, North-eastern Region, Upper Gangetic Plains, Indus Plains (North-western Plains), Malabar/ Western Peninsular Region/ Western Ghats, Deccan/ Eastern Peninsular Region/ Eastern Ghats and Andaman and Nicobar Islands and Lakshadweep (Fig.1). There are several micro centres of India have been identified in these Phyto-geographical regions. As far as fruit-yielding species are concerned, more than 500 species are found in different regions which are being used as fruit and various other purposes. The prominent fruit crop are Table 1.

Table 1: Prominent fruits found in different phyto-geographical regions

Phyto-geographical region	Prominent wild fruits
Western Himalaya	*Elaeagnus hortensis, Ficus palmata, Fragaria indica, Moms spp., Prunus acuminata, P. cerasiodes, P. cornuta, P. napaulensis, P. prostrata, P. tomentosa, Pyrus baccata, P. communis, P. kumaoni, P. pashia, Ribes graciale, R. nigrum, Rubus ellipticus, R. moluccanus, R. fruticosus, R. lasiocarpus, R. lanatus, R. niveus, R. reticulatus, Zizyphus vulgaris. Myrica esculenta,Elaegnus parviflora, Rubus paniculatus, Debregeasiavelutina Aesandrabutyracea, Aesculus india, Artocarpus lacucha, Berberis lyceum Castanea sativa, Cordiadichotoma, Ficus racemosa, Morus alba, Ziziphus glabrerrima, Ziziphus nummularia* Burm
Eastern Himalaya	*Fragaria indica, Morus* spp., *Myrica esculenta, Prunus acuminata, P. cerasiodes, P. cornuta, P. jenkinsii, P. napaulensis, Pyrus pashia, Ribes graciale, Rubus lineatus, R. ellipticus, R. lasiocarpus, R. moluccanus, R. reticulatus.*
North-eastern Region	*Citrus assamensis, C. ichangensis, C. indica, C. latipes, Coixlacryma-jobi var. ma-yuen, Eryobotryaangustissima,Citrus assamensis, C. ichangensis, C. indica, C. jambiri, C. latipes, C. macroptera, C. media, C. aurantium, Docynia indica, D. hookeriana, Eriobotrya angustifolia, Mangifera sylvatica, Musa accuminata/M. balbisiana complex, M. manii, M. nagensium, M. sikkimensis, M. superba, M. velutina, Pyrus pyrifolia, P. pashia, Prunus cerasiodes, P. cornuta, P. jenkinsii, Ribes graciale, Rubus ellipticus, R. moluccanus, R. reticulatus, R. lasiocarpus, Myrica esculenta*
Upper Gangetic Plains	*Syzygiumcumini, Artocarpus heterophyllus, Carissa congesta, Manilkara hexandra, Grewia sub-inaequalis/G. asiatica, Aegle marmelos, Feronia limonia, Syzygium, Cordia myxa, C. rothii,Emblica officinalis, Grewia asiatica, Morus* spp.; *Phoenix* spp.; *Syzygium* spp.; *Zizyphusnummularia and other* spp.
Indus Plains (North-western Plains)	*Zizyphus nummularia and other* spp. *Manilkara hexandra Mimusops sp., Grewia spp, Syzygium spp , Carissa congesta.*
Malabar/ Western Peninsular Region/ Western Ghats	*Diospyros afinis, D. barberi, Eugenia mabaeoides, E. zeylanica var. lineare, Garcinia imbertii, G. rubroechinata, G. travancorica, Madhucadeplostemon, Syzygiumbeddomei, S. coutallense, S. gambleanum, S. param eswaran ii, S. rama-varmae, Vitis gardneri, V. grandis, Eugeniarottleriana, Garcinia wightji, Madhucabourdillonii, Syzygiumchandrsekharanii, Syzygiumutilis, Madhuca insignis, Syzygiumkanarensis. Artocarpus heterophyllus, A. lakoocha, Garcinia indica, Diospyros spp., Ensetesuperba, Mimosopselengii, Spondias pinnata, Vitis spp., Zizyphusoenoplia, Z. rugosa, Rubus ellipticus, R. lasiocarpus, R. moluccanus.*
Deccan/ Eastern Peninsular Region/ Eastern Ghats	*Musa spp., Syzygiumspp., Garcinia spp., Artocarpus spp. Feronia, Madhuca spp. Diospyros afinis, Cordia myxa, Delonia spp.,*
Andaman & Nicobar Islands and Lakshadweep islands	*Ficusandamanica, Garcinia andamanica, G. candelliana, Mangifera andamanica, Mimusopsandamanica, Syzygiumandamanicum, S. kurzii, S. manii, S. viminalis var. fasciculatus, Jasminum multiflorum*

Major Wild Fruit Species

Rich genetic diversity of wild and semi-domesticated fruits is found in India. Some fruits found in different regions and their characters, variability and use and described in Table 2.

Table 2: Major wild fruit species of different regions of India

Fruit/Family	Description	Occurrence	Variability	Uses
Kaphal *(Myrica esculenta* Hook.) Family-Myricaceae	Small, moderate-sized evergreen tree ,may attain 3-15m height. It is Flowering - February to Fruit maturity- April to June.	Found abundantlyat an altitude of 900-2100m.throughout the hills of Uttarakhand, Himachal Pradesh and Jammu and Kashmir	Large variability with respect to growth and flowering and fruitcharacters	Fresh fruit and also processed as jam,juice .popularwild fruits sold at 300-500/kg
Myrica farquhariana Family-Myricaceae	A moderate sized and ever green tree. Fruits are about 2.5 cm ellipsoidal or ovoid reddish or cheese colour at ripening. Fruit is available during April – July in.	Belongs to family *Myricaceae*. It is quite common in Khasi and Jaintia hills and locally known as *Soh-phie-nam* in Khasi and *Saphai* in Pnar	Large variability with respect to growth and flowering and	Fruits are edible and used for making refreshing drink and pickle.
Hisalu *(Rubus ellipticus* Sm.) Family-Rosaceae	Anevergreen, large, spreading shrub, flowering - February to April. fruits mature in April to June.	Found abundantly throughout the hills of Uttarakhand, Himachal Pradesh and Jammu and Kashmir in open slopes and scrub forestsfrom 1000-2200m.	Large variability with respect to growth and flowering and fruit characters	Fresh fruit and jam . popular wild fruits sold at 300-500/kg
Kala Hisalu (*Rubus biflorus* Buch.) Family- Rosaceae	A deciduous small shrub, fruits mature- June to July.	Found in in moist-shaded forests at an altitude of 1600 to 2800m.	Variability with respect to growth and flowering and fruit characters	Fresh fruit and jam
Chook/ Dhur-chuk sea-buckthorn (*Hippophae rhamnoides* L. Family-Elaeagnacae	A dioecious, usually spinescent deciduous shrub or a small tree of. flowering -April-May fruit mature - September to October.	Found in river beds of the drier ranges of the state at an altitude of 1600-3300m	Variability with respect to growth and flowering and fruit characters	Fresh and is made into a jelly with sugar. eruptions
Diploknema butyracea H J Larn (Chiura, phuwara, chura). Family-sapotaceae	Medium to large deciduous tree attaining a height of 10-25m. flowering January-March fruits mature - June-July.	Commonly found growing in ravines and river banks up to 1500m.in Eastern part of Uttarakhand and Nepal	Variability availed in Eastern part of Uttarakhand and Nepal	The fruit is a berry, blackish in colour, with a thick soft, sugary pericarp used to make butter and sugar

Fruit/Family	Description	Occurrence	Variability	Uses
Ficus auriculata Lour Family-Moraceae	A small, evergreen or sub-deciduous, spreading tree. flowering - March-April and fruits mature in June-July.	Commonly found growing in ravines and river banks up to 1800m.	Variability availed in Lower hills of Uttarakhand and Nepal	Nectar type edible substance at the time of maturity which is sweet and tasty. The inner core of the fruit with nectar is edible.
Bedu (*Ficus palmate* Forsk.) Family-Moraceae	Small deciduous or sub-deciduous tree. Flowering - March-June and fruits mature - May to October. The fruits sub-globose or pyriform, dark purple or pink when ripe..	It is commonly found in open places particularly along thebanks of streams and rivers up to 1800m	Variability availed in Lower hills of Uttarakhand and Nepal	The ripen fruits are edible though the flavour is strong and dis-agreeable. rarely available in market
Ban-nimboo (*Glycosmis arborea* (Roxb.) Family-Rutaceace	An evergreen, glabrous shrub of It flowers and fruits almost round the year. The fruits are depressed, globose, dirty-yellowish or pinkish.	Found in abundant in forest throughout the sub-Himalayan tract up to 600m.	Restricted to some areas of Lower hills of Kumaon, Western Ghats and NEH regions	Fruit edible but rarely available in market
*Viburnum continifolium*D.) Family-Caprifoliaceae	large deciduous shrub. cflowering - April-June, fruits mature September to October.	Grows in the Himalayas at an altitude of 1200–3300m	Variability restricted d to higher hills	Ripe fruit is sweet and edible which is used as fresh and dried.
Kilmora *(Berberis aristata* (Fam.) Family-Berberidaceae	A large, erect deciduous spiny shrub of, which reaches up to a height of 1-2.5m.flowering takes place in the month of April-June and the fruits mature in month of June -July.	Commonly found in forested areas at an altitude of 1800-3000m	Variability availed in Lower hills	Fruit is edible and is sometimes given to children as a mild laxative. The dried berries are also eaten.
Ghingaru (*Crataegus oxyacantha* L.) (English Hawthorn)	Thorny, ever green shrub flowering -April-May fruits mature - July-August.	Found in abundance in forests and open places at an altitude of 1500- 2500 m above msl	Variability available in Lower hills	Fruits edible but rarely available in market

Fruit/Family	Description	Occurrence	Variability	Uses
Mehal (*Pyrus pashia*Buch.) Family-Rosaceae	a small, deciduous tree, flowering - February-March, fruits mature -October to November	in open field borders and forest margins from 1000-2400mthroughout the sub- Himalayan tract up to 600m	Restricted to some areas of Lower hills of HP, Utarakhand and NEH regions	fruits are sour -sweet with some astringency. This is widely used as rootstock for pear.
Chilgoza *(Pinus geradiana* L.) Pine nut Family-Pineceae	Large tree, gymnosperm	Grows wild in the higher regions (3000-4000m) of cold and arid Himalayas in HP and J&K	Restricted to some areas of higher hills	A nut fruit, described as low volume having high protein. The fruits are auctioned to the local people.
Indian Hazelnut, Kabasi, *(Corylus colurna*L./*Corylus jacquemontii*Decne.)	tree is small or medium-sized and deciduous tree /shrub, Flowering - April-May, fruits mature -August to September. Fruits are compressed, globose and hard	Mid and high hills (2000-3500m) of the Himalayas	Restricted to some areas of higher hills	Nuts are collected from fruits and sold in the market
Elaeagnus latifolia- (Syn. Elaeagnus infundibularis – Momiy)Family-*Elaeagnaceae*	Large, evergreen, spreading type, woody shrub with rusty-shiny scales that are often thorny Flowering –September Fruit maturity- April	Mostly grown in semi-wild condition in the back yard gardenin Khasi and *Slangi* in Jaintia Hills	Restricted to some areas of Khasi and Jaintia Hills	Fruits are used for making chutney. jam, jelly and refreshing drink
*Prunus undulata*Buch.-Ham. Family-Rosaceae	A small deciduous tree, upright and moderately branchedFlowering– September-March,Fruits are ovoid, pointed and ripen in December- May	a rare species growing wild in Khasi and Jaintia Hills occurring in mountainous or hilly forests or river valleys	Restricted to some areas of Khasi and Jaintia Hills	This species show resistance to powdery mildew, and is incompatible as rootstock on grafting with peach and plum.

Fruit/Family	Description	Occurrence	Variability	Uses
Prunus nepalensis Family-Rosaceae	Trees are large with grey bark, open, upright branches and oblong-lanceolate leaves. Flowering – October, Fruits mature during August–September	Mid hills of the Eastern Himalayas and NEH regions	Variability available in Lower hills	Fruits are eaten raw and used for wine making due to its imparting purple colour to the wine
*Baccaurea sapida*Muell. Arg.Family-Phyllanthaceae	Evergreen tree up to 5 m height, with long branches from near the ground, particularly when the stem is laden with fruits	Khasi and Jaintia Hills. It is native to Southeast Asia.	Variability available in Lower hills	Ripened fruits are eaten and taste delicious. The rind of fruits is used for making chutney
Calamus meghalayensis Family-Arecaceae	Tree grows up to 2 m height and 0.5-0.6 m diameter with leaf sheaths. Fruiting- February-April	native of Meghalaya. It grows in warmer parts of the Khasi, Jaintia and Garo Hills.	Variability available in Lower hills	Fruit – edible, very delicious
Docynia indica Family-- Rosaceae	Tree is deciduouswith medium to tall in height	It grows in parts of the Khasi, Jaintia and Garo Hills	Variability available in Lower hills	Fruits are eaten as fresh or pickles. used as rootstock for imparting semi-dwarf in apple.
Artocarpus hirsutus Lamk Family- Moraceae	A tree may reach a height of up to 35 m. The fruits ripe in May- June	native to western Ghats, prefers moist, deciduous forests up to 1000 mmsl	Variability found in Western Ghats and Eastern India	The mature carpels, seeds are consumed fresh .Tender fruit used in pickle
Chrysophyllum roxburghii Family -Sapotaceae	A tree up to 30 metres tall, with a trunk diameter of up to 40 cm	native to western Ghats, prefers moist, forests	Variability found in Western Ghats	Ripe fruits are consumed fresh, Can be use as rootstocks for Sapota
Cordia dichotoma	small to moderate-sized deciduous tree, fruiting in April-June	native to India	Variability found all over India	Fruits y used as a fresh fruits .Immature fruits are used in pickling

Fruit/Family	Description	Occurrence	Variability	Uses
Garcinia gummigutta	medium tree found in Western Ghats	native to western Ghats, prefers moist, forests	Variability found in Western Ghats	fruit rind is used as spice in curries and its juice is used for reducing fat
Garcinia xanthochymus	medium tree found in Western Ghats, eastern India and NEH region	native to western Ghats, prefers moist, forests	Variability found in Western Ghats, eastern India and NEH region	Fruits are used fresh and fruit rind is used as spice in curries
Xeromphis spinosa Family- Rubiaceae	Small tree or shrub fruits matures in June-July	found in entire of Western Ghats	Variability found in Western Ghats	whole fruit expect the seed is consumed at ripe stage
Zizyphus rugosa LamK Family- Rhamnaceae	Small tree or shrub, spines. Flowering- December-January fruits mature in February-March	found in entire of arid regions	Variability found in arid regions	small tree or shrub, spines.
*Garuga pinnata*Roxb. Family- Burseraceae	Garuga is a deciduous tree of moist tropics growing up to 13 metres tall	Foundin Indian sub-continent,	Variability found in entire India	The fruits are eaten raw or pickled
Meyna laxiflora Roby Family-Rubiaceae	Large shrub or a small tree and commonly found in evergreen forests	Native of India may be in western Ghats. The plants are found in Western UP, West Bengal, North-east India and Deccan peninsula.	Variability found in entire regions	The fruits are used preparation of pick
Salvadora oleoides Decne Family - Salvadoraceae	Evergreen shrub or tree with a dense crown of numerous, drooping branches	Indigenous to India. It is found growing naturally mostly in Rajasthan and Gujarat and also to some extent in Punjab	Variability found in entire regions	Fruits are used raw or dried
Capparis decidua (Forssk.) Edgew. Family - Capparaceae	Straggling glabrous shrub. Branches are zigzag and green (Phylloclades) with a pair of straight spines	Kair is native to India and western Asia. grows wild in Rajasthan and Gujarat.	Low Variability found in entire regions	green immature fruits are preferred for preparation of pickle and vegetable

Fruit/Family	Description	Occurrence	Variability	Uses
Machilus edulis King Family – Lauraceae	Pumsi is an evergreen species (tree) raise upto a height of about 15-20 m, with spreading branches	It grows from Nepal to Sikkim,Bhutan, Arunachal Pradesh and the whole north-eastern region	Variability found in entire regions	Ripe fruits are consumed fresh, Can be use as rootstocks for avocado
Choerospondia saxillaris Roxb. Family-Anacardiaceae	An evergreen tree that can grow up to 10 metres tall, though is usually smaller.	The fruit tree is inhabitant to Nepal to North Eastern Hill states of India. found growing from 850–1900 m msl.	Variability found in entire regions	Fruits are consumed fresh, pickled and processed for making various products locally called as Mada and candy
Borassus flabellifer L Family - palmae(araeceae)	Robust tree.,may reach a height of 30 metres, unbranched stem,	Native to India subcontinent and south east asia, grows in wide range of conditions in dry to moist tropical and subtropical climate.	Variability found in entire regions	Fruits are used raw or cooked. The immature fruits are pickled

Conservation

Exploration and taxonomical aspects of wild fruit species have been studied done by different botanists and taxonomists. Lot of work on pharmalogical properties of these fruits have also been done in different regions of India. But there is no systematic conservation of wild fruits is flowed in India. Some of the wild species have been conversed by National Bureau of Plant Genetic Resources and it's regional stations in different regions of the country. Some of ICAR institutes, State Agricultural Universities, Research stations, and forest department have also conserved some species of the wild fruits grow in their jurisdiction (Table 3).

Table 3: Wild fruit species conserved at different organizations

Institute	Number of species	Major species and Accessions
IIHR, Bangalore	25	*Alangium savifolium, Flacourtia montana (11), Flacourtia indica (2), Garcinia xanthochymus (13), Garcinia gummigutta, Canthium dicoccum, Antidesmabunius, Artocarpus laucha (1) Azimatetra cantha (2), Grewia subinaeqalis, Phyllanthus acidus (2), Salacia chinensis (2), Artocarpus hirsutus*
CHES ,Chettalli	3	*Garcinia xanthochyma (13), Garcinian gummigutta, Garcinia morella,*
College of Forestry, Ponnampet, Kodagu	12 spp	*Alangium savifolium, Flacourtia montana, Flacourtia indica, Garcinia xanthochymus, Garcinia gummigutta,), Grewia subinaeqalis,*
RPRC, Bhubaneswar	51	*Alangium savifolium, Flacourtia jangomas, Flacourtia indica, Garcinia xanthochymas, Garuga pinnata, Canthiumdicoccum, AntidesmabuniusArtocarpus lauchaAzimatetra cantha, Grewia subinaeqalis, Phyllanthus acidus, Rubus niveus, Salacia chinensis, Capparis sepiaria*
NBPGR and it's regional station	>50 spp.	*Malus baccata, Diospyros lotus; Actinidia callosa, Rubus sp., Meetha Reetha/Sweet soap nut tree (Sapindus mukorossi Gaertner) Pistacia integerrima (wild pistachio), Zanthoxylum armatum (Timoor),* Capparis decidua, Prosopis cineraria, Carissa *spp.,* Cordia spp., Morus *spp.,* Grewia subinaequalis, *Buchnania lanzan, Cordia rohtii, Diospyros tomentosa, Garcinai sp, Elaeocarpus floribundus,Artocarpus hirsutus, A. chaplasha, A. gomezianus, Bouea oppositifolia (1), Garcinia hombroniana (1), G. kydia (1), G. speciosa (6), Mangifera andamanica (1), M. sylvatica (1), Mimusops andamanica (1), Musa acuminata*
CAZRI	7	*SaJvadora oJeoides (Pilu). Prosopis jufiflora, Ker (*Capparis decidua*), Lasora (*Cordia spp.), *Ziziphus nummularia, Prosopis cineraria*

Institute	Number of species	Major species and Accessions
JLN-TBGRI, Trivandrum	>25 spp	*Garcinia imberti, Cycas annaikalensis, Goniothalamus malayanus, G. wayanadensis,., Canthium travancoricum, Grewia palodensis,, Calamus andamanicus, Syzygium parameswaranii, Hydnocarpus pentandrus, Garcinia talbotii, Baccaurea spp.*
Forest dept. and SAU	-	-
ICARCNEHR, Umiam	5	*Elaeagnus latifolia, Prunus nepalensis, Pyrus pashia, Baccaureasapida, Doynia indica*
CIARI, Port Blair	21	*Baccaurea ramiflora, Baccaurea sapida Dillenia indica, Dillenia pentagyna, Diospyros discolor, Ficus racemosa, Flacourtia montana, Garcinia andamanica, Garcinia celebica, Garcinia cowa, Garcinia dhanikhariensis, Garcinia dulcis, Garcinia gummi-gutta, Garcinia kydia, Garcinia xanthochymus, Haematocarpus validus, Malpighia punicifolia, Mangifera camptosperma, Mangifera nicobarica, Semecarpus kurzii, Semecarpus prainii, Syzygium claviflorum*

***Source*:** Websites of Institutes and personal communications

The NBPGR Regional Station, Bhowali (Nainital) is maintaining 132 accessions of 25 wild fruit species collected from different parts of Uttarakhand (Table 4).

Table 4: Wild fruit species conserved at NBPGR Regional Station, Bhowali

Genus	Species	State	District	No. of Accessions
Malus	*baccata*	Uttarakhand	Nainital	14
Chaenomales	*japonica*	Uttarakhand	Nainital	02
Pyrus	*pashia*	Uttarakhand	-	12
Prunus	*armeniaca*	Uttarakhand	-	20
Prunus	*mira*	Utarakhand	Chamoli	02
Prunus	*cornuta*	Uttarakhand	Chamoli	01
Prunus	*jacquemontii*	Uttarakhand	Pithoragarh	01
Prunus	*cerasoides*	Uttarakhand	Nainital	01
Punica	*Granatum*(Darim)	Uttarakhand	-	39
Rubus	*ellipticus*	Uttarakhand	-	03
Rubus	*foliolous*	Uttarakhand	-	01
Rubus	*nepalensis*	Uttarakhand	Pithoragarh	01
Rubus	*nivens*	Uttarakhand	Pithoragarh	01
Rubus	*Species (unknown)*	Uttarakhand	-	02
Rubus	*peniculatus*	Uttarakhand	Pauri	03
Rubus	*mecilentus*	Uttarakhand	Pauri	02
Rubus	*rosaefolius*	J&K	Ramban	01
Vitis	*jacquomontii*	Uttarakhand	-	01
Myrica	*esculenta*	Uttarakhand	-	01

Genus	Species	State	District	No. of Accessions
Embalica	*officinalis*	Uttarakhand	-	04
Cotoneaster	*microphylla*	Uttarakhand	-	01
Citrus	*jambhiri*	Uttarakhand	-	05
Citrus	*medica*	Uttarakhand		09
Citrus	*decumana*	Uttarakhand		02
Citrus	*regulosa*	Uttarakhand		03

***Source*:** NBPGR RS, Bhowali

Utilization

The wild fruits species are source of food for several tribes and forest dweller since time immortal. Some of these species are used as rootstocks for cultivated species. Some species have potential to use in breeding programmes. Some examples are described:

Rootstocks

Wild pear (*mehal*, *Pyrus pashia*) is used as rootstocks cultivated pear in Uttarakhand , Himachal Pradesh and Jamun Kashmir since long. Similar *Kaith/ pareu* (*Malus baccatta*) is used as rootstocks for apple. Similarly wild peaches and wild apricot are used as rootstocks for peaches and pear. There are some species as *Pistacia integerrima* (wild pistachio, kakarsingi) , Wild avocado (*Machilus edulis*), Palepan (*Chrysophyllum roxburghii)* have potential to be good rootstocks for Pistachio, avocado and sapota ,respectively.

Breeding

Some the wild species have resistance to disease pests, cold hardiness and other characters. *Malus baccatta*has resistance to collar rot, root rot and woolly aphid. Diospyros lotus is prolific bearer, some are seedless, very good rootstock and breeding material for persimmon (*D. Kaki).Actinidia callosa*is prolific bearer, promising breeding material for cultivated kiwi. Wild apricot is good source for cold hardiness. These can be unitized in breeding programmes.

New Species

Some of the wild species can be used as new species for crop diversification with standardization of production technologies for these species. Wild jack, wild avocado, native hog plum, Chalta, Burmesegrapes and Khirani are some examples.

Source of nutraceuticals: The fruits of Kokum, Jamun are well known from ancient times for their medicinal as well as nutritional properties. Laboratory and field research suggests that substances may have potential for various future applications, since they have shown cytotoxic, anti-inflammatory, anti-diabetic and anticancer effects in laboratory experiments. Similarly, the other fruits can also be explored for the medicinal and nutritional properties.

Processing and value-addition

Some of wild fruits are used to prepare juices, pickles and other products. The wild hog plum pickle is very popular in Nepal. A lot of hog plum pickle is prepared in Nepal and Eastern hills are exported. Similarly The butter and oil produced form Chiura is widely used in Eastern Uttarakhand and Nepal. The pickles of ker, Bhokar, losada are common in some part of India. These are sources of livelihood for the people of these regions. The processing and value addition of wild fruits may helpful in employment generation in these regions.

Prospects

Agroforestry

It is observed that many of these crops usually grow in wild areas without any much care and special attention. Hence the good way to promote, these fruits by introducing these plants in Farming system-. The thorny plants of Karonda, *Flacoutia*, Azima, Ziziphus can be introduced along the borders, so that it can also act as hedge protecting being a source of additional income.

Not to neglect: Though these crops are grown in wild and have less attention, they have their own unique characters and qualities. Hence there is immense need to promote these fruits by creating awareness among the locals and popularisation of these fruits in the form of value added products.

Future Thrust

- Although some of these fruits are abundantly available in their habitat, it is required for collecting the core Germplasm collection for their conservation, characterization and characterization to mitigate the climate change driven problems in future. Therefore, collection and conservation of Germplasm resources needs special thrust for future benefits. In vivo and in situ conservation approaches may be explored for conservation.
- Promising germplasm accessions need to be evaluated for short listing new varieties possessing commercial attributes and processing qualities.
- Production technology and propagation techniques need to be standardization for commercial production and production and supply of quality planting material of newly developed varieties.
- Storage and handling studies are imperative for effective secondary agriculture activities.
- Keeping the unique Nutraceutical qualities of the fruits, value added products like RTS of health drinks and herbal drinks specialty fruits need to be explored and products popularized through appropriate production and marketing linkage chains.

- Since this group of fruits is an important source for health protective properties, like antioxidants, vitamins and secondary metabolites, etc., there is need to give special attention to probably explore them for addressing new outbreak of diseases and strange health problems.

References

Abdullah A.T.M., Hossain M.A., Bhuiyanv M.K., P2005. Propagation of Latkan (*Baccaurea sapida* Muell.Arg.) by Mature Stem Cutting. Research Journal of Agricultural and Biology Sciences 1: 129 – 134.

Agrahar-Murugkar D., Subbulakshmi G. 2005. Nutritive values of wild edible fruits, berries, nuts, roots and spices consumed by the Khasi tribes of India. Ecology of Food and Nutrition 44: 207 – 223.

Arora R. K., Nayar E. R. 1984. Wild Relatives of Crop Plants in India. National Bureau of Plant Genetic Resources. Science Monograph No. 7, New Delhi, 115 p

Arora, R.K., E.R. Nayar and A. Pandey. 1990. Plant genetic resources and their conservation, pp. 42-66. In: Charter for Nature (Dwivedi S.N. and Bhatt V.S. (Eds.)). Department of Ocean Development, Govt. of India, New Delhi.

Ghora C., Panigrahi G. 1995. The family Rosaceae in India. In: Bishen Singh, Mahendra Pal Singh (Eds.). Revisory studies on six genera (Prunus, Prinsepia, Maddenia, Rosa, Malus and Pyrus). Dehradun, India pp. 481.

Jaisal, V. 2010. Culture and ethenobotany of Jaintia tribal community of Meghalaya, NE, India. A mini-review. Indian Journal of Traditional Knowledge 9: 38 – 44.

Kayang H. 2007. Tribal knowledge in wild edible plants of Meghalaya, North East. Indian Journal of traditional Knowledge 6: 177 – 181.

Murthy, S.R. and S. Pandey, 1978. Delineation of agro-ecological regions of India, 11th Congress, International Society of Soil Science, Edmonton, Canada, 17-27 June 1978.

Nayar, M.P. 1980. Endemism and patterns of distribution of endemic genera (*Angiosperms*). J. Econ. Tax. Bot. 1: 99-110.

Nayar, M.P. 1989. In-situ conservation of wild flora resources. National Symposium on Conservation and Sustainable Management of India's Genetic Estate. WWF, New Delhi, 3-4 November 1989.

Sehgal, J.C., Mondal, D.K. Mondal C. and Vadi S. 1990. Agro-ecological regions of India. NBSS Tech. Bull. No. 24. NBSS & UP, Nagpur. 73 p.

Tripathi, P.C. 2020. Wild fruits of India. NIPA, New Delhi, pp174.

Tripathi, P. C., Karunakaran, G., Sankar, V. and Senthil Kumar, R. 2015. Survey and conservation of indigenous fruits of Western Ghats. J. Agric. Sc. and Tech. A.5 (7):608-15.

Uthaiah, B.C. 1995. Wild edible fruits of Western Ghats – A Survey. Higher plants of Indian sub continent. Vol. III. 67.99. Bishen Singh Mahendra Pal Singh, Dehra Dun.

9

Advances in Precision Viticulture

Sharmistha Naik and R.G. Somkuwar

ICAR-National Research Centre for Grapes, Manjari Farm, Solapur Road Pune 412 307, Maharashtra, India

Precision viticulture aims to optimize vineyard management, reduce use of resources, reduce environmental impact and maximize yield and quality of grapes. New technologies as UAVs, satellites, proximal sensors, variable rate machines (VRT) and robots are being developed and used more frequently in some parts of the world in recent years. Developments and abilities of computers, software and informatic systems to read, analyze, process and transfer a huge amount of data are major milestones in precision viticulture. In addition, different decision support systems (DSSs) for making better crop management decisions at the right time also assist vine growers. In India having fragmented small vineyards, relatively cheaper technologies like UAV, proximal monitoring through various tools, and DSSs developed by ICAR-NRCG for Grapes, Pune, can be used by individual grape grower or through farmers' cooperatives/groups to make grape cultivation technologically, economically and environmentally viable. Therefore, current status of precision viticulture technologies and their potential applications in viticulture have been discussed.

Grape (*Vitis* spp. L.), is most important fruit crop grown in the world. Total world grape production was 78.03 million tonnes from an area of 6.950 million ha during 2022 (FAO STAT, 2022). In India, grape is cultivated in almost all parts having diverse climatic conditions ranging from Himachal Pradesh to South India. India produced 2.94 million tonnes of grapes from 0.147 million ha in 2019-2020 (DA&FW, 2022). Grape is a high input crop needing application of several expensive inputs including pesticides. As such cost of grape production is more and indiscriminate use of inputs leads to food safety issues and environmental degradation. Precision viticulture aims at reducing the input costs by following need-based cultural/crop protection practices; applying need-based inputs; increase grape yield and quality while minimizing environmental impact. (Gebbers and Adamchuk, 2010; Urretavizcaya *et al.*, 2017). In simple words, precision farming is "doing the right thing, in the right place at the right time" (Pierce and

Novak, 1999). Vineyards are generally spatially variable and heterogeneous with regard to their location, soil quality, cropping practices and weather conditions (Bramley, 2003). They, therefore, require specific agronomic and crop protection managements to address the real needs of the crop, in relation to these variabilities (Proffit *et al.*, 2006). Recent developments in new precision farming technologies for supporting vineyard management allows improved productivity, quality and food safety, and reducing environmental impact. The essential steps in precision viticulture are assessing variation and its management. Components that are responsible for vineyard performance in terms of yield and quality vary in space and time. The spatial variability in vineyards can be assessed and mapped using surveys, high resolution satellite/aerial data and modeling. Once the variation is adequately assessed, accurate cultural (fertilizers, irrigations etc) and crop protection (pesticides) inputs are applied in site-specific manner to reduce cost of cultivation and environmental impacts. Precision viticulture is mainly used for optimization of inputs, differential grape harvesting, yield forecasting, and accuracy of canopy and soil sampling (Bramley, 1999; Bramley and Lamb, 2003). In comparison to other crops, initiation of precision viticulture is of recent origin. This was due to the complexity of vineyard data, requirement of high-resolution images to differentiate canopy from soil and big data processing capacity to manage spatial information.

There are mainly two aspects of precision viticulture. The monitoring technologies, that are used for mapping spatial variability and the technologies that are utilized to provide site-specific cultural inputs known as variable-rate technologies (VRTs) and robotics. In addition, supplementary technologies such as disease forecasting and Decision Support Systems (DSSs) also assist grape growers to take prophylactic actions. Available precision tools and their applications in viticulture are described below.

Information and Communication Technology (ICT)

Computers, mobile computing systems, internet and mobiles are major components for information gathering, processing and transferring. For collection of huge data from the fields mobile computing systems having high speed microprocessors that could operate at very high speeds were developed to store and transfer massive amounts of data. In addition to the computer hardware, there had been significant progress in development of precision farming software. Development of software for precision agriculture is more an experience than an application. The most important computer application in precision agriculture is Computer Vision (CV). It is a technology that acquires, processes, analyses and extracts data of images to provide numerical or symbolic information such as the estimation or prediction of key traits of the targeted object, in a fast, contactless, reproducible and accurate manner (Vidal *et al.*, 2013). The CV comprises a set of techniques associated with artificial intelligence, that allows a computer to understand and read an image to derive precise information (Ballard and Brown, 1982). Such type of electronic integration has played important role in furtherance of precision farming during the last few decades.

Monitoring Technologies

Several technologies and sensors are deployed for acquiring intra-vineyard and inter-vineyard georeferenced information. Mainly two types of technologies, viz. remote sensing technologies and proximal sensing technologies are used to monitor vineyards.

Remote Monitoring

Satellites

The Global Positioning System (GPS) is a space-based satellite navigation system that provides highly accurate, rapid and timely information. GPS receiver calculates position of the vineyard on earth (up to 15 m accuracy) based on the information it receives from more than 4 satellites. However, a network of fixed ground-based reference stations can correct the positions indicated by satellite systems and provide location accuracy in centimeters. The first satellite, Landsat-1 launched in 1972 was equipped with multispectral sensor and provided a spatial resolution of 80 m with a revisit time of approximately 18 days. The last launched Landsat satellite, Landsat-8 (Ridwan *et al.*, 2018) operates in visible, near-infrared, short wave infrared and thermal infrared spectrums. Other high-resolution satellites that are being used in remote sensing are:

Sentinel-2 (https://sentinels.copernicus.eu/web/sentinel/user-guides/sentinel-2-msi),

RapidEye (https://earth.esa.int/eogateway/missions/rapideye?text=rapideye)

WorldView-1, -2 and -3 (https://earthdata.nasa.gov/worldview)

Sentinel-2 data are open-source and freely downloadable. RapidEye has been used to evaluate Normalized Differential Vegetation Index (NDVI) in order to characterize the vine vigor and some phenolic parameters. Simple linear relationships between NDVI at berry set, pre-veraison and ripening has been found to evaluate sugar content and anthocyanins at harvesting (Santangelo *et al.*, 2013). RapidEye has also been used to evaluate the Leaf Area Index (LAI) in vineyards demonstrating a good correlation with in-field estimation of evapotranspiration (Vanino *et al.*, 2015). WorldView has been used to detect vineyards, canopy estimation and discrimination of varieties (Karakizi *et al.*, 2016).

Aircrafts

Aircrafts monitor large areas with a long flight range and can carry heavy and multiple sensors at a time. Aircrafts provide better ground resolution (up to 10 cm) depending upon flying altitude. However, aircrafts are economically feasible only on areas bigger than 10 ha (Matese and Di Gennaro, 2015).

UAV

UAVs fly autonomously and can be controlled remotely by a ground pilot. UAVs are fixed with several flight control sensors (gyroscopes, compass, GPS, pressure sensor and accelerometers) which are controlled by a microprocessor. They can be equipped with a variety of sensors for performing a wide range of monitoring. UAVs provide very high spatial resolution on the ground (down to cm) and they are ideal for small-to-medium fields (1–20 ha), characterized by high fragmentation and high heterogeneity. There are two types of UAVs having rotary wings and fixed-wings. Rotary-wings UAVs having less flight time (up to 30 minutes) are generally used for monitoring small fragmented areas (up to 20 hectares), while fixed-wings UAVs having more flight time (up to 1 hour) allow monitoring large unfragmented flat areas (Ammoniaci *et al.*, 2021). In comparison to satellites, UAVs have much higher resolution, therefore, use of UAVs with high-resolution sensors is suggested for assessing vine structure and inter-row spacing of vineyards (Khaliq *et al.* 2019). UAVs with true colour cameras (RGB) and multispectral or thermal sensors have been used in viticulture to detect vineyards/vine rows (Comba *et al.*, 2015); estimation of grape yield (Di Gennaro *et al.*, 2019); assessing vegetative vigor and detecting missing plants (Matese *et al.*, 2018); and estimating water stress and assessing grape maturity (Soubry *et al.*, 2018). UAV technology has shown to be economically compatible with the agronomic costs in vineyards with more than 40 ha having lots of fragmented vineyards (Mondino *et al.*, 2017).

Proximal Monitoring

In addition to aerial monitoring, several tools having different types of sensers are available for proximal monitoring. Radiometric and fluorometric sensors and some application programmes such as VitiCanopy are used to assess canopy vigour, stress, chlorophyll content, nitrogen concentration and leaf area index. Geophysical and spectro-radiometric sensors are employed for assessing soil composition and structure. While, optical sensors (fluorometric and spectrophotometric sensors) determine grape quality and berry ripening. These sensors are mounted on farm machinery or some of them such as optical sensors are fixed on handheld devices and various measurements are carried by moving vehicles or manually for getting precise ground information. In addition, wireless sensor network (WSN) technologies configured in vineyards provide a highly useful and efficient tool for remote and real-time monitoring of important variables of vineyard.

Canopy Assessment by Proximal Sensors

Radiometric sensors measure electromagnetic radiation reflected by vegetation and it can allow the rapid collection of important information for indirect measurement of the canopy state. The spectral response of canopy is mainly related to the radiation absorption by the pigments, which allows the assessment of vegetation cover but also the characterization of its health (Casa, 2017). If solar radiations or external light source is used, radiometric sensors are passive and results are affected by variations in lighting conditions. But, if they use their own artificial light source, they are active and less affected by lighting variations. Also, use of inbuilt lighting by active sensors permits recording of information in night. Two commercially available radiometric sensors are OptRx (Ag Leader Technology, Ames, IA, USA) used to assess the vigor of the crop and CropSpec (Topcon, Livermore, CA, USA) used to determine chlorophyll content that can be correlated to nitrogen concentration in leaves.

Another optical sensor is fluorometer that measures fluorescence emitted by the chlorophyll. Based on this information onset of stress conditions in the vineyard can be predicted. The information derived from the fluorescence measurement represents functional state of the plant, and in turn photosynthetic potential of canopy. From this photosynthetic potential expressed as electronic transport rate (ETR) and two other parameters, viz. leaf and air temperatures, Losciale *et al.* (2015) developed an index (I_{PL}) capable of estimating the rate of net photosynthesis of the leaves (2015).

VitiCanopy developed by De Bei *et al.* (2016) is one of the most used Apps to measure the LAI in vineyards. This is done through images acquired by a smartphone/tablet. The image is then analyzed by the App to calculate both the LAI and the porosity of the canopy. The results obtained through VitiCanopy are georeferenced to know the variation in the entire vineyard. Such information can be extrapolated to estimate the vigour, productivity and quality of produce from a vineyard.

Soil Assessment by Proximal Sensors

Monitoring of soil health is one of the most important applications of precision viticulture. Wide range of sensors are used for assessing soil variability. Electrical conductivity (EC) of the soil is directly related with several soil properties such as texture, depth, water retention capacity, organic matter content and salinity. EC of the soil an be measured using mobile platforms equipped with sensors and GPS (Corwin and Lesch, 2005). Two types of sensors viz. Electrical Resistivity Sensors and Electromagnetic Induction Sensors are used to measure soil EC. Commercially available electrical resistivity sensors are Veris 3100 (Veris Technologies Inc. Salina, USA) and Automatic Resistivity Profiling System (Geocarta Ltd, Paris, France). While, DualEM (DualEM, Milton, Canada) and EM-31 and EM-38 (Geonics Ltd, Mississauga, Canada) are commercially available electromagnetic induction sensors. Knowledge about spatial variability of soil characteristics is important for understanding variability in the physiological response of the vines (Priori *et al.*, 2019).

Other type of sensors used for soil assessment are Geophysical sensors. These sensors measure potential drop in introduced current in the soil. Such drop is related with EC. The EC of the soil is related to the soil texture, humidity, salinity, degree of compaction and the presence of gravel and pebbles in the soil. Invasive geophysical sensors (mobile soil resistance-meters) measure the apparent resistivity of the soil by using direct contact electrodes and non-invasive sensors (electromagnetic induction sensors, ground penetrating radars) assess soil properties by using electromagnetic fields. Another type of non-invasive sensor is the ground penetrating radars. These radars gather information of the soil through emission of electromagnetic pulses and phenomena of reflection and refraction of the different materials of the soil.

Third type of sensor used for assessing soil is Spectroradiometer. Each soil has its own spectral signature which is the intensity of the reflected radiation depending on the intrinsic spectral behavior of soil constituents (minerals, organic matter, water etc). Gamma-ray based spectroradiometers consist of a scintillator crystal, generally, made of cesium iodide (Csl) or sodium iodide (Nal), which emits photons when hit by gamma rays. Reflected photons by the soil are measured to characterize soil properties. Vis-nir reflectance spectroradiometer is rapid and relatively cheap alternative to gamma-ray based spectroradiometer. This sensor can measure several soil properties in a single scan.

Grape Quality Assessment by Proximal Sensors

Non-destructive monitoring of grape quality parameters is based on optical sensors. Manual devices or tools fixed with optical sensors are used for non-destructive monitoring of grape quality parameters. Some of the tools that are in common use are: Multiplex (Force-A, Orsay, France) and Spectron (Pellenc SA, Pertuis, France). Multiplex is mainly used to estimate the nitrogen status of grapevine leaves. In addition to nitrogen status, Multiplex can also assess contents of chlorophyll, flavanols and anthocyanins in the leaves and grapes (Cerovic *et al*., 2008). Another sensor used to assess the quality of the grapes is the Spectron (Pellenc SA, Pertuis, France). This sensor is integrated with GPS and can be used for non-destructive measurements of quality-related parameters, like sugar, acidity, anthocyanins and water content (Matese and Di Gennaro, 2015).

Yield Assessment by Proximal Sensors

Many systems are available to obtain georeferenced yield information. These are: Harvest Master Sensor System HM570 (Juniper Systems Inc., Logan, USA), Canlink Grape Yield Monitor 3000GRM (Farmscan, Bentley, WA, Australia), and Advanced Technology and Viticulture (ATV) (Advanced Technology Viticulture, Joslin, SA, Australia). These systems are mounted on mechanical harvesters and for yield assessment they use volumetric grape measurement on the discharge conveyor belt of the harvester and/or direct measurement of the transported grape weight by means of load cells.

Assessment of Microclimate and other Parameters in Vineyard using Wireless Sensor Network

Wireless sensor network (WSN) technologies provide a useful and efficient tool for remote and real-time monitoring of important variables of a vineyard. A WSN consists of wireless peripheral nodes and a sensor board equipped with sensors and a wireless module for data transmission from nodes to a base station. At base station the data are stored and these data accessible to the end user for taking appropriate vineyard management decisions. Applications and configuration of WSN in vineyards has been adequately described (Burrell *et al.*,2004) and WSN has been successfully used for prolonged temperature and microclimate measurements (Beckwith *et al.*, 2004; Matese *et al.*, 2013). The primary application of WSN is the acquisition of micrometeorological parameters at vine canopy and soil level. Recent developments of new kinds of sensors such as dendrometers and sap-flow sensors made it possible to continuously measure plant water status for irrigation scheduling.

Forecasting and Decision Support Systems (DSSs)

Precision viticulture aims at reducing input costs; enhancing productivity and quality of the produce; and protecting environment. Reliable and timely forecasts provide useful information for planning in advance. Viticulture is highly input and cost intensive. In crops, production and attack of insect pests and diseases and production estimates are the two major aspects that need attention. Insect pests and diseases are major causes of reduction in productivity and their appearance and intensity can be forecasted based on weather parameters. Timely application of remedial measures reduces the yield loss. Forecasts of crop production before harvest are required for different policy decisions related to storage, value addition, pricing, marketing, import-export, etc. Other tools for timely application of inputs are Decision Support Systems (DSSs). Decision support systems (Bidgoli, 1989; Power, 2002) are interactive, computer-based systems which help users to accurately identify specific problem (mainly nutrient deficiency and incidence of insect pest/disease) based on symptoms in the vineyard. Once the problem is identified, DSS suggests management strategies for it. In viticulture, DSSs have been of great help in making appropriate decisions at appropriate time, thereby, reducing crop losses to a greater extent. Several DSSs are available for guiding the vine growers to take suitable management strategies to address macro- and micro-fertilizer deficiencies, incidence of diseases and infestations by inset pests. Metos® (Pessl Instruments GmbH, Werksweg, Weiz, Austria) is commercially available decision support system for grape production.

ICAR-National Research Centre for Grapes, Pune, India (https://nrcgrapes.icar.gov.in) has developed some DSSs on irrigation, soil nutrition, diseases and insect pests for assisting Indian grape growers for judicious use of inputs.

Variable Rate Technology (VRT)

Farm machines fitted with VRT technology (sensors) and GPS make precise operations based on prescription maps prepared using remotely and proximally recorded data of the vineyard. Based on these prescriptions, these VRT machines apply various inputs (fertilizers, pesticides and other inputs) at right doses and at right places without manually changing rate settings. This is possible due to GPS module and electronics, consisting of control units and proportional servo-valves mounted on VRT machine. In addition to fertilizers and pesticide applications, such thematic prescription maps can be developed for yield and quality parameters like acidity, polyphenols and anthocyanin. Another application of VRT machines is selective harvesting. Selective harvesting is split picking of fruit according to a vineyard vigor mapping, grape composition, and quality as per market demand. For example, grapes grown for juice purpose, sugar content is an important target for harvest, while, combinations of sugar content, anthocyanin content and acidity could be targets for grapes grown for wine purpose.

It has been reported that a variable rate fertilization can save up to 30% of fertilizers (Donna *et al.*, 2013) and a variable rate application of pesticides can save up to 30% of pesticides and increase the profit up to 20% (Casa, 2017).

Robots

The use of robotics in agriculture is still in infancy, however, the agricultural robotics is poised to change agricultural scenario in the world by 2050 (Blackmore, 2014). In viticulture, The VineRobot project at the University of La Rioja, Spain is aimed to develop a new agricultural robot, equipped with noninvasive monitoring technologies and GPS. Such robots are expected to perform a proximal monitoring of various critical parameters such as yield, vigor, water stress, quality of the grapes and assist vine growers to improve vineyard management. The Commercially available Wall-Ye robot (http://www.wall-ye.com/) can move along vine rows and acquire data on each vine, thereby, producing detailed vineyard map. Wall-Ye can prune about 600 plants per day and can be remotely controlled by iPad. Another robot called VineGuard developed by Ben-Gurion University, Israel is designed for foliar applications. This robot can move within the vineyard using a complex set of sensors. Another commercially available robot is "Vitirover" (https://www.vitirover.fr/en-home). This robot was developed by Chateau Coutet (Saint Emilion, France) and it can cut the weeds up to a distance of 2–3 cm from the base of the vine. Solar power system is fixed on this robot. Several other robots and robotic tractors are under various stages of development and testing.

Precision Technologies in Viticulture

The introduction of new technologies for vineyard management facilitate enhancement in efficiency, productivity and quality and reduction in cost of cultivation and environmental impact. Precision technologies have been used for various purposes in vineyards.

Soil Properties

Variations in soil fertility impact vineyard performance. Mobile platforms fitted with proximal soil sensors can be moved over the field to acquire geo-referenced soil data. High-resolution maps developed using this soil data and GPS provide valuable soil information on spatial variability of soil properties and topography which are relevant when establishing new vineyards and/or redeveloping existing vineyards (Bramley, 2010). Data acquired from Recently, data acquired from electromagnetic induction sensors and multispectral imaging were combined to estimate vineyard soil and vine vigour variability (Hubbard *et al.*, 2021). Use of electrical resistivity sensors is limited to determination of soil nutrients, pH and organic matter; however, optical and electrochemical sensors can be used for assessing patterns of chemical fertility parameters of the soil (Joseph *et al.*, 2010). Performance of mobile near infrared spectrometry for in situ soil mapping and gamma-ray spectrometers for detecting the presence of particular minerals have been explored in vineyards by Schirrmann *et al.* (2013) and Simone *et al.* (2014), respectively.

Vegetative Growth, Nutritional Status and Canopy Architecture

Remote and Proximal Sensing-Derived Spectral Indices and Biophysical Variables have been widely employed to evaluate vine canopy growth and health in commercial vineyards (Darra *et al.*, 2021). Nutritional status of leaf nitrogen can either be assessed using fluorescence-based portable sensors. Such sensors can also be mounted on mobile platforms (Diago *et al.*, 2016a) Another technology known as light detection and ranging (LiDAR) has been shown to be a powerful technology for the rapid and non-destructive assessment of canopy and leaf parameters in vineyards (Arno *et al.*, 2013). Some other precision viticulture applications are RGB camera imagery acquired by UAV for estimating canopy biomass and detecting missing plants (Di Gennaro and Matese, 2020) and UAV-based point cloud analysis to detect the severity in canopy decline caused by dieback-like disease symptoms (Ouyang *et al.* 2021). Canopy architecture, including fruit and leaf exposure and canopy porosity can assessed using machine vision technologies (Diago *et al.*, 2016c) or on-the-go assessment using RGB image analysis (Diago *et al.*, 2019). "VitiCanopy" App uses smartphones as imaging devices to measure vine performance attributes such as canopy vigour, LAI and porosity (De Bei *et al.*, 2016).

Pests and Diseases

Grape cultivation in India faces serious threats from several diseases and insect pests. Major fungal diseases are downy mildew, powdery mildew and anthracnose, whereas, mealybug and thrips are major insect pests that cause enormous economic losses to grape cultivation. Use of appropriate pesticide in right dose at right time and right place holds the key for effective pest management. Visual inspection of diseases and insect pests is time consuming, subjective, risky and expensive. Use of new sensors and monitoring tools has been shown to provide objective, rapid, cheap and reliable diagnosis of pests and diseases in vineyards (Lee and Tardaguila, 2021). These technologies have opened new frontiers to map disease/pest across vineyards in order to apply fungicides differentially using VRT technology (Chen *et al.*, 2020). Using computer vision and deep learning, Gutierrez *et al.* (2021b) could detect and differentiate downy mildew and spider mite in commercial vineyards. Computer vision, hyperspectral imaging and machine learning have been applied for assessing downy mildew in grapevines (Rose *et al.*, 2016). Use of hand-held UV-LED fluorimeter for early detection of stilbenoid phytoalexins associated with Downey mildew infections on grape leaves was reported by Latouche *et al.* (2015). Wine producers set tolerance levels for diseases such as powdery mildew and Botrytis bunch rot. Two apps have been released (PMapp® and RotBot®) which allow users to quickly assess the severity of these diseases on clusters and calculate both the incidence and severity of the disease and also record other parameters such as date, time and geo-reference position (Hill *et al.*, 2014; Birchmore *et al.*, 2015).

Forecasting models and DSSs are valuable tools to manage biotic stresses in viticulture. During the last few years several computer-based disease and insect pest prediction models and decision support systems (DSS) have been developed in many crop plants including grapes. Rosa *et al.* (1993) developed PLASMO (*Plasmopara* Simulation Model) model for forecasting downy mildew in *Vitis vinifera*. The model simulates the development of downy mildew on the basis of climatic conditions. Chen *et al.* (2020) developed an efficient and accurate machine learning algorithms for predicting Downey mildew that reduced at least 50% of fungicide use in Bordeaux region of France. Brischetto *et al.* (2021) developed a mechanistic model to predict secondary infections of *Plasmopara viticola* and their severity as influenced by environmental conditions. Powdery mildew caused by *Uncinula necator* fungus is another important disease of grapes. Fungal growth, conidia formation and germ tube formation are mainly influenced by temperature. Management strategy of Powdery Mildew disease on grapes by a decision support system based on weather and image processing was developed by Mundankar *et al.* (2007). In warm tropical and sub-tropical conditions, anthracnose disease caused by *Colletotrichum gloeosporioides* affects tender shoots and young fruits reducing vine productivity and fruit quality. Disease incidence and severity have been shown to be dependent on weather parameters. Ghule *et al.* (2015) reported favourable weather conditions for development and progression of anthracnose. These were rainfall with minimum temperature between 22.33 to 23.12 ^{0}C, maximum temperature between 30.12 to 31.88 ^{0}C, RH-1 more than 67% and RH-2 more than 51%.

In India, major insect pests in grapes are mealybugs, thrips, fleabeetle, leafhoppers, stem borer and mites in order of their economic damage to the crop. Indiscriminate use of pesticides not only increases cost of cultivation but also is harmful to the environment and human health. Therefore, pest management in viticulture should follow an integrated approach, including best agronomic practices, advance forecasting using models, decision support systems (DSSs), biological control agents and chemical sprays for reducing pesticide use (Pertot *et al.* 2017). Chougule *et al.* (2019) developed a grape crop protection decision support system named as "PDMGrapes" using ontology, semantic web rule language and image processing techniques for management of insect pests and diseases on grapes in hot tropical region of India. Lessio *et al.* (2021) have reviewed mathematical models and DSSs developed to predict key aspects of insect pests. These models are used for forecasting seasonal occurrence and spread of insect pests. Under integrated pest management (IPM), one of the most important components is the reduction in number of pesticide sprays and pesticide doses. Roman *et al.* (2022) developed DOSA3D decision support system that allows the dose to be adjusted to the specific scenario. DOSA3D calculates the optimal application volume rate by estimating the leaf area index and takes into account the overall spraying efficiency and the pest or disease to be controlled. DOSA3D could achieve pesticide savings up to 53% in fruit trees and 60% in vineyards.

Water Status in the Vineyard

Current changing climates are characterized by water scarcity and higher temperatures; therefore, assessment of vineyard water status and irrigation management are becoming increasingly important. Thermal imaging technology has been used extensively to determine vineyard water status manually (Grant *et al.*, 2016) or remotely using aerial platforms (Bellvert *et al.*, 2014). Thermographic instrument can also be mounted to a ground-based vehicle for on-the-go mapping of vine water status in commercial vineyards (Gutierrez *et al.*, 2021a). Small thermal camera attached to a smartphone has also been used for assessing vine water status (Petri *et al.*, 2019). Recently, near infrared spectrometers mounted on mobile platforms have been used to assess vine water status in a stop-and-go mode (Diago *et al.*, 2017). A mobile phone application (ApeX-Vigne) has also been developed for monitoring vine water status in vineyards (Pichon *et al.*, 2021).

Yield Components and Crop Forecasting

During harvesting of grapes, yield can be easily monitored by measuring the weight of berries flowing across load cells fitted to mechanical harvesters (Taylor *et al.*, 2019). Computer vision systems have recently been used to assess grape yields based on cluster compactness (Palacios *et al.*, 2019), number of berries per cluster (Aquino *et al.*, 2018), cluster weight (Liu *et al.*, 2020b) and berry size (Roscher *et al.*, 2014). Three android based Apps, viz., vitisFlower® (Aquino *et al.*, 2015), vitisBerry® (Aquino *et al.*, 2015) and 3DBunch® (Liu *et al.*, 2020a) have been developed for measurements of flower, berry and bunch parameters. For assessing large commercial vineyards, automatic RGB image capturing gadgets are mounted on mobile vehicles and yield predictions are made using computer vision technology (Palacios *et al.*, 2020).

Fruit Composition and Quality Attributes

Near infrared spectral analyzers are capable of monitoring dynamic changes in berry composition during the ripening period and, therefore, provide an alternative option to destructive quality testing procedures. Portable near infrared spectrometers have been used for determining total soluble solids and other compositions in grape berries under both laboratory and field conditions (Barnaba *et al.*, 2013) This technology has also been successfully implemented on-the-go from a moving vehicle to monitor the dynamics of berry ripening in the vineyard (Fernandez-Novales *et al.*, 2019). Another non-destructive technology consisting of hyperspectral imaging (HIS) to fingerprint the colour pigments of whole grape berries has also been developed for laboratory testing (Diago *et al.*, 2016b) and on-the-go from a mobile platform in the vineyard (Benelli *et al.*, 2021). A chlorophyll fluorescence-based sensor Multiplex® (FORCE-A, Orsay, France) has been developed for contact-free measurement of anthocyanin content in grape berries under laboratory (Ben Ghozlen *et al.*, 2010). These sensors can be mounted on harvesters for data acquisition on-the-go (Bramley *et al.*, 2011) Computer vision technology is also being used for sorting berries in many wineries before crushing. Vegetation indices derived from remote and proximal sensors were also used to evaluate quality characteristics of table grapes (Anastasiou *et al.*, 2018). Proximal sensing proved to be more accurate in terms of table grape yield and quality characteristics compared to satellite-based monitoring.

Conclusion

In general, these technologies are aimed at cost reduction in crop management, improving yields and crop quality, and environmental sustainability with a rational and judicious use of chemical inputs. Choosing appropriate technologies for different type of application is important in precision viticulture. For example, though satellites and aircrafts are excellent tools for developing prescription maps for VRT machines, satellites have low resolutions and the operational cost of aircrafts is high. In this regard, UAV platforms give high resolution, flexibility of use and economic feasibility. However, they can only monitor small areas. The VRTs are well developed and widely used, especially in chemical applications. Remote and proximal monitoring technologies and VRT machinery are being extensively used in some parts of the world, while robotics is in an experimental stage. In India, where majority of vineyards are small and fragmented, use of expensive precision technologies may not be feasible. However, at village/block/ district level, some of these technologies like UAV and proximal sensors can be adopted by farmers' cooperatives to make viticulture technologically, economically and environmentally sustainable.

References

Ammoniaci M, Kartsiotis SP, Perria R, and Storchi P. 2021. State of the Art of Monitoring Technologies and Data Processing for Precision Viticulture. Agriculture 11:201. https://doi.org/10.3390/agriculture 11030201

Anastasiou E, Balafoutis A, Darra N, Psiroukis V, Biniari A, Xanthopoulos G and Fountas S. 2018. Satellite and Proximal Sensing to Estimate the Yield and Quality of Table Grapes. Agriculture 8: 94. https://doi:10.3390/agriculture8070094

Aquino A, Barrio I, Diago MP, Millan B ad TardaguilaJ. 2018. VitisBerry®: an Android-smartphone application to early evaluate the number of grapevine berries by means of image analysis. Computers and Electronics in Agriculture 148:19–28.

Aquino A, Millan B, Gaston D, Diago MP and Tardaguila J. 2015. vitisFlower® development and testing of a novel android-smartphone application for assessing the number of grapevine flowers per inflorescence using artificial vision techniques. Sensors 15: 21204–218.

Arno J, Escola A, Valles JM, Liorens J, Saniz R, Masip J, Palacin J, and Rossel-Pola JR. 2013. Leaf area index estimation in vineyards using a ground-based LiDAR scanner. Precision Agric 14, 290–306.

Ballard DH and Brown CM. 1982. Computer Vision. Prentice Hall, Englewood Cliffs, New Jersey. 539 P

Barnaba FE, Bellincontro A and Mencarelli F. 2013. Portable NIR-AOTF spectroscopy combined with winery FTIR spectroscopy for an easy, rapid, in-field monitoring of Sangiovese grape quality. Journal of the Science of Food and Agriculture 94: 1071–77.

Beckwith R, Teibel D, and Bowen P. 2004. Report from the field: results from an agricultural wireless sensor network. In: Proceedings of the 29th Annual IEEE International Conference on Local Computer Networks, Tampa, FL, USA, pp. 471–78.

Bellvert J, Zarco-Tejada PJ, Girona J and Fereres E. 2014. Mapping crop water stress index in a 'Pinot-noir' vineyard: comparing ground measurements with thermal remote sensing imagery from an unmanned aerial vehicle. Precision Agriculture 15:361–376.

Ben Ghozlen N, Cerovic ZG, Germain C, Toutain S and Latouche G. 2010. Non-destructive optical monitoring of grape maturation by proximal sensing. Sensors 10:10040–68.

Benelli A, Cevoli C, Ragni L and Fabbri A. 2021. In-field and non-destructive monitoring of grapes maturity by hyperspectral imaging. Biosystems Engineering 207:59-67.

Berry JK. 1995. Spatial Reasoning for Effective GIS. CIS World Books, Ft. Collins, Co.

Bidgoli H.1989. Decision support systems: Principles and practice. West Publishing Company. 368 pp.

Birchmore W, Scott E, Zanker T, Emmett B and Perry W. 2015. Smart-phone app field assessment of powdery mildew. Australian and New Zealand Grapegrower and Winemaker 622: 46-47.

Blackmore S. 2014. A vision of farming with robots in 2050. Address to the Oxford Farming Conference. January 8, 2014. Oxford, UK. Book Gallery, Greenwood Publishing Group. 251 P.

Bramley R. 2003. Smarter thinking on soil survey. Australian and New Zealand Wine Industry Journal 18:88–94.

Bramley RGV and Lamb DW. 2003. Making sense of vineyard variability in Australia. In: Proceedings of an International Symposium, Held as Part of the IX Congreso Latinoamericano de Viticultura y Enologia, Chile; Centro de Agricultura de Precision, Pontificia Universidad Catolica de Chile, Facultad de Agronomia e Ingeneria Forestal: Santiago, Chile. pp. 35–54.

Bramley RGV, Le Moigne M, Evans S, Ouzman J, Florin L, Fadaili EM, Hinze CJ and Cerovic ZG. 2011. On-the-go sensing of grape berry anthocyanins during commercial harvest: development and prospects. Australian Journal of Grape and Wine Research 17: 316–26.

Bramley RGV. 1999. Precision Viticulture: Managing vineyard variability for improved quality outcomes. In: Viticulture and Wine Quality. Woodhead Publishing Series in Food Science, Technology and Nutrition; Woodhead Publishing: Cambridge, UK, pp. 445–80

Bramley RGV. 2010. Precision Viticulture: managing vineyard variability for improved quality outcomes. In: Managing Wine Quality Vol-I [Reynold AG (Ed)]. Woodhead Publishing, UK. pp 445-80.

Brischetto C, Bove F, Fedele G and Rossi V. 2021. A Weather-Driven Model for Predicting Infections of Grapevines by Sporangia of Plasmopara viticola. Front. Plant Sci. 12:636607.

Burrell J, Brooke T, and Beckwith R. 2004. Vineyard computing: sensor networks in agricultural production. IEEE Pervasive Comput. 3:38–45.

Casa R. 2017. Agricoltura di precisione. Metodi e tecnologie per migliorare l'efficienza e la sostenibilita dei sistemi colturali. Edagricole Calderini. 450 P.

Cerovic ZC, Goutouly JP, Hilbert G, and Moise N. 2008. Mapping wine grape quality attributes using portable fluorescence-based sensors. Frutic 9: 301–10.

Chen P, Xiao Q, Zhang J, Xie C and Wang B. 2020. Occurrence prediction of cotton pests and diseases by bidirectional long short-term memory networks with climate and atmosphere circulation. Computers and Electronics in Agriculture 176: 105612. https://doi:10.1016/j.compag.2020.105612

Chougule Archana and Jha VK. 2019. Decision support for grape crop protection using ontology. Int. J. Reasoning-based Intelligent Systems 11:24-37.

Comba L, Gay P, Primicerio J, and Aimonino DR. 2015. Vineyard detection from unmanned aerial systems images. Comput. Electron. Agric. 114: 78–87.

Corwin DL and Lesch SM. 2005. Characterizing soil spatial variability with apparent soil electrical conductivity: Part II. Case study. Comput. Electron. Agric. 46: 135–152.

DA&FW. 2022. Department of Agriculture and Farmer's Welfare, Govt. of India. https://www.agricoop.gov.in/en/statistics/horticulture (accessed on 19 November 2022).

Darra N, Psomiadis E, Kasimati A, Anastasiou A, Anastasiou E, and Fountas S. 2021. Remote and Proximal Sensing-Derived Spectral Indices and Biophysical Variables for Spatial Variation Determination in Vineyards. Agronomy 11: 741. https://doi:10.3390/agronomy11040741

De Bei R, Fuentes S, Gilliham M, Tyerman S, Edwards E, Bianchini N, Smith J, and Collins C. 2016. VitiCanopy: A Free Computer App to Estimate Canopy Vigor and Porosity for Grapevine. Sensors 16: 585. https://doi.org/10.3390/s16040585

Di Gennaro SF and Matese A. 2020. Evaluation of novel precision viticulture tool for canopy biomass estimation and missing plant detection based on 2.5D and 3D approaches using RGB images acquired by UAV platform. Plant Methods 16:91. https://doi:10.1186/s13007-020-00632-2

Di Gennaro SF, Toscano P, Cinat P, Berton A, and Matese A. 2019. A Low-Cost and Unsupervised Image Recognition Methodology for Yield Estimation in a Vineyard. Front. Plant Sci.10. https://doi:10.3389/fpls.2019.00559

Diago M P, Rey-Carames C, Le Moigne M, Fadaili EM, Tardaguila J, and Cerovic Z G. 2016a. Calibration of non-invasive fluorescence-based sensors for the manual and on-the-go assessment of grapevine vegetative status in the field. Australian Journal of Grape and Wine Research 22: 438–49.

Diago MP, Aquino A, Millan B, Palacios F, and Tardaguila J. 2019. On-the-go assessment of vineyard canopy porosity, bunch and leaf exposure by image analysis, Australian J. Grape Wine Res. 25: 363–74.

Diago MP, Bellincontro A, Scheidweiler M, Tardaguila J, Tittmann S and Stoll M. 2017. Future opportunities of proximal near infrared spectroscopy approaches to determine the variability of vineyard water status. Aust. J. Grape Wine Res. 23: 409–14.

Diago MP, Fernández-Novales J, Fernandes AM, Melo-Pinto P and Tardaguila J. 2016b. Use of Visible and Short-Wave Near-Infrared Hyperspectral Imaging to Fingerprint Anthocyanins in Intact Grape Berries. Journal of Agricultural and Food Chemistry 64: 7658–66.

Diago MP, Krasnow M, Bubola M, Millan B, and Tardaguila J. 2016c. Assessment of vineyard canopy porosity using machine vision. Am. J. Enol. Vitic. 67:229–38.

Donna P, Tonni M, Divittini A, and Valenti L. Concimare. 2013. il vigneto guidati dal satellite. L'Informatore Agrar. 22: 42–46.

FAOSTAT. 2022. FAO Web Page. Available online: http://www.fao.org/faostat (accessed on 19 November 2022).

Fernandez-Novales J, Tardaguila J, Gutierrez S and Diago MP. 2019. On-The-Go VIS + SW – NIR Spectroscopy as a Reliable Monitoring Tool for Grape Composition within the Vineyard. Molecules 24:2795. https://doi.org/10.3390/molecules

Gebbers R and Adamchuk V. 2010. Precision Agriculture and Food Security. Science 327: 828–31.

Ghule S, Sawant IS, Shetty DS and Sawant SD. 2015. Epidemiology and weather-based forecasting model for anthracnose of grape under the semi-arid tropical region of Maharashtra. J. Agrometeorology 17: 265-67.

Grant O, Baluja J, Ochagavía H, Diago MP and Tardaguila J. 2016. Thermal imaging to detect spatial and temporal variation in the water status of grapevine (Vitis vinifera L.). The Journal of Horticultural Science and Biotechnology 91:44–55.

Gutierrez S, Fernandez-Novales J, Diago MP, Iniguez R and Tardaguila J. 2021a. Assessing and mapping vineyard water status using a ground mobile thermal imaging platform. Irrigation Science 39:457–68.

Gutierrez S, Hernandez I, Ceballos S, Barrio I, Diez-Navajas AM and Tardaguila J. 2021b. Deep learning for the differentiation of downy mildew and spider mite in grapevine under field conditions, Computers and Electronics in Agriculture 182:105991. https://doi:10.1016/j.compag.2021.105991

Hill GN, Evans KJ, Beresford RM and Dambergs RG. 2014. Comparison of methods for the quantification of botrytis bunch rot in white wine grapes. Aust. J. Grape Wine Res. 20: 432–41.

Hubbard SS, Schmutz M, Balde A, Falco N, Peruzzo L, Dafflon B, Leger E, and Wu Y. 2021. Estimation of soil classes and their relationship to grapevine vigour in a Bordeaux vineyard: advancing the practical joint use of electromagnetic induction (EMI) and NDVI datasets for precision viticulture. Precision Agric 22: 1353–76.

Joseph V, Sinfield DF, and Oliver C. 2010. Evaluation of sensing technologies for on-the-go detection of macro-nutrients in cultivated soils. Computers and Electronics in Agriculture 70:1-18.

Karakizi C, Oikonomou M, and Karantzalos K. 2016. Vineyard Detection and Vine Variety Discrimination from Very High-Resolution Satellite Data. Remote Sens. 8: 235. https://doi.org/10.3390/rs8030235

Khaliq A, Comba L, Biglia A, Aimonino DR, Chiaberge M, and Gay P. 2019. Comparison of Satellite and UAV-Based Multispectral Imagery for Vineyard Variability Assessment. Remote Sens. 11: 436. https://doi.org/10.3390/rs11040436

Latouche G, Debord C, Raynal M, Milhade C and Cerovic Z. 2015. First detection of the presence of naturally occurring grapevine downy mildew in the field by a fluorescence-based method. Photochemical and Photobiological Sciences 14: 807–1813.

Lee WS and Tardaguila J. 2021. Pests and diseases management, Advanced Automation for Tree Fruit Orchards and Vineyards. Springer, New York.

Lessio F and Alma A. 2021. Models Applied to Grapevine Pests: A Review. Insects 12: 169-81.

Liu S, Zeng X and Whitty M. 2020a. 3DBunch: a Novel iOS-Smartphone application to evaluate the number of grape berries per bunch using image analysis techniques, IEEE Access 8:114663–674.

Liu S, Zeng X and Whitty M. 2020b. A vision-based robust grape berry counting algorithm for fast calibration-free bunch weight estimation in the field. Computers and Electronics in Agriculture 173:105360. https://doi:10.1016/j.compag.2020.105360

Losciale P, Manfrini L, Morandi B, Pierpaoli E, Zibordi M, Stellacci AM, Salvati L, and Corelli Grappadelli L. 2015. A multivariate approach for assessing leaf photo-assimilation performance using the IPL index. Physiol. Plant. 154: 609–20.

Matese A, and Di Gennaro S. 2018. Practical Applications of a Multisensor UAV Platform Based on Multispectral, Thermal and RGB High Resolution Images in Precision Viticulture. Agriculture 8: 116. https://doi:10.3390/agriculture8070116

Matese A, and Di Gennaro SF. 2015. Technology in precision viticulture: A state of the art review. Int. J. Wine Res.7: 69–81.

Matese A, Vaccari FP, Tomasi D, Di Gennaro SF, Primicerio J, Sabatini F, and Guidoni S. 2013. CrossVit: enhancing canopy monitoring management practices in viticulture. Sensors (Basel) 13:7652-67.

Mondino EB, and Gajetti M. 2017. Preliminary considerations about costs and potential market of remote sensing from UAV in the Italian viticulture context. European J. Remote Sens. 50: 310–19.

Mundankar KY, Sawant SD, Sawant IS, Sharma J and Adsule PG. 2007. Knowledge based decision support system for management of powdery mildew disease in grapes. In: Proceedings of 3rd Indian International Conference on Artificial Intelligence (IICAI-07). Pp 1563–1571.

Ouyang J, De Bei R, and Collins C. 2021. Assessment of canopy size using UAV-based point cloud analysis to detect the severity and spatial distribution of canopy decline. OENO One 55: 253–66.

Palacios F, Bueno G, Salido J, Diago MP, Hernandez I and Tardaguila J. 2020. Automated grapevine flower detection and quantification method based on computer vision and deep learning from on-the-go imaging using a mobile sensing platform under field conditions. Computers and Electronics in Agriculture 178: 105796. https://doi.org/10.1016/j.compag.2020.105796

Palacios F, Diago MP and Tardaguila J. 2019. A non-invasive method based on computer vision for grapevine cluster compactness assessment using a mobile sensing platform under field conditions. Sensors (Basel) 19: 3799. https://doi:10.3390/s19173799

Pertot TC, Rossi V, Mugnai L, Hoffmann C, Grando MS, Gary C, Lafond D, Duso C, Thiery D, Mazzoni V and Anfora G. 2017. A critical review of plant protection tools for reducing pesticide use on grapevine and new perspectives for the implementation of IPM in viticulture. Crop Protection 97:70-84.

Petrie PR, Wang Y, Liu S, Lam S, Whitty MA and Skewes MA. 2019. The accuracy and utility of a low-cost thermal camera and smartphone-based system to assess grapevine water status. Biosystems Engineering 179:126–39.

Pichon L, Brunel G, Payan JC, Taylor J, Bellon-Maurel V and Tisseyre B. 2021. ApeX-Vigne: experiences in monitoring vine water status from within-field to regional scales using crowdsourcing data from a free mobile phone application. Precision Agriculture 22: 608–62.

Pierce FJ and Novak P. 1999. Aspects of Precision Agriculture. Advances in Agronomy 67: 1–85.

Power DJ. 2002. Decision support systems: concepts and resources for managers. Faculty

Priori S, Pellegrini S, Perria R, Puccioni S, Storchi P, Valboa G, and Costantini EAC. 2019. Scale effect of terroir under three contrasting vintages in the Chianti Classico area (Tuscany, Italy). Geoderma 334: 99–112.

Proffit T, Bramley R, Lamb D and Winter E. 2006. Precision Viticulture – A New Era in Vineyard Management and Wine Production. Published by Winetitles Pty Ltd., Ashford, South Australia. Pp 1–90.

Ridwan MA, Radzi NAM, Ahmad WSHMW, Mustafa IS, Din NM, Jalil YE, Isa AM, Othman NS, and Zaki WMDW. 2018. Applications of Landsat-8 Data: A Survey. Int. J. Eng. Technol. 7: 436–41.

Roman C, Peris M, Esteve J, Tejerina M, Cambray J, Vilardell P and Planas S. 2022. Pesticide dose adjustment in fruit and grapevine orchards by DOSA3D: Fundamentals of the system and on-farm validation. Science Total Environment 808:152-58.

Rosa M, Genesio R, Gozzini B, Maracchi G and Orlandini S. 1993. PLASMO: a computer program for grapevine downy mildew development forecasting. Computers and Electronics in Agriculture 9: 205-15.

Roscher R, Herzog K, Kunkel A, Kicherer A, Topfer R and Forstner W. 2014. Automated image analysis framework for high-throughput determination of grapevine berry sizes using conditional random fields. Computers and Electronics in Agriculture 100:148–58.

Rose JC, Kicherer A, Wieland M, Klingbeil L, Topfer R and Kuhlmann H. 2016. Towards automated large-scale 3D phenotyping of vineyards under field conditions. Sensors 16: 2136–2161.

Santangelo T, Di Lorenzo R, La Loggia G, and Maltese A. 2013. On the relationship between some production parameters and a vegetation index in viticulture. Remote Sensing for Agriculture, Ecosystems, and Hydrology XV. https://doi:10.1117/12.2029805

Schirrmann M, Gebbers R, and Kramer E. 2013. Performance of automated near-infrared reflectance spectrometry for continuous in situ mapping of soil fertility at field scale. Vadose Zone Journal 12:1-14.

Simone P, Bianconi N, and Costantini EAC. 2014. Can γ-radiometrics predict soil textural data and stoniness in different parent materials? A comparison of two machine-learning methods. Geoderma 226–227:354-64.

Soubry I, Patias P, and Tsioukas V. 2017. Monitoring vineyards with UAV and multi-sensors for the assessment of water stress and grape maturity. J. Unmanned Veh. Syst. 5:37–50.

Taylor JA, Tisseyre B and Leroux C. 2019. A simple index to determine if within-field spatial production variation exhibits potential management effects: application in vineyards using yield monitor data. Precision Agriculture 20: 880–895.

Urretavizcaya I, Royo JB, Miranda C, Guillaume TS and Santesteban LG. 2017. Relevance of sink-size estimation for within-field zone delineation in vineyards. Precision Agric 18: 133–44.

Vanino S, Pulighe G, Nino P, De Michele C, Falanga Bolognesi S, and D'Urso G. 2015. Estimation of Evapotranspiration and Crop Coefficients of Tendone Vineyards Using Multi-Sensor Remote Sensing Data in a Mediterranean Environment. Remote Sens. 7: 14708–30.

Vidal A, Talens P, Prats-Montalban JM, Cubero S, Albert F and Blasco J. 2013. In-line estimation of the standard colour index of citrus fruits using a computer vision system developed for a mobile platform. Food Bioproc. Tech 6: 3412–19.

10

Harnessing Indigenous and Exotic Fruits for Nutritional and Livelihood Security

G. Karunakaran, M. Arivalagan, P. Singh, Kanupriya T. Sakthivel, A. Thirugnanavel and P.C. Tripathi

ICAR-Indian Institute of Horticultural Research, Hessaraghatta Lake Post Bangalore 560 089, Karnataka

Underexploited

Unexploited fruit crops, jackfruit, tamarind, jamun, karonda, star fruit, etc having high nutritive value, are gaining popularly among the people. Similarly, many exotic fruits, Dragon fruit (*Hylocereus* spp), Avocado (*Persea Americana* Miller), Rambutan (*Nephelium lappaceum* L.), Mangosteen (*Garcinia mangostana* L), Passion fruit (*Passiflora edulis* Sims), Durian (*Durio zibethinus* L.,) Longan (*Dimocarpus longan* Lour) have been introduced in India. Exploitation of their genetic diversity could provide nutritional and livelihood security. There is a need to know about nutritional composition and health significance of these crops. Therefore, efforts have been made to know their potential nutritional and health benefits.

Indigenous Fruit Crops

Jackfruit (*Artocarpus heterophyllus* Lam.)

Jackfruit belonging to family Moraceae, is native to tropical rainforests of the Western Ghats, India (Baliga *et al*., 2011; Wangchu *et al*., 2013; Reddy *et al*., 2004). An evergreen tree (Fig. 1), it grows well in warm humid climate and soil rich in nutrients and good drainage system. It can grow up to the elevation of up to 1500 MSL, and tolerant to both drought and frost to some extent. But the crop is sensitive to prolonged heavy frost and drought. India, Bangladesh, Malaysia, Sri Lanka, the Philippines, Myanmar, Indonesia, South China, Vietnam, and Thailand are jackfruit growing countries (Lin *et al*., 2009; Balqis and Rosma, 2011; Baliga *et al.*, 2011; Dutta *et al.*, 2011). In India, it is commercially grown in Kerala, Tamil Nadu, Karnataka, Andhra Pradesh, Assam, Bihar, Orissa, West Bengal, Chhattisgarh, Jharkhand, and Uttar Pradesh and on a limited scale in Arunachal Pradesh, Nagaland, Maharashtra, and Himachal Pradesh (Vazhacharickal *et al.*, 2017).

Fig. 1: Jackfruit tree at yielding stage

Jackfruit is a multi-purpose tree and provides healthy food, and also timber, fodder and fuel. Some of the bioactive compounds isolated for the jackfruit have medicinal value and it is a rich source of pectin. Immature and mature flakes (fleshy aril) are the prime economic part and tender fruits are used as vegetables and the ripe ones as table fruits (Baliga *et al.*, 2011). The ripen fruit significantly contributes to the nutrition of lower income families as it is a good source carbohydrate, total soluble solids (TSS), and carotene and it has a considerable amount of protein, fat, fibre, calcium, phosphorous, iron and vitamins. The fruit, a gigantic syncarp, is the largest of all cultivated fruits. Different shapes (obloid, spheroid, high spheroid, ellipsoid, clavate, oblong, broadly ellipsoid, etc) of fruits with different coloured flakes are found in jackfruit (Fig. 2). The fruits can have 10-500 flakes and seeds. In India it is commonly called as "Poor man's Food" and it is mainly due to easy availability and nutritive value (Samaddar, 1990; Prakash *et al.*, 2009; Swami *et al.*, 2012). Various value added products like chips, papad are prepared out of unripe flakes. In some of the Indian states, the immature tender fruits are used as a vegetable.

Fig. 2: Jackfruit diversity in terms of size and flake colour

Jackfruit seeds are a rich source of starch, minerals, fat, and potassium and it is one of the popular ingredients in several culinary preparations, particularly in Kerala and Tamil Nadu. Due to the presence of about 70-75% of starch, jackfruit seed is one of the potential sources of starch for commercial purpose (Mukprasirt and Sajjaanantakul, 2004; Odoemelam, 2005). Jackfruit is rich in antioxidants which may provide health-promoting and disease-preventing benefits. Resveratrol present in the jackfruit skin has antioxidant and anti-inflammatory properties, due to which it protects human health against diseases like cancer, diabetes, and Alzheimer's disease (Akshatha *et al.*, 2015). The seeds are boiled or roasted and used in many culinary preparations. The peel of fruit and its leaves are excellent cattle feed. Its timber is valued for furniture making since it is rarely attacked by white ants. The latex obtained from the bark is used as a resin. Pickles and dehydrated leather are its preserved delicacies.

Farmers are custodians of jackfruit diversity. Identification unique varieties from such farmers will help sustainable livelihood of the farmers and also ensure the nutritional security of the community. The scientists of ICAR- IIHR, Benagaluru, Karnataka, undertook survey and identified three superior jackfruit genotypes (i) Siddu, a novel variety with coppery red, sweet and firm flakes with average fruit weight of 2.44 kg (ii) Shankara, a unique variety with aromatic, crispy and coppery red flakes with average fruit weight of 2.73 kg, and iii) Shree Halasu, 12-year-old seedling identified from of Karnataka having coppery red coloured flakes rich in carotenoids and lycopene (Fig. 3) (Karunakaran *et al.*, 2017; Karunakaran and Singh., 2019). All these varieties are rich in carotenoids, contain about 2-2.5% dietary fibre and total phenols and antioxidant activity in both varieties are more when compared to other varieties with yellow and white coloured flakes. Custodian farmers were recognised as the "Custodians of Genetic Diversity" for conserving these unique jackfruit varieties.

Siddu Shankara Shree Halasu

Fig. 3: Jackfruit varieties identified and released as farmer's varieties (Siddu, Shankara and Shree Halasu)

Tamarind (*Tamarindus indica*)

Tamarind is an evergreen tree legume, distributed all over the world in tropical and subtropical countries (Fig. 4). The tree produces fruits in pods which consist of a brittle outer shell encapsulating the pulp and enclosed seeds. Once established, its tree develops a large tap root which protects it from strong winds and cyclones, making it well suited to the region prone to such weather phenomena. It is also considered to be a suitable tree for inter-planting with other commercial forest species. Tamarind starts bearing from 6-8 years and has productive life of 50- 70 years after which it declines. The normal life span of the tree is 150 years. A typical established tree yields between 50-100 Kg of fruit which is harvested during multiple picks over an 8-10-week period between February and April. Apart from tamarind pulp other by-products such as seed, shell, fiber is also useful for various purposes. Tamarind comes in two main types; sweet and sour. Sweet tamarind is harvested ripe and usually consumed fresh, while the sour tamarind is processed into a range of value-added products (Joshua and Dudhade, 2006.).

Fig. 4: Tamarind tree

Tamarind seeds are gaining importance as a rich source of proteins and valuable amino acids. India is the world's largest producer of tamarind about 300,000 tons of tamarinds are produced annually and exported mainly to Europe and Arab countries ((El-Siddig, 2006). Tamarind has a wide range of genetic variation in India.

A unique tamarind variety (Fig. 5) having broad pods with good pulp colour, recovery and sour taste (18.61 mg/100 g pulp) was identified in Nandihalli village of Tumkur district. It is a 40-year-old seedling tree superior with better yield and pod characters compared to local and registered mean annual yield (4 years from 2016-2020) of 251.4 kg/tree as against 165.0 kg/tree in local trees. The pods are long, broad and curved having mean pod length of 25.4 cm. Pulp is light brown with less fibre. Pulp recovery is high (43%) as against 28% in local tamarind trees. The variety was released in the name of "Lakshmana tamarind" as farmer's variety and submitted for registration under farmer's variety under PPVFRA (Kanupriya *et al.*, 2020).

Fig. 5: Lakshmana tamarind – a farmer's variety

Dragon fruit (*Hylocereus* spp.)

Dragon fruit is an herbaceous perennial climbing cactus. Widely known as Pitaya, it has recently drawn much attention among the Indian growers due to its economic value as well as numerous health benefits. Five major types of *Hylocereus* species are available which are mainly differentiated based on their fruit characteristics. The *Hylocereus undatus* is characterized with white pulped fruits and pink skin, *H. polyrhizus* has red pulped fruits with pink skin, *Hylocereus costaricencis*, which have violet-red pulp and pink skin, *H. guatemalensis*, has red pulp with reddish-orange skin, and *H. megalanthus* have white pulp and yellow skin (Arivalagan *et al.*, 2021). It prefers dry tropical and subtropical climate with average temperature of 20-29°C, but can withstand up to 40°C, and as low as 0°C for short periods. The crop provides fast return with economic production during first year after planting, and full production is attained after 3-4 years. Although, the initial investment (11.95 lakhs/ha) is relatively high, profit is substantial within 2-3 years. Once planted, it can continuously yield for 20 years.

Presently, dragon fruit is mainly imported to India from Vietnam, Sri Lanka, Malaysia, and Thailand. Its import was noticed mainly during January, April, October, November and December. In India, dragon fruit cultivation was initiated during 2008's and the area under cultivation is gradually increasing since 2015. Cultivation of dragon fruit has already started in different parts of the country with many success stories of farmers from different regions. Karnataka, Gujarat, Maharashtra, Telangana, Andhra Pradesh, West Bengal, Bihar, Kerala, Tamil Nadu, Orissa, NEHs, and Andaman & Nicobar Islands are the major states cultivating dragon fruit, and now cultivation extended to Rajasthan, Punjab, Haryana, Madhya Pradesh, and Uttar Pradesh. At present, the area expansion under dragon fruit cultivation is increasing every year, thus, India has great scope in a near future for self-sustainability. Due to high demand both in domestic and international markets, fruit production could be an economical avocation to both backyard growers as well as entrepreneurs of medium and large-scale plantations (Karunakaran *et al.*, 2020).

Hylocereus undatus (white pulp and pink skin)

Hylocereus polyrhizus (red pulp and pink skin)

Fig. 6: Dragon fruit species commonly available in India

Dragon fruit is considered to be one of the 'super foods' due to its nutrient richness. It is rich in various nutrients and low in calories. The TSS of the fruit is about 9-11°Brix, and sugar content is about 5.0 to 7.0 g/ 100g. Glucose and fructose are the major sugars present in the fruit. Fruit contain about 1.0 g of protein and dietary fibre. It is good source for essential amino acids such as histidine, lysine, methionine, and phenyl alanine. Malonic acid and citric acid are the major organic acids identified. Presence of organic acids ensures the increased shelf life, stability, and microbial safety. Fruits with red pulp have significantly high quantum of phenolics and antioxidant potential than white pulped ones. 100g of Fruit contained 120-200 mg K, 30-45 mg Mg, 20-45 mg Ca, 20-35 mg P, 0.70-1.5 mg Fe and 0.20-0.40 mg Zn. Vitamin C was found maximum (~6 mg), followed by vitamin E (~150 µg), pantothenic acid (~50 µg), vitamin K1 (~25 µg) and niacin (~15 µg). Due to its low caloric value dragon fruit is ideal option for weight loss treatment, control of diabetes and lowering cholesterol, etc. Reports suggest that it's consumption could help in the control of chronic illnesses, improves the health of alimentary canal and boosting the body's immunity. All these beneficial factors make dragon fruit a suitable candidate for inclusion in diet aimed at weight loss treatment, control of diabetes, lowering the cholesterol level, etc (Arivalagan *et al.*, 2021).

Since the dragon fruit is introduced to India, the availability of diverse germplasm is limited, there is a need to enrich varietal wealth by widening the genetic base. Location specific complete package of nutrient management needs to be developed. Post-harvest handling, cold storage, processing and value addition needs to strengthened. Technical know-how for farmers and awareness campaign for consumers about health benefits to be imparted.

Avocado *(Persea americana)*

Commonly known as butter fruit in India, it is also called as avocado, alligator pear, butter pear and vegetable butter. It is important exotic fruit crop, belonging to family of Lauraceae, originated in Mexico and Central America, possibly from more than one wild species. Avocado includes approximately 150 species among which Guatemalan, Mexican, and West Indian races are very important from horticultural point of view. In India, avocado was introduced from Sri Lanka in early part of the twentieth century. In a very limited scale and in a scattered way, it is grown in Tamil Nadu, Kerala, Maharashtra, Karnataka and Sikkim (Ghosh, 2000). It cannot tolerate the hot dry winds and frosts of northern India. Climatically, it is grown in tropical or semitropical areas experiencing some rainfall in summer, and in humid, subtropical summer rainfall areas.

Fig. 7: Avocado

All three horticultural races i.e. West Indian, Guatemalan and Mexican races adapted to tropical and sub-tropical conditions, have been grown different parts of India and the variety preferred is depends upon the prevailing agroclimatic conditions. The cultivars of West Indian race are grown in localized pockets in Maharashtra, Tamil Nadu and Karnataka. In tropical and near-tropical areas, only West Indian race is well-adapted but its hybrids with Guatemalan (e.g. Booth selection) perform well and are considered valuable for extending the harvest season. In less tropical regions, hybrids of Guatemalan with Mexican race predominate since they combine the cold hardiness of the latter with the superior horticultural traits of both and also bridge the two seasons of maturity. In the eastern Himalayan state of Sikkim, avocado has been introduced successfully in hill ranges with an altitude of 800- 1,600 metres. Both the Mexican and Gautemalan races are grown successfully in Sikkim. The varieties that are cultivated in India go by several names, such as Purple, Green, Fuerte, Pollock, Peradeniya Purple Hybrid, Trapp, Round and Long. Among the several existing varieties, perhaps Fuerte is the most widely grown, but it is regarded as unsuitable for the tropics. The Purple and Green varieties were introduced into India from Ceylon in 1941. In Kodagu, avocado is grown as one of the mixed crops in coffee-based cropping system. Almost each house is maintaining few plants of avocado, and lots of variability of avocado is available in Coorg and adjoining areas.

Avocadoes are the good source of bioactive compounds, especially lipids, fibres, carotenoids, polyphenols, tocopherols, and phytosterols (Bhuyan *et al.*, 2019). Most of these compounds have numerous biological activities, such as antimicrobial, anticancer, anti-inflammatory, antidiabetic and anti-hypertensive. Presence of high energy and low sugar makes avocado as one of the superior food for diabetic patients. It is widely used in cosmetic industries for manufacturing of shampoos, lipsticks, face creams and body lotions due to the presence of high lipids. It is also used in the preparation of many food products such as ice creams, sandwich, avocado pizza, salad, avocado toast, guacamole. In India, avocado is mostly used for preparation of milkshakes and consumed by mixing pulp with sugar. Avocado pulp contains up to 30 per cent oil depending on the variety and orchard location. The composition of avocado oil is similar to olive oil, highly digestible and mainly consists of unsaturated fatty acids especially oleic acid which contributes for taste and consistency of the pulp. Presence of high monounsaturated fatty acids helps to reduce the low density lipids (LDL) and increases the high density lipids (HDL) which reduces the risk of cardio vascular diseases by reducing the glucose tolerance and triglycerides in the blood (Ledesma *et al.*, 1996). Several reports showed that the carotenoids especially lutein, violaxanthin, and neoxanthin present in the avocado pulp can be used to control the age related macular degeneration and improves the vision.

Rambutan (*Nephelium lappaceum L.*)

Rambutan is a tropical fruit belonging to the Sapindaceae family; it is a fruit, originally from Malaysia (Muhamed *et al.*, 2019), whose name is derived from the Malay word "*Rambut*", meaning "*Hair*", concerning the soft thorns that cover the surface. It is widely cultivated in Thailand, Indonesia, Malaysia and Sri Lanka. In India, it is primarily introduced and grown in backyards of some parts of Kerala, Karnataka and Tamil Nadu (Tripathi *et al.*, 2014). Fruit resembles to litchi in appearance. Its hairy outgrowth has eye catching yellow or pink and red colours and it imparts a distinctive exotic appearance to fruits. The translucent pulp is sweet, juicy and delicious with a pleasant flavour. Recently, cultivation of rambutan has gained substantial momentum due to steadily increasing market demand and a lot of planting material is being produced, and plantings have increased appreciably in southern India (Karunakaran *et al.*, 2021).

Fig. 8: Rambutan

The crop is adapted to warm tropical and subtropical climate and requires around 22-30 °C for optimum growth and it is very sensitive to low temperature. The high (>40 °C) and low (less than 10 °C) temperatures affect the growth and development of plants. The tree grows well up to 700 m above sea-level and requires well-distributed rainfall of 200-500 cm throughout the year, but a short spell of the dry period is essential for initiation of floral bud (Ravishankar *et al.,* 2007). Rambutan seedlings grow into male, female, or hermaphrodite trees which can be identified only after flowering. Hence, rambutan plant should be vegetatively propagated by approach grafting, patch budding and air-layering. Patch budding on 8 - 12-month-old rootstocks during May or July give the best results. Several commercial varieties of rambutan are grown in South-Eastern Asian countries. CHES (IIHR), Chettalli, has recently released two varieties namely Arka Coorg Arun and Arka Coorg Peetabh (Figure 9). Arka Coorg Arun is an early maturing variety with red coloured fruits. Fruits mature in September - October and average fruit weight is 40 - 45 g. Arka Coorg Peetabh is a high-yielding, yellow coloured variety with the average fruit weight of 25-30 g (Tripathi *et al.,* 2011).

Fig. 9: Rambutan - Arka Coorg Arun and Arka Coorg Peetabh

Rambutan pulp, peel and seed have numerous health promoting substances. The pulp is edible portion of fruits and is consumed as fresh. It contained about 83 % moisture, 15 % carbohydrates, about 1 % protein, and contained about 25-50 mg vitamin C and have caloric value is about 65 cal. The pulp is one of the good sources of mineral and contained about 75 mg potassium, 25 mg calcium, 15 mg phosphorous, 0.3 mg iron, etc. Rambutan peel has a high content of phenolic compounds and antioxidant activity, thus it could be a potential source of antioxidants for food products, cosmetics, and pharmaceuticals due to its antioxidant activity content and nontoxic capacity for normal cells. Rambutan seed have about 34.4% moisture, 1.2% ash, 7.8% protein, 11.6% crude fiber, 46% carbohydrates, and 33.4% fat on a dry basis. Among fatty acids, oleic acid is major (40%), followed arachidonic acid (35%), palmitic acid (6.0%), stearic acid (7.1%), gondoic acid (6.1%), behenic acid (3%) and palmitoleic acid (1.5%). Both pulp, seed and peel of rambutan is reported to have various biological properties which includes antioxidant, antibacterial, antidiabetic, antiproliferative, anti-inflammatory, and antiviral activities (Manaf *et al.*, 2013; Hernandez *et al.*, 2019).

Mangosteen (*Garcinia mangostana* L.)

Mangosteen is an evergreen tropical tree having medium size with broad-leaves. It belonging to the Clusiaceae family that grows in Southeast Asia, and is cultivated mainly as a source of its highly palatable fruit, consisting of a fragrant white internal pulp divided in septa, contained in a dark purple rind (Figure 10). It is considered by many to be the most delicious fruit of the tropics with a universal appeal, 'the finest fruit of the world' or 'queen of fruits' (Moongkarni *et al.*, 2014). The fruit was introduced more than a century ago in India, but currently it is successfully grown only in selected places on slopes of Nilgiris, Malabar and Waynad. It is a slow growing tree, and the mature tree reaches a height of 10 to 15m with a dense pyramidal crown and glossy bright leaves, but it has a long juvenile period. Mangosteen requires humid tropics with 180-250 cm rainfall. It is found growing in an altitude of 400-900 m in south India. It requires good soil moisture and higher humidity for better growth and yield. Trees produce flowers and fruits on the terminals of new growth on small twigs that develop from the main stem or large branches. A large number of strong healthy and productive branches and terminal shoots should be left on the tree. The trees begin to bear fruits 10 to 12 years after seeding. However, if the plant is well cared, fruiting can begin at 7 - 8 years after seeding.

Fig. 10: Mangosteen

The fruit is rich in various health promoting substances. Xanthones and anthocyanins are the major bioactive compounds present in the fruit pericarp. Xanthones have a unique chemical structure composed of a tricyclic aromatic system (C_6–C_3–C_6). Isoprene, methoxyl and hydroxyl groups located at various locations on the A and B rings, resulting in a diverse array of xanthone compounds. At least 68 distinct xanthones have been identified in different parts of the mangosteen plant with 50 being present in the fruit's pericarp at higher concentrations than in the aril or edible portion of the fruit. α- and γ-mangostin are the most abundant xanthones present in the pericarp and β-mangostin, gartanin, 8-deoxygartanin, garcinones A, B, C, D and E, mangostinone, 9-hydroxycalabaxanthone and isomangostin are the minor once. Along with xanthones and anthocyanins, mangosteen tissues like pericarp, seeds, leaves, and plantlets, are known to contain bioactive compounds, such as phenolics and flavonoids which are comparable with jamun polyphenolics and gallic acid, protocatechuic acid, chlorogenic acid, quercetin, epicatechin, rutin, catechin, and cyanidin-3-sophoroside are the major polyphenolic compounds reported (Aizat *et al.*, 2019; Azima *et al.*, 2017). Various studied showed that mangosteen possesses high antioxidant and anti-inflammatory properties, and fruits are widely used to treat various diseases such as tumors, diabetes, bacterial infections, hypertension, and arthritis (Aizat *et al.*, 2019; Tousian *et al*, 2017).

Durian (*Durio zibethnus* Murr.)

Durian, also named Shexiangmaoguo, belongs to the family Bombacaceae (Figure 11). It is one of the most important tropical fruits in Southeast Asia and adjacent islands. It originated in the Malay Peninsula and now is mainly distributed in Thailand, Malaysia, Indonesia, Cambodia, Laos, Myanmar, Sri Lanka, the Philippines and Singapore (Wang, 2020). Durian is not only used as fresh fruit and processed food, but also as a traditional folk medicine in Southeast Asia. Southeast Asian folk believe that the fruit is an aphrodisiac and abortion drug, which can improve menstruation and treat infertility (Lim, 2012; Reshma, 2016).

Fig. 11: Durian

Various works demonstrated that durian fruit have high potential as a therapeutic indicator that could help patients suffering from certain diseases. The presence of different types of phytochemicals like carotenoids, polyphenols, flavonoids, and anthocyanins import medicinal and therapeutic value to durian. A significant amount of fiber (7.5–9.1 g/100 g dry matter), carbohydrate (62.9–70.7 g/100 g DM), and sugar (47.9–56.4 g/100 g DM) were reported in durian fruit varieties. Pulp also reported to have linoleic acid (2.20%), myristic acid (2.52%), oleic acid (4.68%), 10-octadecenoic acid (4.86%), palmitoleic acid (9.50%), palmitic acid (32.91%), and stearic acid (35.93%). Phytochemical contents of durian had a strong correlation with the antioxidant capacities based on different biological activities that could be used as an alternative use in medicinal foods (Zhan *et al.*, 2021).

Blueberry (*Vaccinium* sect. *Cyanococcus*)

Blueberry belongs to the genus *Vaccinium* and the subgenus *Cyanococcus* (Figure 12). In general, highbush blueberries (HB, *V. corymbosum* L.) and rabbiteye blueberries (RB, *V. ashei Reade*) are considered commercially important blueberry types (Yang *et al.*, 2022). The presence of high amount flavones and other poly phenolic compounds contribute to strong antioxidant properties that provide health benefits and made the blueberry 'a functional food' (Zeng *et al.*, 2021). Anthocyanins, the major part of total phenolic compounds and the most important subclass of flavonoids, are ubiquitous in the world's fruits and vegetables (Koh, Xu, & Wicker, 2020b). Additionally, compared to other nutrients in fruit, the anthocyanin content of blueberries is high (Kalt & Dufour, 1997), which is an indication that the nutrient quality of the fruit as a functional food is also high. Dietary blueberries allegedly enhance vision and brain function, by enacting an anti-inflammatory role to attenuate projected chronic diseases (such as obesity and diabetes), simply through a prebiotic role that favorably regulates the gut microbial population (Hidalgo *et al.*, 2012, Morais *et al.*, 2016, Rodriguez-Daza *et al.*, 2020, Tsuda *et al.*, 2012). Blueberries express a diverse selection of anthocyanin types. Although the identities of anthocyanins among the cultivars are similar, proportion of each compound is typically different. Creation of awareness among the producer, consumers and health experts about the importance of blueberry anthocyanins in human health will help in wide spread of the crop cultivation and consumption.

Fig. 12: Blueberry

References

Aizat WM, Ahmad-Hashim FH, Syed Jaafar SN. 2019. Valorization of mangosteen, "The Queen of Fruits," and new advances in postharvest and in food and engineering applications: A review. J Adv Res. 20:61-70.

Aizat WM, Jamil IN, Ahmad-Hashim FH, Normah MN. 2019. Recent updates on metabolite composition and medicinal benefits of mangosteen plant. Peer J. 7: e6324.

Akshatha S, Anbarasu K, Vijayalakshmi, G. 2015. Resveratrol content and antioxidant properties of underutilized fruits. J. Food Sci and Technol. 52(1): 383-90.

Arivalagan M, Karunakaran G, Roy TK, Dinsha M. Sindhu BC, Shilpashree VM, Satisha GC, Shivashankara KS. 2021. Biochemical and nutritional characterization of dragon fruit (Hylocereus species). Food Chem. 353, 129426.

Azima AS, Noriham A, Manshoor N. 2017. Phenolics antioxidants and color properties of aqueous pigmented plant extracts: Ardisia colorata var. elliptica, Clitoria ternatea, Garcinia mangostana and Syzygium cumini. J Funct Foods. 38:232–41.

Baliga, M.S., Shivashankara, A.R., Haniadka, D'Souza, J. Jason and Harshith P. Bhat. 2011. Phytochemistry, nutritional and pharmacological properties of Artocarpus heterophyllus Lam (jackfruit): A review. Food Res Int. 44: 1800–11.

Balqis ZS. Rosma, A. 2011. Artocarpus integer leaf protease: Purification and characterization. Food Chem. 129(4): 1523-29.

Bhuyan DJ, Alsherbiny MA, Perera S, Low M, Basu A, Devi OA, Barooah MS, Li CG, Papoutsis K. (2019). The Odyssey of bioactive compounds in avocado (*persea americana*) and their health benefits. Antioxidants (Basel). 8(10):426.

Dutta H, Paul SK, Kalita D, Mahanta CL. 2011. Effect of acid concentration and treatment time on acid-alcohol modified jackfruit seed starch properties. Food Chem. 128(2): 284-291.

El-Siddig K, Gunasena HPM, Prasad BA, Pushpakumara DKNP, Ramana KVR, Vijayanand P, Williams JT. 2006. Tamarind, Tamarindus indica. Southampton Centre for Underutilised Crops, Southampton, UK.

Ghosh SP. 2000. Avocado Production in India. In Avocado production in Asia and the Pacific. FAO Corporate Document Repository. Bangkok, Thailand: FAO Regional Office for Asia and the Pacific.

Hernandez C, Aguilar C, Rodriguez R, Flores C, Morlett J, Govea SM, Ascacio-Valdés, J. 2019. Rambutan (*Nephelium lappaceum* L.). Nutritional and functional properties. Trends in Food Sci Technol. 85. 10.1016/j.tifs.2019.01.018.

Hidalgo M, Oruna-Concha MJ, Kolida S, Walton GE, Kallithraka S, Spencer JPE, de Pascual-Teresa S. 2012. Metabolism of anthocyanins by human gut microflora and their influence on gut bacterial growth. J Agri Food Chem. 60(15):3882-90.

Joshua ND, and Dudhade, PA. 2006. Analysis of economic characteristics of value chains of three underutilised fruits of India. Research Report No. 3. The International Centre for Underutilized crops, Sri Lanka.

Kalt W, Dufour D. 1997. Health functionality of blueberries. HortTechnology, 7(3); 216-221

Kanupriya C, Karunakaran G, and Singh P. 2021. Lakshamana: New tamarind selection for improving rural livelihood. Ind Hort. 66(1): 21-23.

Karunakaran G and Singh P. 2019. Jackfruit varieties serves niche for growers. Ind Hort. 64(2); 18-19

Karunakaran G, Arivalagan M and Sriram S. 2020. Dragon Fruit - A potentially healthy niche fruit crop in waiting. Ind Hort. 65(4):41-44.

Karunakaran G, Singh P and Ravishankar H. 2017. New jackfruit for homesteads. Ind Hort. 62(4): 46-48.

Karunakaran G, Tripathi PC, Singh P, Arivalagan M, Sriram S. 2021. Smart Fruits for 21st Century. In: Current Horticulture: Improvement, Production, Plant Health Management and Value-addition. pp. 423-437.

Koh J, Xu Z, Wicker L. 2020. Blueberry pectin and increased anthocyanins stability under in vitro digestion. Food Chem. 302: 125343-50.

Ledesma LR, Munari FAC, Domínguez HBC, Montalvo CS, Luna HMH, Juárez C, Lira MS. 1996. Monounsaturated fatty acid (avocado) rich diet for mild hypercholesterolemia. Arch Med Res. 27(4):519-23.

Lim TK. 2012. Fruits, Edible Medicinal and Non-Medicinal Plants. Springer; The Netherlands: p. 580.

Lin KW, Liu CH, Tu HY, Ko HH, Wei BL. 2009. Antioxidant prenylflavonoids from Artocarpus communis and Artocarpus elasticus. Food Chem. 115(2):558-62.

Manaf Y, Marikkar N, Long K, Ghazali MH. (2013). Physico-Chemical Characterisation of the Fat from Red-Skin Rambutan (*Nephellium lappaceum* L.) Seed. J oleo Sci. 62: 335-343. 10.5650/jos.62.335.

Moongkarndi P, Jaisupa N, Samer J, Kosem N, Konlata J, Rodpai E, Pongpan N. 2014. Comparison of the biological activity of two different isolates from mangosteen. J Pharm Pharmacol. 66: 1171–9.

Morais CA, de Rosso VV, Estadella D, Pisani LP. 2016. Anthocyanins as inflammatory modulators and the role of the gut microbiota. J Nutr Biochem. 33: 1-7.

Muhamed S, Kurien S, Iyer KS, Remzeena A, Thomas S, 2019. Natural diversity of rambutan (*Nephelium lappaceum* L.) in Kerala, India. Genet Resour Crop Evol. 66(5): 1073-90.

Mukprasirt A. Sajjaanantakul K. 2004. Physico-chemical properties of flour and starch from jackfruit seeds. International J. Food Sci Technol. 39(3): 271-76.

Odoemelam SA. 2005. Functional properties of raw and heat processed jackfruit (Artocarpus heterophyllus) flour. Pakistan J Nutr. 4(6): 366-370.

Prakash O, Kumar R, Mishra A, Gupta, R. 2009. Artocarpus heterophyllus (jackfruit): an overview. Pharmacogn Rev. 3(6):353–58.

Reddy BMC, Patil P, Shashikumar S, Govindaraju L. R. 2004. Studies on physico-chemical characteristics of jackfruit clones of south Karnataka. Karnataka J Agril Sci. 17(2): 279-282.

Reshma MA. 2016. Potential use of durian fruit (*Durio Zibenthinus* Linn) as an adjunct to treat infertility in polycystic ovarian syndrome. J Integr Med. 14(1):22-28.

Rodriguez-Daza MC, Roquim M, Dudonne S, Pilon G, Levy E, Marette A. 2020. Desjardins. Berry polyphenols and fibers modulate distinct microbial metabolic functions and gut microbiota enterotype-like clustering in obese mice. Front Microbiol. 11:1-7.

Samaddar, H.N. 1990. In: Jackfruit. p.638-649. Bose T.K. and Mitra S.K. (Eds.), Naya Prokash, Calcutta.

Swami, S.B., Thakor, N.J., Haldankar, P.M. and Kalse. S.B. 2012. Jackfruit and its many functional components as related to human health: a review. Compr Rev Food Sci Food Saf. 11(6): 565-76.

Tousian SH, Razavi BM, Hosseinzadeh H. 2017. Review of Garcinia mangostana and its xanthones in metabolic syndrome and related complications. Phytother Res. 31:1173–82.

Tripathi PC, Karunakaran G, Sakthivel T, Sankar V, Senthilkumar R. 2014. Rambutan Cultivation in India. Tech Bull., pp. 1-2.

Tsuda K, Yamanaka K, Kondo M, Matsubara K, Sasaki R, Tomimoto H, Mizutani H. 2012. Ustekinumab improves psoriasis without altering t cell cytokine production, differentiation, and T cell receptor repertoire diversity. PLoS ONE, 7(12): 1-8.

Vazhacharickal P, Sajeshkumar NK, Mathew J, Albin A. 2017. Morphological diversity of jackfruit (*Artocarpus heterophyllus*) varieties in Kerala: An Overview.

Wang QJ. 2020. The Effect of Drying Methods on the Flavor Compounds of Durian. Shenyang Agricultural University; Shenyang. p. 5.

Wangchu L, Singh D, Mitra SK. 2013. Studies on the diversity and selection of superior types in jackfruit (*Artocarpus heterophyllus* Lam.). Genet Resour Crop Evol. 60(5):1749-1762.

Yang W, Guo Y, Liu M, Chen X, Xiao X, Wang S, Gong P, Ma Y, Chen F. 2022. Structure and function of blueberry anthocyanins: A review of recent advances. J Funct Foods. 88: 104864.

Zeng F, Zeng H, Ye Y, Zheng S, Zhuang Y, Liu J, Fei P. 2021. Preparation of acylated blueberry anthocyanins through an enzymatic method in an aqueous/organic phase: Effects on their colour stability and pH-response characteristics. Food Funct. 12(15): 6821-29.

Zhan YF, Hou XT, Fan LL, Du ZC, Ch'ng SE, Ng SM, Thepkaysone K, Hao EW, Deng JG. 2021. Chemical constituents and pharmacological effects of durian shells in ASEAN countries: A review. Chin Herb Med. 13(4):461-71.

Section 02: Vegetables

11

Advances in Vegetable Nursery Raising

S.N.S. Chaurasia, Anant Bahadur and Swati Sharma

ICAR - Indian Institute of Vegetable Research, Post Bag. No.01 Shahanshahpur (Jakkhini), Varanasi-221 305, Uttar Pradesh, India

Nursery is a place where planting material, such as seedlings, saplings, cuttings, etc., are raised, propagated and multiplied under favourable conditions for transplanting in prepared beds. In other words, "A vegetable nursery is a place or an establishment for raising or handling of young vegetable seedlings until they are ready for more permanent planting." There are some vegetables whose seed need to be sown directly in the main plot such as cowpea, okra, bottle gourd, bitter gourd, pumpkin, sponge gourd, ridge gourd, peas, beans, and similar others. Similarly, certain vegetables cannot grow by seeds directly and need to multiply by vegetative means like pointed gourd, ivy gourd, kakrol, kartoli, garlic, zinger, colocasia, Aravi, water chestnut etc. But there are several other vegetable crops which seeds cannot grow directly by sowing the seed into main plot and needs to raise nursery first and then transplanted in the main plots are tomato, eggplant, chili, capsicum, cabbage, cauliflower, broccoli, Brussels' sprouts, knol-khol, Chinese cabbage, red cabbage, savoy cabbage, lettuce, leek, kale, collard, parsley, celery, Bak-choy etc.

Traditional method of vegetable nursery raising suffers from a number of climatic factors and stresses. The shortage of agricultural laborers nears the city and out of the city areas, technical know-how etc. ultimately increases the production cost and deteriorate the quality of seedlings. The automation in vegetable nursery produces sufficient quantities of high quality seedlings to satisfy the needs of users in very short span of time at cheaper rate. Healthy vegetable seedlings up to a large extent determine the productivity and profitability from vegetable cultivation. The recent trend of automated seedlings raising technology enhanced cropping intensity, change in cropping pattern and quantitative and qualitative improvement in the marketable yields. Now-a-day's vegetable seedlings are being raised in plug treys by automation under controlled conditions round the year. The plug tray raised transplants are uniform, bold and establish better in the field because of less root damage.

Direct seed sowing under open field conditions

The main features for direct sowing the seeds in nursery are:

- For direct seed sowing the total sown area need to be prepared just like nursery bed for better germination.
- Needs more time, labour and capital for caring the growing tiny seedlings in irrigation, weed control and protection form biotic and a-biotic stresses in quite a large area.
- There is need to sprinkled water both the time daily with the help of Rose can for better germination and growth as sprayed in nursery bed takes more time and labour.
- Seeds are very small as compared to direct sown vegetables so it needs special care.
- Less time will be available for field preparation when some other crop is standing in the same field.
- 100% plant populations can't be maintained by direct sowing.
- Seed sowing in the field is impossible during heavy rains and un-even weather conditions when going to sow the seeds directly in the field.
- Impossible to take early crop as it will totally depend on weather conditions.
    - Apart from the direct seed sowing in the field if we go for nursery and after attaining 5-7 cms height the ready seedling planted in the main field have certain advantages as it has certain importance which are as follows:

Importance of Nursery Raising

- There is ample scope of saving land, labour and capital.
- Better care of younger seedlings as it is easy to look after nursery in small area against pathogenic infection, pests and weeds.
- More intensive crop rotations can be followed.
- Crop grown by nursery raising is quite early and fetch higher price in the market, so economically more profitable.
- Nursery raising helps in taking early crops.
- 100% plant populations may be maintained.
- More time is available for the preparation of main field because nursery is grown separately.
- Vegetable seeds are very expensive particularly hybrids, so we can economize the seed by sowing them in the nursery.
- Favourable growing condition can easily be provided for raising seedlings successfully when there is uneven weather condition.

- Helpful to the entrepreneurs to establish a vegetable nursery for self-employment and future prospects.

Thus, nursery raising is more economical, especially when seeds are very small, costly (e.g. hybrids of tomato, capsicum, parthenocarpic cucumber etc.).

Nursery raising under field conditions

Selection of Site

While selecting the site for vegetable nursery the following points should be considered in mind.

- Nursery area should be free from water logging particularly during rainy season.
- It should be much away from shade so that tree or building shade should be come over nursery beds.
- There should arrangement of water supply for irrigating the nursery beds.
- The nursery area should be protected from the wild and pet animals.
- Such places should be discarding where in previous season/ year nursery beds were there. It means nursery places should be changed.
- Such area should also be discarded where weed problems are there.

Otherwise, the seedlings will not be of appropriate quality.

Alignment of Beds

The direction of nursery bed is very important for raising healthy seedlings. Because the moisture near the root zone is very injurious for causing damping off diseases. Once damping off caused in the bed the majority of the seedling within the 10-15 days time goes die. Size and type of nursery beds depends on the place, season and time. The size of the nursery beds should be of 3 to 5 meter in length (towards North-South direction) and width restricted to 1 m (East-West direction) and raised up to 15 - 20 cm from the ground level to facilitate proper aeration and light in the root zone to reduce moisture (harmful for damping off diseases). A space of 30 - 40 cm between two rows is preferred for easy intercultural operations and also to drain excess rain/ irrigation water. However, during unfavorable weather conditions, the flower pots, polythene bags, potting plugs, wooden treys, earthen pots etc. and low tunnel poly house, poly shed and covered rows may be used for nursery raising. Soil, sand and compost mixture in the ratio of 1:1:2 are filled in seedling raising pots. Arrangement should be made to drain excess water from these structures by making a hole at the bottom. During too low and too high temperature conditions the artificial temperature regimes can be created in the glass house or poly house for seed germination and raising the seedlings successfully to catch the early market by producing the early crop of cucurbits.

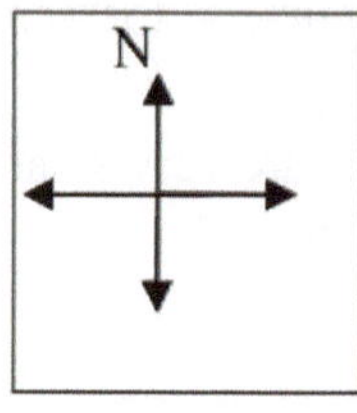

Soil Preparation

The growth and development of vegetable seedlings depends on the development of seedlings root. So it requires proper field preparation. The nursery soil should be loose and friable to prepare the beds and sowing the seeds. The field preparation requires the following points

- Loose and friable soil is needed for vegetable nursery so remove clots, stones and weeds from the field should be removed and land should be levelled.
- Mix 2 kg well rotten and fine Farm Yard Manure/compost or leaf compost or 500 g vermi- compost per square meter and mix in the soil. If the soil is heavy mix 2-3 kg sand per square meter so that the seed emergence may not be hampered.
- Raising of vegetable seedlings requires fertile and healthy soil.
- Preferably, the soil for nursery should be loam to sandy loam, loose and friable, rich in organic matter and well drained.
- The soil p^H should be close to the neutral i.e. about 7.0
- It needs a deep cultivation of the nursery land either by soil turning plough or by spade and subsequent 2-3 hoeing with cultivator.

Soil and Seed Treatment

***Soil treatment*:** The nursery soil must be treated to protect the young tender seedlings from harmful diseases and insect-pests. The chemical fungicides like captan, thirum, *Trichoderma* etc. are helpful in controlling the pathogens. The insect-pest may be controlled with application of insecticides. The weeds in the nursery may be controlled either by hand using Khurpi or by chemical weedicides. While covering of nursery beds with transparent polythene sheet (200 gauge thickness) for a period of 5-6 weeks and margin of the polythene sheet should be buried with wet soil (compressed mud) in order to stop the air from outside to inside and vice-versa. It also cares all weed seeds in top 5-6 inches soil, kills all the insect-pests, their eggs, larva and pupa, kills all the damping off causing pathogens, their spores and all other stages. For immediate nursery raising bio-fungicides @10g per square meter area of nursery, save the seedlings against harmful soil borne pathogens.

***Seed treatment*:** The seed treatment is very essential because it carry pathogen of certain diseases and insect pest. Seeds treated with Captan or Thiram @ 5 g/kg seeds or with *Trichoderma viridi* @ 10 g per kg seeds in order to protect seedlings from the seed borne diseases gave better performance at the Institute compared to the other chemical fungicides tried.

Sowing of Seeds in Nursery: Broadcasting method of seed sowing is practiced by majority of the nursery growers which causes hues loss of the seed mortality in the process of nursery raising. Moreover, the seed cost of the most of the vegetables particularly hybrids are very costly.

There are two methods so seed sowing in nursery i.e. broadcasting method and line sowing method. But there are some major disadvantages of broadcasting method of seed sowing, are:

- Uneven distribution of seeds in the nursery beds.
- Growth and development of seedlings is poor.
- Sometimes nursery becomes so dense to look like as patches of grasses.

- There is more possibility of damping off disease occurrence.

The seed sowing in the nursery beds revealed that the line sowing is the best method of sowing the vegetable seedlings. Lines are made 0.5 cm deep parallel to the width keeping 5 cm distance between two rows. The seeds are sown or placed singly at a distance of about 1-2 cm apart. Cover the seeds with fine mixture of well rotten and sieved FYM or leaf compost, sand, and soil etc. in the ratio of (2:1:1). As experienced it has certain advantages. Those are:

- Conservation of moisture in the root zone will be less. Hence, spread of nursery diseases will be less and slower.
- Weed management of seed beds will be easy
- Each and every seedling will get proper air, water, light and nutrients so all will be healthy, bold and uniform.
- Less seeds are required compared to the broadcasting method.
- Less amount of seed bed covering mixture will be required.

Seedbed Covering Materials

There is a lot of controversy among the vegetable grower's that what material should be used for covering the seeds after sowing. Among 36 different materials and combinations used the combination of compost: soil : sand in the ratio of (2:1:1) was found superior for covering the seeds after sowing in the nursery beds. The data reveals that vermicompost shows their superiority over all the other treatments. The smallest plants were found under saw dust.

In general the loam to sandy loam, loose and friable, organic matter rich and well drained soil with preferably close to neutral pH (7.0) is considered as most suitable soil for seedling raising. A deep cultivation of the nursery land followed by 2-3 hoeing is recommended. Well rotten Farm Yard Manure (FYM) @ 2 kg or compost or leaf compost or 500 g Vermicompost per square meter should be mixed with the nursery bed.

Seed Requirement

The seeds and area for raising the seedlings may vary according to the soil, crop (variety), season and method of nursery raising.

An experiment on different vegetable crops was conducted to educate the nursery man and farmers for the appropriate seed rate and area required for raising the bold and uniform seedlings (Table 1).

Table 1: Vegetable, seed rate/ha and area required for sowing of seeds

Vegetables	Seed rate (g)	Area (m^2)
Tomato (OP)	250-300	100-125
Tomato (hybrid)	150-200	75-100
Brinjal (OP)	300	100-150
Brinjal (hybrid)	200	75-100
Chilli (OP)	500-600	75-100
Chilli (hybrid)	300-350	100-150
Cauliflower (early)	700	100-150
Capsicum (hybrid)	400-600	100-150
Cauliflower (mid-late)	400-500	150-200
Cabbage	450-500	100-150
Knol-khol	700-750	150-200
Onion	8000-10000	400-500

Mulching of Beds

Table 2: Vegetables and approximate period of mulch removal from beds

Vegetables	Seed germination (days)	Vegetables	Seed germination (days)
Cauliflower	3-4	Chicory	3-4
Cabbage	3-4	Endive	3-4
Knol-Khol	3-4	Kale	3-4
Chinese cabbage	3-4	Lettuce	4-5
Savoy Cabbage	3-4	Tomato	4-6
Broccoli	3-4	Onion	7-10
Brussels sprouts	3-4	Sorrel	8-10
Sweet Marjoram	10-12	Celery	10-15

Mulching of nursery beds and their removal is utmost important aspect in nursery raising to maintain the proper soil moisture for better seed germination. Seed bed cover with a thin layer of mulch materials either by paddy straw or sugar cane trash, or Sarkanda leaves, or dry grasses or any organic mulch / plastic mulch is ideal for uniform seed emergence. Because It maintains the soil moisture and temperature for better seed germination, It suppresses the weeds and their growth and gives protection from direct sunlight and raindrops and also protects against bird damage.

To verify the period of mulching the results obtained from the different experiments shows that length and thickness of the emerging plumule is very important to decide the health of the seedlings. It was observed that due attention should be given to remove the covered mulch from the seedbed. As and when the white thread like structure seen from about 50% seeds above to the ground, remove the mulch carefully to avoid any damage to emerging plumules. Always remove mulch in the evening hours to avoid harmful effect of bright sun on newly emerging seedlings. There are certain waiting duration's in different vegetables is given in Table- 2.

Curing Vegetable Nursery: In hot / cool weather conditions seedlings got damage with unfavourable weather conditions. Studies conducted at IIVR, Varanasi on different protectants screen in different weather conditions proved that seedlings may be protected from the thatched screen or shading nets during warm weather to protect the nursery from the direct sunlight. Similarly polythene structure/cover should be used to protect the seedlings against rain water and low temperature conditions for proper germination of the seeds and after growth of the seedlings.

Leaf curl is very harmful disease of tomato, chilies, capsicum etc. It is transmitted through white fly. Infestation starts from seedling stage and continue till harvest of the crop. The leaves of affected plants show curling, mottling, rolling, puckering etc. The disease is specially seen in the tomato, chilli and capsicum and causes great loss of the crop. To overcome with these problems an experiment on seedlings protection from different screen, net, Agronet and muslin cloth net was tried at the institute and observed that nursery bed covered with the muslin cloth net performed better than other covered screen. The other precaution should follow to control this disease. First treat the soil of the nursery by Carbofuran 5 g/m^2. Seed treatment with Gaucho @ 5 g/kg seed. Cover the seed bed after seed sowing by Agronet/ Markin cloth net making a tunnel like structure about 1-1.5 feet above to the ground, and spray the nursery beds with Imidachloprid 0.5 ml/litre of water at 15 days after seed germination at 7 days interval. Last spray of Imidachloprid @1 ml/ 3 liter of water 1- 2 days before transplanting.

Nursery raising of direct sown vegetables

The vegetables like bottle gourd, pumpkin, bitter gourd, cucumber, sponge gourd nursery are raised to get 1-2 month early crop compared to main season. To get higher price of their produce. For that only alternative is to raise the nursery of direct sown crop too. But care should be taken that direct sown crop nursery can't be uprooted unlike to tomato, brinjal and chilles etc. as corky cell "Suberine" in the cell of roots is found which is impervious to water resulting the death of the plants. There are certain other combinations such as damage of growing tips and poor root regeneration which also hinder plant establishment after transplanting. But one can pla nt such vegetable crops without disturbing its root zone (earth ball). Generally seedlings of direct sown crops are raised in the adverse situation when there is too low temperature for seed germination or the field preparation is delayed due to some standing crop in the field. Experiments conducted at the Institute proved that cucurbitaceous vegetable nursery can be successfully raised in small poly bags (15 x 10 cm size) filled with FYM, soil, and sand mixture (2:1:1). Sowing of seeds is done in the month of December kept in poly house where temperature is above 20° C. Seedlings become ready upto15th of Feb. and then it is transplanted in the well fertilized field by removing the polythene or by cutting lower portion of the polythene bag for further root development. In this way an early crop of Bottle gourd, bitter gourd, muskmelon, watermelon, cucumber etc. can be taken and one can advance the crops for about 30-45 days than normal crop and earn more income.

Nursery raising of Vegetative Propagated Vegetables

Vegetables like pointed gourd and Ivy gourd, stem cuttings of one year old stems are used for propagation. The cutting is planted generally in the open field and causes great loss due to improper emergence of buds in the transplanted cuttings. Studies conducted at the Institute reveals that sprouting in the transplanted cutting may be enhanced 2-3 fold by putting them in the polyethylene bags of 15 x 10 cm size filled with mixture of sand, soil and well rotten FYM (1:1:2) and kept in the polyhouse/low tunnel poly house. About 30 cm long vines of pointed gourd of one year old vines are to be selected for plants raising in Oct.-Nov. While in the Ivy gourd 15 cm long and 2-3 cm thick stem cuttings are selected and planted in the polyethylene bags leaving 1.5-2.0 cm one side exposed in Jan-Feb. The cutting becomes ready for transplanting in about 6 weeks after planting.

Advances in Vegetable Nursery Raising

The production of good quality seedlings is essential for getting higher yields and improving crop quality. In the past, the farmers themselves produced the seedlings required for transplanting at a lower cost, as most of the vegetable varieties were open pollinated types. Now, most commercial farmers are going for intensive vegetable cultivation using high yielding F1 hybrids to augment productivity. As these hybrid seeds are expensive, converting every individual seed into a healthy seedling becomes essential and this requires intensive nursery management. Vegetable seedling production is taken up by specialized farmers/companies or as a specialized activity in most advanced countries. In India too, the production of vegetable seedlings is gradually changing from open field nurseries to protected raised bed or seedling tray production in some of the intensive vegetable growing areas. Seedling production as a specialized practice is also rapidly catching up. However, establishment of a shade net nursery by every individual farmer owning a small piece of land under vegetable cultivation is not practically feasible and economically viable. Such farmers have to depend on commercial nurseries for hybrid vegetable seedlings to meet their requirements. Many seedlings are still grown in the ground, pulled up when mature and sold as bare-root plants, or grown in plastic tubes in a mixture of soil, sand and compost. Many such seedlings suffered from soil-borne diseases or transplant shock when they are moved into the open field. Many nursery growers do not use proven seedling raising technologies such as proper media preparation, using seedling trays, maintaining hygienic conditions, using quality insect mesh, shade net materials, or a double door system.

A lot of advancement in the field of vegetable nursery has been made either in terms of automation or in terms of growing medium. A number machine has been developed in India and in abroad to make nursery raising easy and convenient by which the raising of seedlings under controlled environment become easier and labour intensive. Because vegetable nursery production has become a highly commercialized business, wherein most farmers buy their plugs from professional growers. The growth chambers are usually an insulated room where temperature and relative humidity can be maintained which ease to germinate the seedlings very less time compared to open field conditions. Air circulation is important to ensure uniform temperature and humidity throughout the chamber. Cell trays are used by commercial growers to produce seedlings for planting out. The seedlings are easily removed from the tray for transplanting and the growth check to transplants from cell trays is minimal when planted in the field compared to the use of other types of transplants. Seedlings from cell trays may be used in manual or automatic trans-planters. Soil-less mixes are usually used for commercial seedling production. Commercial sterilized soil-less mixtures are available with added fertilizer and this simplifies the production process. Different plant species require differing amounts of space, nutrients and water. Accordingly larger (12 cell) to smaller (300 cell) sizes containers/ treys are used for raising the nursery for bolder to smaller vegetable seeds. They may have cell sizes as small as 0.8 inch square or as large as 6 inches square. Therefore, transplants growing in larger cells require less frequent watering and fertilizing. Plastic and polystyrene containers most often come in straight row arrangements. Polystyrene containers normally have inverted pyramid-shaped cells that taper toward the bottom. The number of cells in a container depends on the cell size. So, there are a series of advancement made in span of time in the field of nursery raising where one can raise healthy seedlings in any part of the year successfully with automation. This may be helpful to the entrepreneurs to establish a vegetable nursery for self-employment and future prospects.

Advantages of Mechanization in Nursery Raising

Apart from the traditional method of nursery raising mechanization has certain additional advantages which help in raising nursery faster and healthy. The main advantages are follows:

- Less time needed in seed emergence than conventional method.
- Less risk in raising nursery under controlled conditions.
- Multiplication is very fast as compared to old method.
- Less risk of protection from biotic and a-biotic stress.
- Seedlings/ transplants become more uniform, bold and healthy with more root system.
- Can be transported at distant places.
- Easy in transplanting at any time in the day.
- Establishment in the field is very early.
- Easy to look after the small seedling.
- Easy to weed control.
- Reduced field management cost.
- Less growing season and higher yield.
- More efficient use of surface area.
- Seedling can be raised in off season.
- Mortality percentage is less during transplanting as compared to traditional nursery.
- Easy to handle from one place to another.

Selection of Nursery-Growing Area and Potting Material

In the present scenario in place of selection of site for nursery there is need of nursery raising protected structure like, net house, shade net house, poly house, low tunnel , poly shades etc. to raise the nursery. Avoid areas with intensive fruit or vegetable cultivation around the nursery to prevent infestation of seedlings with insects, fungal and bacterial spores (e.g. mites, thrips, whitefly, or blight). Pests and diseases present on the crops in surrounding areas may be incidentally carried onto the seedlings in the nursery Seedlings. Locating a nursery in a cereal cropping area is preferable as cereals do not harbour as many pests and diseases as vegetable. At such places climatic condition may be manipulated according to the need. During summer season screen net may be stretched for shade as per the requirement. Need not to worry about soil type as seedlings are grown in soilless media in germination trays. Since the nursery is being raised in potting trays so there is no need of making beds. Trays may be kept in any direction under controlled conditions.

Media Treatment

In traditional method of nursery raising we treat the nursery soil by many ways but while we raise the nursery in potting trays the media used for filling in the trays need to be treated against harmful insect pest and pathogens. For that we put the heap of soil/compost/perlite/cocopit or any other with hot air vapour continuously for 6-8 hours. A vapour generating machine is now available for treating such pot filling materials. We have make a heap of that material and perforated pipes must be buried beneath the heap and whole material must be covered with Tarpaulin sheet to protect outgoing of hot vapour. The perforated pipes must be connected with the vapour generating machine. In this way the material is treated within 6-8 hours as compared to the 1-2 month by solarisation method.

Growing Structures for Seedling Raising

***Polyhouse*:** A polyhouse structure will be with galvanized nuts and bolts so that it will be stronger and easier to repair if it is damaged. The specifications for the polyhouse are as follows:

- A double door system that closes automatically is recommended with 6 feet (2 m) spacing between the doors.
- Doors should be sliding, not swinging to ensure there is no gap while closing.
- To ensure there are no holes or gaps in doors to allow insect entry, insert rubber flaps or brushes in the gaps.
- The side walls should be made of insect-proof net and the top to be covered with plastic to prevent rain water entering.
- The width of a single span polyhouse should be about 8 m.
- The roof structure should be extended at least 0.75 m on all four sides of a polyhouse to drain off rain water in heavy rainfall areas.

- The complete structure should be made of 2 mm thick galvanized steel tubular pipes.
- The gutter height should be at least 15 feet above the floor area.
- The gable needs to be covered with insect net instead of polythene to prevent insects entering the polyhouse from the roof.
- The height of the curtain wall or apron should not be beyond 1 foot from the ground level.
- A truss height of 1.5-1.8 m is recommended for high rainfall and windy areas.
- Suggested direction of the windward side for good ventilation: Southwest monsoon region: southwest direction. Northeast monsoon region: northeast direction.
- Preferred orientation of the structure is north-south to capture the best sunlight.

The potting trays should be kept on tables inside the polyhouse. Movable iron tables, arranged contiguously with 1 m working space in between two rows. The length and breadth of each table could be 3.5 m x 1.75 m so that two rows of tables could be arranged in a single span of the polyhouse. Apart from the wooden tables Bamboo benches that can last for at least 1.5-2 years may be used wherever bamboo is readily available.

Net house: A nethouse is a structure made with the help of iron pipes as stated in the polyhouse but inplace of polyethene 40 mesh net is stretched all around the structure to protect the seedlings from white flies. Mostly it is recommended where high rainfall and humidity is not a problem and the investor requires low cost structures which fully provides protection against the environment and complete protection from pests and diseases. The net house structure has certain feature for manufacturing which are as follows:

The net house structure should not be less than 13 feet in height.

- Galvanized steel pillars 2.0-2.2 mm thick and 2.8-2.9 inches in diameter will usually used for making net houses which able to withstand a wind velocity up to 140 km/h.
- UV stabilized insect-proof net with 40 mesh is generally recommended.
- For smaller structures and in cooler areas, 50-60 mesh can be used which provides better insect exclusion.
- Red, silver or white shade nets can be used, but not green.
- Green coloured net cuts off more light.

When seedlings are grown in net houses, it can be important to protect the seedlings from rainwater. Metal hoops have to be fixed over the raised beds and a polythene sheet is fitted over the hoops and fixed at one end of the raised bed. It is spread over the beds during rain

***Types of shade net*:** There are different types of shade nets preferred for different needs.

***Green shade net*:** Produces lanky seedlings and is preferable for ornamental plants. It is not preferred for vegetable nursery seedling production as a 50% shade net cuts off more light and looks like 80% shade is provided to the seedlings, hindering the seedling growth. The common fact that green color in the plants is important for photosynthesis does not apply to the shade net.

***Red shade net*:** Not preferred in cloudy areas or during cloudy times of the year. If too bright double net can be used. It will cut infrared light and this can help produce higher yields in Capsicum.

***White shade net*:** This will give compact seedlings, and works well for nurseries growing sun-loving plants such as tomato, eggplant, or chili.

***Silver shade net*:** Ideal for sun loving plants such as tomato, eggplant and chili.

***Black shade net*:** Produces lanky seedlings and is preferred for ornamental plants but not for vegetable seedlings.

Alumin shade net (thermal screens): Thermal screens can be used as a secondary layer in polyhouses and shade nethouses to reduce the temperature 4 degrees. In winter, thermal screens will retain the heat inside the nethouse/polyhouse.

***Types of polythene sheets* :** Polythene sheeting that can be used to cover the roof of a polyhouse also plays an important role in regulating the amount of light that penetrates into the polyhouse. UV (Ultraviolet) stabilised polythene is recommended to ensure the longevity and durability of the sheets and prevent degradation from UV light.

There are different types of polythene sheets available for different needs. For polyhouses, 200 GSM (grams per square meter), 5 layered, UV stabilised, anti-dust plastic sheets of different types can be used.

***Transparent*:** More light penetration. This can be used in regions where the light intensity is low in general and often cloudy.

Translucent: Gives diffused light. Appropriate location is Kerala.

Yellow: Less light penetration. Can be used in dry places such as Pune. In general, a yellow color increases the heat and humidity inside the polyhouse.

Use of Sensors in Polyhouse

There are two kinds of sensors i.e. Temperature dependant which trigger the functioning of foggers when the temperature crosses the set limits. The second one is Relative Humidity dependant which Trigger the fogger when humidity levels decrease. In South and Central India, temperature-dependent sensors are preferred, as the temperatures inside polynet houses can rise dramatically in summer. One sensor in the centre of the polyhouse hanging from the roof, one to two feet below the foggers will provide for the requirement of a 600 sq m polyhouse.

Humidity Control

The preferred humidity level for vegetable production is 60% to 65% and this can be maintained by providing vents around the polyhouse to expel excessive humid air. Ideally, there should be at least one large vent placed

The preferred humidity level for vegetable production is 60% to 65% and this can be maintained by providing vents around the polyhouse to expel excessive humid air. Ideally, there should be at least one large vent placed near or on the roof of the polyhouse to allow heat to escape. There should also be several vents placed around the perimeter of the polyhouse, near the base. This allows for the best, most natural type of ventilation: cross-ventilation.

Growing Medium

Growing seedlings in an artificial medium without soil or compost is healthier as it prevents contact with soil-borne diseases. Drainage also can be easily varied in artificial media. Perlite, vermiculite and peat moss are expensive. Cocopeat is a cheaper and effective alternative base for an artificial growing medium, but it is important to get the pH right and avoid any salt contamination. Cocopeat is the pith derived from coconut husks after the removal of fibre by the coir industry. The recommended pH of cocopeat is 5.8-7. Highly acidic or alkaline cocopeat will hinder seedling growth. It is suggested to test the pH of coco peat with a pH meter before using it. It is always better to sieve the coco peat. Instead of raw cocopeat alone as a medium, a mixture of cocopeat, vermiculite and perlite in the ratio of 3:1:1 by volume is better for the growth of vegetable seedlings. Vermiculite or perlite not only reduces the weight of the medium but also provides better drainage and porosity to the medium, enhancing the growth of young roots.

Salinity is measured as Electrical Conductivity (EC). The EC of cocopeat should be less than 1mS/cm. If the EC is higher, the seedlings grown in this media will be weak and lanky, and the EC needs to be reduced by either adding buffering chemicals (may be done by the supplier) or simply by repeated washing at the nursery level. Soak the cocopeat in good quality water, and then remove the water by squeezing it two to three times. Repeat the process two or three times to remove the salts and bring down the EC. It is usually sterilized before sale, but if sterilizing is not done properly, every type of fungal infection may also appear on cocopeat. To prevent from the diseases treat the cocopeat before use. It is better to use decomposed cocopeat as it can provide the required moisture for seedling growth immediately after sowing. Un-decomposed cocopeat needs to be decomposed by wetting it regularly until it is ready. Cocopeat or any similar substrate can be enriched to reduce pest and disease problems in seedlings by mixing Neem cake @ 50 kg + carbofuran @ 5 kg + Trichoderma harzianum @ 2 kg + Pseudomonas fluorescens @2 kg. These additives either act as insecticides or bactericides or can be particularly useful for producing healthy seedlings of tomato, eggplant, capsicum, cauliflower, cabbage, chilies and onion.

Selection of Potting Trays

Potting trays or seedling raising trays are of different sizes are available in the market. The selection of the seedling tray may depend on the type and nature of the vegetables. There are two types of cavities in the trays available in the market: inverted cone shaped and inverted pyramid shaped. The experiences of nursery entrepreneurs indicate that a tray with an inverted pyramid shape is ideal for a strong stem and good root growth as it provides a wider space for the root system to grow. In contrast, the inverted cone shape narrows down the space for roots to spread, resulting in a cluster. Raising cucurbits seedlings require wider cell size trays while tomato brinjal, chillies, cabbage, cauliflower requires medium cell size trays. For raising the onion seedlings in potting trays needs more than 300 cell sizes trays. A seedling that is 25-26 days old with 4-5 true leaves is at the right stage for transplanting.

Pot Filling with Growing Medium

At smaller scale the filling of pot takes place manually i.e. by hand, but at commercial scale considering the problem of farm laborers mechanized methods were followed to economize the production cost of seedlings. In case of manual method about 20-25 trays may be filled per hour but in case of mechanized method about 300-400 trays are filled in an hour. Before filling the growing medium in trays perlite and vermiculite may be mixed with cocopeat. The perlite and vermiculite are the light weight material which boost the seed germination.

Seed Sowing in Trays

Seed sowing in the trays is done either by hand or by automatic machine. By hand one can sow 8-10 trays per hour but by automatic machine more than 100 trays may be sown in an hour. To maintain uniform sowing depth and speed, sow the seeds with the help of a dibbler that pokes uniform holes in the medium. The recommended dibbling depth is three times the seed diameter. After sowing, a layer of vermiculite or finely sieved cocopeat is spread on the sown seeds to a depth of 1-1.5 times the diameter of the seed. In large scale nurseries a seeder assembly can be used for automated tray filling, sowing, covering and watering. This enables about 90-100 trays to be filled in an hour.

Placing the Trays in Germination Chamber

Normally in old method of seedling raising 4-6 days' time is taken in germination of tomato, brinjal and chillies seeds and 3-4 days in cauliflower, cabbage, broccoli and other crucifers. With the mechanization the germination may be takes place in one or two days. Care to be taken after sowing the seeds it is important to transfer the sown trays to a darkened germination room / closed chamber that is kept at a mild temperature for a good germination percentage. Stacking many trays on top of each other is generally not recommended, as the top ones will compress the trays beneath. If the germinating trays need to be stacked due to space constraints, it is recommended to use a ziz-zag arrangement where the upper tray does not directly touch the media of the lower tray. Before germination the trays may be kept on one another but after germination it may be spread. If the germinating chamber is not available the trays may be wrapped with black polythene as shown in the figure to maintain darkness and humidity later. After germination it may be uncovered and then spread for uniform germination.

Watering and Aftercare

Any plant species needs water for their survival. Similarly young growing tender seedlings need proper spraying water at regular interval for proper growth and development. Water quality is important and can have a big impact on the health and growth of seedlings. The most common water quality problems are from salts, particularly when bore water is used. Water quality is measured by Electrical Conductivity (EC) and pH. The ideal ranges for these are: • EC should be below 1 mS/cm. Seedling growth may be affected badly if the water EC is high. The pH of water should be 6.5-8.4. Higher pH levels are associated with higher levels of salts that will damage seedlings. An EC meter and pH meter need to be installed in all nurseries. Always lower EC (less than 1 mS/cm) is preferable. If the EC of irrigation water is too high then there are several options to reduce the problem. Rain water are having lower EC. A rainwater harvesting structure may be constructed to mix rainwater with ground water to reduce the EC. Water softeners (such as potassium chloride) with bactericidal properties may be used to reduce water EC. Sodium chloride should not be used to soften water, as it is toxic to seedlings. Good drainage in the seedling trays can prevent the build-up of salts, even when the EC is not initially ideal. Application of irrigation water in the trays in the morning is ideal. If seedlings are irrigated in the evening, water droplets will remain on the leaves and lead to fungal infections. If the growing medium is porous, less water is required. Depending on weather conditions spraying of the water in trays may be made two or three times a day. During the bad weather, irrigation can be reduced to once a day. Reduce watering in the last week of seedling to become ready for hardening of seedlings prior to transplanting form more growth to establish in main field.

Application of Manures and Fertilizers

In the nursery never add fertilizers in the growing medium, but nutrients are supplied to the growing seedling in the artificial medium through fertigation every day. There are several different ways this can be applied: 8.1 Basic fertilizer application Daily application of 19:19:19 containing micronutrients @ 5 g per 10 litres of water, starting from the cotyledon stage till 16 days and gradually increasing the dosage every 3-4 days from 16 to 24 days, and then later stopping for hardening

Pest and Disease Control

Young growing tender seedlings are affected by a number of insect-pest and diseases. Pest and disease problems can be minimized by careful construction and maintenance of the protective nursery structure. Some general precautions will help in reducing the number of sprays:

- Avoiding excess irrigation which promotes diseases.
- Repairing holes in the net whenever noticed.

- Closing the doors properly without any gaps to exclude insects.
- Disinfecting the trays, nursery tools and nursery area.
- Sterilizing the growing medium.
- Installing sticky traps in between the two doors to catch any insects that do enter.
- Prophylactic sprays may be needed if seasonal conditions or local outbreaks suggest a potential problem.
- When pests and diseases are noticed, a quick application of mild chemicals is needed.

Grafting of Seedlings

Grafting is newer techniques followed with certain objectives to overcome Soil-borne diseases and nematodes causing serious problems in vegetable cultivation. Secondly in acidic soils and under humid conditions, bacterial wilt are a serious threat to tomato and brinjal production. Thirdly some vegetables are having poor root system like capsicum etc. So if capsicum or tomato grafted on brinjal root stock the uptake of water and nutrients are more as compared to sole capsicum or tomato plant which ultimately increased the growth and yield. Tomato seedlings can be grafted on brinjal wild species plant named to *Solanum torvum* or wilt-resistant brinjal varieties.

For raising the rootstock, seeds of wild brinjal i.e. ***Solanum torvum*** are sown into seedling trays in a soilless medium containing three parts of washed cocopeat and one part each of vermiculite and perlite. Solanum torvum takes at least 15 to 20 days to germinate. Normally 8-10 seeds are sown in a single cell and when the seedlings reach the 3-4 leaf stage, excess seedlings are transplanted into a new tray at one seedling per cell. The rootstock seedlings require daily irrigation and fertigation once every 5-7 days. ***Solanum torvum*** seedlings have an initial slow growth rate and may take around 30-40 days to grow to a height of 5-8 cm to get ready for grafting. Hence, they have to bc sown 20-25 days before sowing the scion seeds Slant/side grafting is also used for vegetable seedlings. The stock is de-topped using a slanting cut 5.0 – 6.0 cm above the base. A small polyethylene sleeve (1.0 cm long) is then fixed on the rootstock where the slanting cut has been made. A matching slanting cut is made on the scion seedling and it is then inserted into the polyethylene sleeve attached to the rootstock so that both the cuts are aligned with each other. The vegetable seedlings after grafting are immediately shifted to a healing chamber that can provide a hot (25 – 30°C), humid (90 - 95% RH) environment for the grafts to heal. High humidity is maintained inside the healing chamber by spraying water or fogging at periodic intervals depending upon the weather. The spraying intensity increases in dry weather. After 4 to 5 days, when the grafts heal, they are transferred to a polyhouse and kept for a week for further hardening. The hardened seedlings are then transplanted. It is important to harden seedlings before they are transplanted. The best way to help strengthen seedlings for the outside environment is to harden them off. It is an easy process and will enable the plants grow better and stronger when transplanted out into the main field. Hardening should not be done outdoors because of the danger of pest and disease infections. The required high light intensity can be provided inside the nursery. The hardening process should start from 16 days after sowing by slowly reducing the fertilizer and water supply. When seedlings are hardened under a shade net in polyhouse, the shade net on the top should be removed and watering should be reduced.

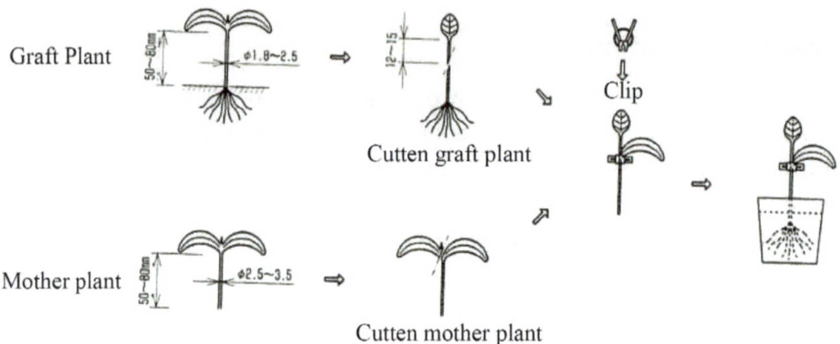

Transportation of Seedlings

The nursery floor has to be cleaned every day and insect netting needs to be cleaned once every three months. To disinfect the seedling trays: Trays can be steam cleaned and then rinsed with water, or treated with water mixed with chlorine. Mechanical damage to the insect mesh/polyfilm or shade net due to wear and tear, high winds or rodents needs to be repaired immediately to avoid entry of insects and pathogens. It is strongly recommended to go for disposable trays to reduce the risk of infections. However, some nursery owners use re-usable trays when they sell seedlings locally around their nursery. While using re-usable trays, they have to make sure to replace them immediately if there is any wear and tear. If the trays are disinfected between the seedling cycles, they may be used 3-4 times before being discarded. The best method for transporting seedling trays is to use plastic crates arranged one above the other in the transport vehicle. The transplanting may be done starting from the morning to evening as while transplanting there is very less shock to roots as compared to traditional method because while lifting the transplants along with whole root comesout from the trays which also help in early establishment in the main field.

References

Bharathi, P.V.L., Ravishankar, M. 2018. Vegetable nursery and tomato seedling management guide for south and central India. World Vegetable, Publication No. 18-829. World Vegetable Center, Taiwan, 30 p.

Bose, T.K. and Som, M.G. 1986. Vegetable Crops in India. Naya Prokash, Bidhan Sarani, Calcutta. P. 128.

Chaurasia, S.N.S, Nirmal De, K.P. Singh, M. K. Banerjee and Mathura Rai 2003. Nursery management in vegetables. IIVR Technical Bulletin (15):1-35.

Chaurasia, S.N.S. and Singh, B. 2016. Quality transplants: a healthy way of raising vegetable nursery, Indian Horticulture. 61 (1):32-35.

De Nirmal, Chaurasia S.N.S., Pandey K.K., Singh K.P. and Rai M. 2004. Sabjiyon ki paudh taiyar karna (Hindi), Technical Bulletin no. 20 :1-34.

Easdown, W. and Ravishankar, M. 2016. Study of tomato nursery production practices inselected districts of Maharashtra and Karnataka. World Vegetable Center 16(807):1-40.

Mohanta, S.D. and Basabadatta Sahu 2022. Nursery Raising for Vegetables and Flowers in Greenhouse, Smaranika, Centre for Smart Agriculture, Centurion University of Technology and Management, Paralakhemundi, Odisha.

Pandiyaraj, P., Kumar, Y.R., Vijayakumar, S. and Das, A. 2017. Modern nursery raising systems in vegetables. International Journal of Agriculture Sciences 9(52): 4889-92.

Prasad, S. and Kumar, U. 2007. Greenhouse management for horticultural crops, Agrobios, India, pp. 476.

Singh, R.R., Meena, L.K. and Singh, P. 2017. High Tech Nursery Management in Horticultural Crops: A Way for Enhancing Income. International Journal of Current Microbiology and Applied Sciences 6(6):3162-172.

12

Hybrid Seed Production of Vegetable Crops

B.S. Tomar[1], Yalamalle V.R.[2] and Chaukhande Paresh[1]

[1]*Division of Vegetable Science, ICAR-IARI, New Delhi - 110 012*
[2]*Division of Seed Science and Technology, ICAR-IARI, New Delhi - 110 012*

India is the second-largest producer of vegetables in the world (FAO, 2021). Its diverse soil and climate condition provides ample opportunity to grow a variety of crops (be it tropical, semi-temperate, or temperate type). Vegetables form a significant part of the total agricultural production of the country. India grows vegetables in an area of 10.86 mha with annual production of 200 million tonnes (NHB, 2021). Vegetables alone contribute 66.97 % of the total horticultural production (341.63 million tonnes). During 2021-22, India exported vegetables worth Rs. 5,745.54 crore. The growth in vegetable production is largely due to an increase in productivity of average productivity is 18.18 tonnes per hectare, The increase in productivity has led to improved per caput availability of over 250g. This impressive improvement can be attributed to the development of improved varieties and hybrids, along with the availability of high-quality seeds, production, and protection technologies through systematic research, and their large-scale adoption by farmers.

The total vegetable seed requirement in India is 51,000 tonnes but the actual availability is around 40,000 tonnes and a large chuck of seeds is still being multiplied by farmers themselves (Dutta, 2004). Vegetable plays an important role in India's rural economy by improving the income of the farmers. Cultivation of vegetables is labour intensive, generating a lot of employment opportunities throughout the year. They fit well in the cropping system either as main crop, intercrop or as a catch crop between the growing seasons.

Most of the farmland in India are small holdings with average size of 1.08 ha.. For small and marginal farmers with assured irrigation facilities, vegetable production is an productive enterprise. Vegetable crops have the potential to provide nutritional security for both urban and rural populations and are more productive than other crops. In vegetables, hybrid cultivars are preferred because of higher yield, uniformity, biotic an abiotic stress tolerance and high shelf life. A consistent increase in productivity can be achieved by using high-quality seeds with built-in inbred and hybrid vigor, along with modern vegetable cultivation technologies and effective government policies. Therefore, adopting hybrid vegetable technology is the better option (Table 1).

Table 1: Comparison of productivity in variety vs hybrid

Crop	Yield (tonnes/ha)	
	Variety	Hybrid
Tomato	25-45	70-85
Brinjal	25-30	50-60
Cucumber	7-10	20-25
Muskmelon	10-20	20-25
Bitter gourd	16-18	22-25
Watermelon	20-25	35-40
Carrot (Tropical)	16-20	20-22
Cabbage	30-35	50-70
Cauliflower (tropical)	18-20	25-27
Broccoli	15	20-22

Vegetable seed production involves two processes: pollination and fertilization. The pollination involves the transfer of pollen grains from the male part of the flower, called the androecium, to the female part, called the gynoecium. There are two main types of pollination: self-pollination, which occurs when pollen is transferred from the anther to the stigma of the same plant. Self-pollination increases homozygosity. Self-pollination in some species causes inbreeding depression, these species have evolved as cross-pollinated species. Cross-pollination refers to the transfer of pollen between different plants (Tomar et al., 2021). Cross-pollination leads to heterozygosity. Species that rely on outbreeding typically develop a balance of heterozygosity but may exhibit significant inbreeding depression if self-pollinated. In addition to self-pollination and cross-pollination, there is another type of species known as often cross-pollinated species, which are basically self-pollinated but outcrossing to an extent of 5% - 30% occurs. Certain mechanisms such as bisexuality, homogamy, cleistogamy, and the position of anthers promote self-pollination, while others such as dicliny (including monoecy and dioecy), dichogamy, heterostyly, herkogamy, self-incompatibility, and male sterility promote cross-pollination (Singh et al., 2020).

Hybrid Seed Production Technology

In recent decades, availability of an affordable technique for mass-producing F_1 seeds has gained momentum, because of which the availability and affordability of seeds of hybrid varieties have increased. Many techniques have been developed to create hybrid vegetable crops, but because of practicality and economic viability only a few of these are used to produce commercial hybrids in vegetables. The most commonly utilized mechanism or methods for hybrid seed production in commercial vegetables are (Table 2).

Table 2: Method/systems of hybrid seed production used in vegetable crops

Crop	Hybrid seed production mechanism commercially exploited
Cucumber, bitter gourd	Gynocecism and natural pollination
Bitter gourd, bottle gourd, Pumpkin	Pinching of staminate flowers and hand pollination
Watermelon and muskmelon	Removal of staminate flower + emasculation + hand pollination
Tomato, brinjal	Functional male sterility and hand pollination
Chilli	GMS + bee pollination
Capsicum, onion, cabbage, carrot, radish	CMS + natural pollination
Cauliflower, broccoli	Self-Incompatibility and natural pollination
Squash's	PGR and natural pollination
Sweet corn, baby corn	Detasseling + wind pollination
Tomato, brinjal, sweet pepper, okra, chilli	Hand emasculation and manual pollination

***Source*:** Tomar *et al.* (2017)

Gynoecious lines

The majority of cucumber hybrids are produced by crossing gynoecious and monoecious cucumber lines. The other systems of producing gynoecious hybrid seeds are gynoecious × gynoecious but gynoecious × monoecious hybrids are still widely grown hybrids because this offers advantages like earliness (Jat *et al.*, 2015), high degree of female sex expression (Jat *et al.*, 2016; Jat *et al.*, 2017), with uniform and concentrated fruit formation, which was especially advantageous for mechanical harvest (Robinson 1999, 2000). Several institutes have released gynoecious hybrids, IARI, New Delhi has released a cucumber hybrid Pusa Sanyog and PAU Ludhiana have released muskmelon hybrid (MH-10). Gynoecious hybrids employ honey bees for crossing in seed block, ideally 4:2 (female: male) is maintained, when the seed is produced under open field condition isolation of 1,000 m is recommended. The maintenance of male parents is done by selfing. Since female plants only produce female flowers male flowers are induced by spraying 200 ppm silver nitrate at 2-4 true to inducing the staminate flowers and plants are selfed. One of the major drawbacks of Gynoecious lines is the stability of lines under varying photo-thermal regimes, particularly in tropical countries like India. Efforts have been done in IARI, New Delhi to develop stable gynoecious lines in cucumber and muskmelon. As a result, F_1 hybrid Pusa Rasraj (Monoecions 3 x Durgapura Madhu) was developed and released by IARI.

Use of Growth Regulators

Gynoecy is most important sex form which has made phenomenal exploitation of hybrid vigour in cucumber, bitter gourd, and muskmelon (Munshi *et al*., 2017). The gynoecious inbreds could self-reproduce if a growth regulator is applied to induce male flowers (Robinson, 1999). The gibberellic acid (1500-2000 ppm) is used for the induction of male flowers in cucumber. (Peterson and Anhder (1960) but different gynoecious lines vary in response to GA application and the number of induced male flowers are not sufficient for hybrid seed production, cause excessive stem elongation or malformed male flowers (Robinson, 2000). Therefore, the application of silver compounds such as silver nitrate (250-400 ppm) is done to induce male flowers. These ions inhibit ethylene action and promote male flower induction in gynoecious cucumber lines (Beyer, 1976). However, due to the phytotoxic effects of silver nitrate such as the burning of plants, now a day's silver thiosulphate (400 ppm) is widely used by seed producers for the maintenance of gynoecious cucumber lines. It induces male flowering of cucumber plants over a longer period and is less phytotoxic compared to silver nitrate.

Fig. 1: Induction of male flowers in gynoecious lines using growth regulators

Hand Emasculation and Hand Pollination

Hand pollination and emasculation are the usual techniques for producing hybrid seeds. This technique is only economically viable for crops like brinjal, tomatoes, and cucurbits where there are more seeds produced per pollination and fewer seeds needed per hectare (Vishwanath *et al*., 2008). Since pinching, pollen collection, and hand pollination require a lot of trained labor, this method is only appropriate for small-scale production.

Fig. 2: Protection of female and male flower and hand pollination

Male Sterile Lines

Male sterility refers to the inability of a plant to create or release functional pollen as a result of failing to form or grow functional gametes, microspores, or stamens. Male sterility is consequently sometimes subdivided into the following categories: (a) "Pollen sterility," in which male sterile individuals differ from normal only in the absence or extreme scarcity of functional pollen grains; (b) "Structural or staminal male sterility,". in which male flowers or stamens are deformed, non-functional, or entirely absent; and (c) "Functional male sterility," in which perfectly good and viable pollen is present but fail to dehisce (Lasa, and Bosemark, 1993). On a genetic basis, male sterility is classified in the following three groups 1) Genetic male sterility (GMS) 2) Cytoplasmic male sterility (CMS) 3) Cytoplasmic genetic male sterility (CGMS).

Table 3: Genes responsible for male sterility in vegetables

Crop	Gene number/condition	Gene
Tomato	Single recessive gene	*ps*-2
Chilli	Single recessive gene	ms-12 & ms-3
Muskmelon	Single recessive gene	*ms-1, ms-2, ms-3, ms-4, ms-5*
Winter squash	Single recessive gene	*ms-1*
Summer squash	Single recessive gene	*ms-2*
Cucumber	Single recessive gene	*ms-2*

Source: Singh and Singh, 2022

The male sterility system has a number of benefits. It lowers the cost of producing hybrid seeds. It expedites hybrid breeding programs, prevents hand pollination and emasculation enables large-scale production of hybrid seeds, and allows for the economic exploitation of hybrid vigour (Sharma *et al.*, 2019). But it also has several limitations which include- the identification and transfer of male sterility in a suitable background is a long process. Adequate cross-pollination should be there between A and R lines for a good seed set. Synchronization of flowering should be there between A and R lines. Fertility restoration should be complete otherwise the F1 seed will be sterile. Cytoplasmic male sterility can be utilized in those vegetables where the economical part is the vegetative part. Male sterility is influenced by environmental factors leading to selfed plants. In GMS sytem, in the hybrid seed production field, 50% male fertile segregants (Ms/ms) need to be identified and removed before they shed pollen. Isolation is needed for the maintenance of parental lines and for producing hybrid seeds which are often difficult to find. The use of the male sterility system has been commercially exploited and several hybrids have been released by public sector organizations which are included in Table (4).

Table 4: Hybrids released by public sector using male sterility system

Crop	Hybrids	Male sterility systems	Institute
Chilli	CH-1, CH-3	GMS	PAU, Ludhiana
	Kashi Surkh, Kashi Tej	CGMS	ICAR-IIVR, Varanasi
	Arka Megha (MSH-172), MSH-149, MSH-96	CGMS	ICAR-IIHR, Bengaluru
Carrot	Pusa Nayanjyoti	CGMS	ICAR-IARI, New Delhi
Onion	Arka Kirtiman and Arka Lalima	CGMS	ICAR-IIHR, Bengaluru
Cabbage	H-64 and KCH-4	CMS	ICAR-IARI, New Delhi
Tomato	Pusa Divya	CGMS	ICAR-IARI, New Delhi

Self-Incompatibility

In vegetable crops, self-incompatibility is a common occurrence that discourages inbreeding and promotes outcrossing. Numerous multi-allelic loci control the genetic reaction to self-incompatibility, which is dependent on complex interactions between self-incompatible pollen and pistil combinations (Hiscock, 2002). It is a genetically controlled phenomenon that prevents self-pollination in cole crops and other vegetables like tomatoes. The two types of self-incompatibility are gametophytic and sporophytic, while in sporophytic technique, pollen phenotype (self-incompatibility response) is identified with the genotype on the female plant on which pollens are developed, self-incompatibility response of pollen and stigma is determined with the genotype of the female plant on which pollens are developed (e.g. tomato) in gametophytic technique. (cole greens). Sporophytic self-incompatibility (SSI), has been effectively used for the development of commercial hybrids in the Brassicaceae (Singh *et al.*, 2020). One of the major advantage of, self-incompatibility is during hybrid seed production with appropriate parent combinations seeds can be harvested from both parents. Self-incompatibility system has some limitations -the purity of hybrid seeds is doubtful due to the presence of sib-seeds produced on the SI line during environmental aberrations, and this makes it unsuitable. Although it is a widely grown crop; unfortunately, no one hybrid from the public sector is in cultivation.

Hybrid Seed Production Under Protected Structures

The lack of sufficient isolation, insect vector, diseases, and a virus-free environment in the production of disease-free, healthy, and genetically pure seed for commercial cultivation are the major challenges in the quality hybrid seed production of vegetables. Compared to open field conditions, protected cultivation can fetch higher seed yields with better quality (Tomar and Jat, 2015). Insect vectors and viral diseases are the most devastating problems for quality seed production in most of the vegetable crops grown under open fields, and if the insect vectors are checked by protected structures the use of pesticides will automatically reduce. Seed production in the summer season is affected by a sudden increase in temperature and severe infestation of mottle mosaic virus and other insect pests in the rainy season; against which still there is no effective and reliable management measure. The change in climatic conditions like unseasonal rains during April-June, and increased temperature drastically reduced the seed yield and quality even in the summer season crop. Raising seed crops in insect-proof net houses can overcome these problems by protecting the crop from various insect vectors and unfavorable climatic conditions. It also provides an option for quality and off-season seed production. The insect-proof net house is the most suitable and low-cost protected structure for quality hybrid seed production of open-pollinated varieties in a large number of vegetables. The major interest is to grow virus-free seed crops and protection against major insects/ pests. Insect-proof net house is suitable for hybrid seed production of tomato, sweet pepper, chilli, okra, brinjal, and cucurbits as compared to open field conditions (Jat *et al*., 2015; Jat *et al*., 2016).

The semi-climate-controlled greenhouse is suitable for hybrid seed production of indeterminate type varieties and hybrids of standard tomato, cherry tomato, sweet pepper, bitter gourd, and parthenocarpic cucumber varieties. The seed yield of such crops can be 3-4 times more compared to their open-field cultivation (Kaddi., 2014; Kalyanrao *et al*., 2012; Jat *et al*., 2017). Similarly, the naturally ventilated greenhouse is also suitable for hybrid seed production, where the seed yield is usually 2-3 times more over an open field, but the cost of seed production is only 1/3 of the seed produced under semi-climate-controlled greenhouse conditions (Kalyanrao *et al*., 2014; Singh and Tomar, 2015). The major advantages of hybrid vegetables seed production under protected conditions are:

- Higher seed yield (generally 2-4 times more) and seed quality as compared to open field
- Requirement of isolation distance in cross-pollinated vegetables can be minimized.
- Problem of synchronization of flowering can be minimized.
- Maximum plant population can be maintained.
- Seed production under adverse climatic conditions is possible.
- Training, pruning and hand pollination practices are very easily manageable under protected conditions compared with to field seed crop.
- Emasculation of female parents is not required as there is no insect pollinators.
- Seed crops will not be damaged by un-seasonal rains at the time of their maturity.
- Seed viability and seed vigour could be extended through better nutrient management in seed crops under protected conditions.

The vegetable seed industry has enormous employment generation potential. Hybrid seed production of tropical particularly solanaceous vegetables, which requires hand emasculation and pollination has led to employment generation of 2.71 million man-days in rural areas. for which seed production of vegetables can be a major role player in doubling the farmers' income. India with its diverse climatic and soil and vast pool of manpower can become the world leader in the export of vegetable seeds and ushering the rural prosperity.

References

Singh, Balraj, Singh A.K. Tomar, B S, Ranjan, J K, and Dutt Som 2021. Advances in Hybrid Seed Production of Vegetable Crops September. In "Current Horticulture: Improvement, Production, Plant Health Management and Value- Addition: Brillion Publishing.

Beyer, EMJr., 1976. Silver ion: a potent anti-ethylene agent in cucumber and tomato. Hort Science 11(3): 195–96.

Dutta, O.P. 2004. Recent innovations in hybrid seed production in vegetables. Proc. Indian Sci. cong. New Delhi. Vol-1 (6-9). pp 217-42.

Hiscock SJ. 2002. Pollen recognition during the self-incompatibility response in plants. Genome Biology. 3(2): reviews1004.1-reviews1004.6.

Jat, G.S, Singh, Tomar, B, Singh, B.S., Ram, J.H. and Kumar, M. 2016. Seed yield and quality as influenced by growing conditions in hybrid seed production of bitter gourd (*Momordica charantia* L.) cv. Pusa Hybrid-1. Journal of Applied and Natural Science,8 (4): 2111-15.

Jat, G.S., Munshi A.D., Behera T.K. and Tomar B.S. (2016). Combining ability estimation of gynoecious and monoecious hybrids for yield and earliness in cucumber (*Cucumis sativus*) Indian Journal of Agricultural Sciences 86 (3): 399–403.

Jat, G.S., Singh, B, Tomar, B.S., Muthukumar, P. and Kumar, M. (2017). Hybrid seed production of bitter gourd is a remunerative venture. Indian Horticulture, March-April 62(2):34-37.

Kaddi G, Tomar B.S, Singh, B. and Kumar, S. (2014). Effect of growing conditions on seed yield and quality of hybrid Cucumber (*Cucumis sativus*). Indian Journal of Agricultural Sciences 84(5): 624–7.

Kalyanrao, Tomar B.S. and Singh B. (2012). Influence of vertical trailing on seed yield and quality during hybrid seed production of bottle gourd (cv. Pusa hybrid-3). Seed Research 40(2): 139– 44.

Kalyanrao, Tomar B.S. and Singh B. (2014). Effect of stage of harvest and post-harvest ripening on hybrid seed yield and quality in bottle gourd. Indian Journal of Horticulture 71(3): 428–32.

Lasa, J. M. and Bosemark, N.O. (1993). Male sterility. Plant Breeding: Principles and Prospects, pp. 213-28.

Lasa, J.M. and Bosemark, N.O. 1993. Male sterility. In: Hayward, M.D., Bosemark, N.O., Romagosa, I., Cerezo, M. (Eds) In: Plant Breeding. Plant Breeding Series. Springer, Dordrecht. https://doi.org/10.1007/978-94-011-1524-7_15

Munshi, A.D., Tomar, B.S., Jat, G.S. and Singh, J. 2017. Quality seed production of open pollinated varieties and F1 hybrids in cucurbitaceous vegetables. In: ICAR sponsored short Course Advances in variety maintenance and quality seed production for entrepreneurship". Kumar et al., (Eds,) February pp.107-125.

NHB. 2022. www.nhb.gov.in

Peterson, C.E. and Anhder, L.D. (1960). Induction of staminate flowers on gynoecious cucumbers with gibberellins A3. Science 131: 1673–76.

Robinson, RW. 2000. Rationale and Methods to Produce Hybrid Cucurbit Seed, p 1-47. In: Hybrid Seed Production in Vegetables, Rationale and Methods in Selected Species, Basra, A.S. (Ed.). Food Products Press, 135 p.

Sharma Pramod, Nair A. Sunil, Sharma Payal (2019). Male Sterility and its Commercial Exploitation in Hybrid Seed Production of Vegetable Crops: A Review. Agricultural Reviews. 40(4): 261-270. doi: 10.18805/ag.R-1880.

Singh B and Tomar B S. 2015. Vegetable seed production under protected and open field conditions in India: A review. Indian Journal of Agricultural Sciences 85 (10): 3-11.

Singh Brar, N., Kumar Saini, D., Kaushik, P., Chauhan, J., & Kumar Kamboj, N. (2020). Directing for Higher Seed Production in Vegetables. IntechOpen. doi: 10.5772/intechopen.90646.

Singh Shubham and Singh Abhilash.2022. Male Sterility in Vegetable Crop: A Mini Review. Annals of Plant Sciences.11 (01): 4561-70.

Tomar, B.S. and Jat, G.S. 2015. Vegetable seed production under protected structure. In: MTC on Entrepreneurship development to ensure quality vegetable seed production for making the country nutritionally secure, 0-17th December 2015, Division of Vegetable Science, pp. 51-57.

Vishwanath, Tomar, B S., and Singh Balraj. 2008. Studies on Methods of Pollination for Hybrid Seed Production of Pumpkin (*Cucurbita moschata* Poir.). Seed Research 36(2):214-17.

13

Vegetable Based Microgreens for Human Health

J.K. Ranjan, Ankita Saha, Gyan P. Mishra, R.K. Yadav, R. Bharadwaj, Ajeet Singh, Pragya, Jogendra Singh and B.S. Tomar

Division of Vegetable Science, ICAR-Indian Agricultural Research Institute New Delhi

Microgreens, popularly known as "vegetable confetti," are tender young greens that are grown from seeds of cereals, vegetables, and herbs. These shoots are typically harvested 7-21 days after germination when they are just (2.5-7.5 cm tall. At this stage, microgreens have a central stalk, two mature cotyledon leaves, and a second pair of tender true leaves. Compared to baby greens and sprouts, they are bigger than sprouts and smaller than baby greens. Their distinctive color, texture, and flavour can be either sweet or spicy.). Microgreens are especially important because of their high amount of micronutrients and bioactive substances. Microgreens of the Brassicaceae family are excellent source of K, Ca, Fe, and Zn. They are employed as dietary supplements and also known as functional foods. The microgreens super foods are considered as a component of space life support systems and are indicated for growth in metropolitan and peri-urban environments (Paradiso *et al.*, 2018). In order to maintain quality, increase nutritional value, and lengthen shelf-life, research has looked at preharvest and postharvest interventions such as calcium treatments, modified atmosphere packaging, temperature management, and light (Turner *et al.*, 2020).

History of Microgreens

Although there is no concrete evidence that who created microgreens and documented the name, some research are claim that Craig Hartman created the phrase "Microgreens". Wheatgrass was first produced, dried, and sold as medicine in the majority of North American pharmacies in during the 1930s. Winter greens, buckwheat, radish, and sunflower were frequently planted in the 1960s. Healthy homegrown grasses gained popularity during 1970s due to their health advantages. Chefs began cultivating Cresses and seedlings for garnishing during 1980s. The regional farmers in North America began supplying their neighborhood retail stores with fresh microgreens during 2000. Then, microgreens emerged in supermarkets so that foodies could consume them at home (Pinto *et al.*, 2015). A few alternatives were first offered, including arugula, basil, beets, kale, and cilantro. At the time, a colorful combination of these veggies was known as a Rainbow Mix. Consumption of microgreens has grown recently, along with consumer appreciation for their delicate texture, distinctive fresh aromas, and concentrated bioactive elements including vitamins, minerals, and antioxidants compared to mature leafy greens. Mr. Sanjeev Kapoor, Vikas Khanna, Ranveer Brar, and Kunal Kapoor, India's most well-known chefs, are now blending different microgreens into Indian dishes.

Advantages of Growing Microgreens

- ***Concentrated nutritional source*:** Microgreens are a concentrated nutritional source since they are a powerhouse of important vitamins, minerals, and antioxidants. They can also be cultivated at home. Due to small size of microgreens, there is no risk from harmful chemicals or pesticides. It is possible to get a concentrated source of food nutrients. They offer freshly made multivitamins as a result.
- ***Health benefits*:** Microgreens are extremely beneficial for health. It can lower blood pressure, regulate blood sugar, and prevent certain inflammatory disorders. The risk of heart disease is decreased. It can stop various malignancies. It enhances skin health as well as digestive health.
- ***Low Risk of food poisoning*:** Microgreens don't require soil to develop, and they don't even require a lot of humidity. Both peat and reusable growth mats are appropriate for growing microgreens. These factors contribute to the minimal risk of foodborne diseases.
- ***Convenient to grow*:** Growing microgreens is more convenient than conventional farming. It spends 10 to 12 hours in the sun and water. The average length of the sowing and harvesting cycle is 15 to 21 days.
- ***Lower costs*:** Microgreens can be produced at a low cost. It's not required to have a huge area of land. Therefore, if high-quality seeds are utilized, there is no need to boost seedlings with fertilizers. Furthermore, pesticides and insecticides wouldn't be necessary if they were grown on the right medium.

Additionally, because of their small, it is not essential to utilize expensive gardening equipment.

- ***Less time for cooking*:** Microgreens heated to a certain temperature will ensure that susceptible infection-causing agents are suppressed. However, cooking them is not necessary. Microgreens can be chopped and eaten right away. The bare minimum would be to wash them in running water. Even raw, they taste good and have a good flavour.
- ***Versatile in their use*:** Microgreens can be used for a variety of purposes. Green dips and smoothies are both possible. They can be added as a garnish to gravies and soups.

Global Market of Microgreens

The market for microgreens worldwide was worth US$1,3385 million in 2021. According to IMARC Group, the market is projected to reach US$ 2,284 Million by 2027, exhibiting a CAGR of 8.58 per cent from 2022 to 2027. The market size for microgreens is examined by country, product type, cultivation, growth rate, and distribution method. The USA, Canada, Mexico in North America, Germany, France, UK, Netherlands, Switzerland, Belgium, Russia, Spain, Turkey in Europe, China Japan, India, South Korea, Singapore, Malaysia, Thailand, Indonesia, the Philippines in the Asia-pacific region, Saudi Arabia, UAE, South Africa, Egypt, Israel, rest of middle east as a part of the middle east and Africa and Brazil, Argentina and rest of South America as a part of South America are main markets of microgreens. Due to rising interest in organic farming, North America is the region with the largest market for microgreens. Asia-Pacific is expected to experience the fastest growth in microgreen market. Dummen Orange (U.S.), Syngenta (Switzerland), Bredekamp Group (Netherlands), Selecta Klemm (Germany), Double H. Nurseries (U.K.), ARKANGELI GIOVANNI (Italy), and Ball Horticultural Company are prominent players in microgreens market (USA).

https://www.kenresearch.com/blog/2022/12/global-microgreens-

According to Sourcing Intelligence Platform 2020, Swedish vertical farming startup Urban Oasis raised USD 1.2 million to construct a new, fully automated facility that will increase productivity by 15-20 times. The main crops included in the worldwide microgreens market include peas, arugula, radish, carrot, broccoli, cabbage, cauliflower, and other kinds. In India, production and consumption of microgreens are rising steadily, and a number of enterprises, mostly online, are active in this market. The size of microgreens market in terms of rupees, however, is not covered.

How to grow Microgreens ?

Microgreens should be protected from rain and other environmental pressures because they are delicate and sensitive to physical harm. They have a short life-cycle. Majority of them are grown hydroponically or semi-hydroponically. The cost of seeds is significant in production of high-quality microgreens and is demanded in large quantities. Precautionary sanitary treatments should be applied to seeds to get rid of harmful pathogens. Depending on whether the microgreens are grown in or out of the soil, shallow plastic seed trays with or without suction holes are typically used to plant microgreens. A tray with dimensions of 10x10 or 10x20 with a depth of 1.2 to 2.5 inches is frequently used. Sterile, loose, soilless germination media can be used to develop microgreens; the pH of the media should be between 5.5 and 6.5. Light, temperature, and water are additional external factors. One of the most energy-efficient and quickly evolving plant lighting solutions is light-emitting diodes (LEDs). Microgreens are said to respond well to wavelength combinations of the red, blue, and far-red LED light.

Uses of Microgreens: Microgreens can easily be included in dishes to fully utilize their nutritional value because they are small and tender. However, it is not advised to cook microgreens due to their small size and high-water content. Here are some tips on how to use it: Add some salad, top cooked fish or meat with a little salt. Eggs or omelets with the mixture smoothie by blending. served as a topping on pasta or soup. For more texture in a sandwich, add. Make hummus or other dips look nice.

Dry microgreens: Dried microgreens can help with nutritional security because they are easy to make at any time. To increase their shelf life to eight years, they can be dehydrated. Dehydration can be accomplished via raw dehydration technology or cold sterilization. Unique dehydrated microgreens can be added to smoothies, soups, salad dressings, eggs, baked products, and other dishes. Additionally, dried microgreens-based energy drinks can be made.

Space food : Space travelers require a healthy diet. Space travelers frequently experience a variety of stress-related side effects, including weight loss, changes to blood composition, and radiation-induced stress. They require a supply of food-based antioxidants to prevent this stress, and microgreens appear to be a great choice in this regard. Future space missions plan to produce carotenoid-rich food as a part of space life support systems since microgreens can be grown on board throughout the mission. Reduced gravity (or microgravity), which affects fluid and gas distribution around the plants and causes increased transpiration rates, is a major obstacle to adapting agricultural practices in space. Low irradiance levels (300 mol/m2/s) necessitate supplemental lighting, which is energy-intensive for the space farm.

Microgreens for Nutritional Security

These are organic substances made by plants or animals. Microgreens are rich in vitamin K, vitamin C and vitamin E, and carotenoids (which are precursors of vitamin A). Till now less work has been done on microgreens in India and as well as in the world.

Vitamin K: The vitamin K known as phylloquinone is essential for bone remodelling and blood coagulation. As previously mentioned, Xiao *et al.* compared 25 types of microgreens. There is a wide range of phylloquinone content, ranging from 0.6 to 4.1 ug/g. In general, green and bright red microgreens have higher levels of vitamin K (2.8-4.1 ug/g), while yellow microgreens have relatively lower levels of phylloquinone (0.7 to 0.9 ug/g). In comparison to mature amaranth, basil, and red cabbage, microgreens have comparatively high levels of phylloquinone (USDA's national nutrient database).

***Vitamin C*:** Vitamin C, commonly referred to as ascorbic acid, is a powerful antioxidant and is necessary for a number of biological processes, including collagen formation, wound healing, and immune system control. Yadav *et al.* discovered that Vitamin C levels of cucumber microgreens were greater than those of mature stages (25 mg/100 g fresh FW and 34.90 mg/100 g FW, respectively). Traditional Sicilian broccoli has a substantially more Vitamin C content in microgreen stage (7.5 mg/g) compared to the baby green stage (6.1 mg/g). The microgreen stages of fenugreek, spinach, and roselle have higher Vitamin C than their mature stages. In 25 commercially available microgreens, There is range of total Vitamin C contents (20.4–147.0 mg/100g FW), and red cabbage had greatest Vitamin C concentration among the samples, whereas sorrel had the lowest.

***Vitamin E*:** There are four isomer versions of each tocopherol and tocotrienol, which are members of the Vitamin E family. The different isomeric forms are α, β, γ, δ. Among four isomer forms of tocopherols, α tocopherol is most active, while – γ tocopherol is the form that is most frequently found in plants. The green daikon radish contains highest levels of tocopherol in both α (87.4 mg/100g) and γ (39.4 mg/100g) forms. In addition, microgreens of cilantro, opal radish, and pepper cress have high levels of α and γ -tocopherols. Although tocopherol levels in garden pea tendrils are the lowest, they are nevertheless higher than those in mature spinach leaves.

B-carotene: Most microgreens are a richer source of β-carotene (Choe *et al.,* 2018). Beta-carotene levels in microgreens is 0.6-12.1 mg/100g.Red sorrel had the highest beta-carotene (12.1 mg/100g), followed by cilantro, red cabbage, and pepper cress. Red cabbage microgreens have 260 times more red cabbage than mature red cabbage leaves (11.5 mg/100g). Wasabi, green basil, pea tendrils, and garnet amaranth microgreens have higher beta-carotene levels. Than carrot and sweet potato. Beta-carotene is especially abundant in coriander.

Lutein/Zeaxanthi: Cilantro had highest lutein / zeaxanthin concentration of 10.1 mg/100g. Red cabbage, red sorrel, and garnet amaranthus have 8.8, 8.6, and 8.4mg/100g respectively. Mature cilantro and red cabbage contain 0.9 and 0.3mg/100g respectively. Compare to mature cilantro and red cabbage lutein/ Zeaxanthin concentrations cilantro and red cabbage microgreens had 11.2 and 28.6 times more lutein/zeaxanthin concentrations.

Violaxanthin: The cilantro microgreens have highest concentration of 7.7 mg/100 g, while others. The have 1.3-4.3 mg/100gm. The cilantro microgreens have violaxanthin five times higher than mature spinach leaves.

Minerals: Potassium, calcium, iron, zinc, magnesium, and copper are most abundant in microgreens. Brassicaceae microgreens are a significant source of minerals, Potassium ranges from 176-387 mg/100g in microelement, followed by P (52-86 mg/100g), Ca (28-66 mg/100g), Mg (28-66 mg/100g), and Na (19-68 mg/100g). On a dry-weight basis, microgreen concentrations are higher at earlier stages. According to research by Pinto *et al.* (2015), microgreen lettuce contain majority of minerals (Ca, Mg, Fe, Mn, Zn, Se, and Mo) than mature lettuce (10-week-old). Basil and swiss chard microgreens and great providers of K and Mg Kyriacou *et al.*, (2019).

Polyphenol and glucosinolates: Red cabbage microgreens contain higher content of polyphenols (71.01miumol/g) than mature red cabbage, (Choe *et al.,* 2018) In five microgreen cultivars of the genus Brassica, Sun *et al.* (2013) analyzed polyphenols and discovered 165 phenolic compounds, including several highly glycosylated and acylated quercetin, kaempferol, cyanidin aglycones, and complicated hydroxycinnamic and benzoic acids micro-greens of *Brassica*. The red cabbage microgreens have higher glucosinolate than mature red cabbage.

Total sugar content: The total sugar content in six microgreens, including red amaranthus, bull's blood beet, China rose radish, Dijon mustard, opal basil, and pepper cress have been analyzed. The highest sugar 10.3g/kg content is found in China rose mustard.

Why Nutritional Profile is High in Microgreens

Seed germination increases nutritive value by activating some enzymes which decrease anti-nutritional substances, during germination. A few days of photosynthesis_typically provides 4-6 times more nutrients, including vitamins, minerals, ascorbic acid, beta-carotene, antioxidants, and ascorbic acid. Germinating embryos possess an abundance of glucose_which is used for synthesis of various carotenoids and vit C. Increase in hexose-p pool_in germinating embryo results in the activation of various pathways that involves the synthesis of different vitamins.

Health Benefits of Microgreens

Seven health benefit of microgreens are presented below and it table-1

Prevention of inflammation, and chronic diseases: Eating red cabbage microgreens reduced liver inflammation, caused by a high-fat diet, including levels of tumor necrosis factor and C-reactive protein (CRP). This result could result from microgreens' capacity to reduce liver cholesterol levels.

Prevention of cancer: A diet high in fruits and vegetables and low in fat and calories can prevent one third of cancer. Carcinogenesis is significantly influenced by inflammation. Microgreens may prevent cancer by acting through inflammatory mechanisms. Brassica crops are rich in glucosinolate, precursors and other compounds that aid in prevention of cancer.

Heart disease: Microgreens have great content of antioxidants, e.g. polyphenols which can reduce the risk of heart disease. Different animal studies showed that microgreens lower the level of triglycerides and bad LDL cholesterol (Huang *et al.*, 2016).

Control blood sugar levels: Research has revealed that microgreens are helpful in regulating blood sugar. The fenugreek microgreens can reduce the activity of enzyme that produces glucose and helps control blood sugar levels.

Lower blood pressure: Microgreens are a good source of fiber and vitamin K, which can help keep blood pressure in a healthy range.

Alzheimer's disease: Consuming foods high in polyphenols and other antioxidants can lower your risk of developing memory-related diseases like Alzheimer's modulation.

Gut microbiome: The bacteria and archaea that inhabit the digestive systems of vertebrates, such as humans and insects, comprise the gut microbiome. They also aid with digestion, immunological protection, and energy harvesting. They have a crucial role in emergence of malignancies, intestinal health, and chronic disorders. Microgreens are high in flavonoids, which regulate the gut microbiome and hence provide defense against a number of diseases linked to changes in the microbiome.

Good for skin health: Microgreens are a great source of polyphenols and antioxidants. This polyphenol aids in the production of collagen and elastin, which maintain skin supple and smooth. When combined with a quality collagen supplement, the results are improved. Microgreens' anti-inflammatory properties reduce the frequency and intensity of breakouts and leave a clear, healthy complexion.

Differences in Bioactive Compounds and Minerals in Microgreens

Knowing the genotypic variability in the phytochemical composition of microgreens can significantly contribute to growing industry given the relative concentration of bioactive chemicals between species and the consequences for their sensory and functional properties (Kyriacou *et. al.*, 2017). Thirteen microgreens representing Brassicaceae, Chenopodiaceae, Lamiaceae, Malvaceae, and Apiaceae had their compositional variation investigated. Na, K, and S concentrations showed significant genotypic variations, and nitrate hyperaccumulators. Brassicaceous microgreens had best antioxidant capacity, and considerable genotypic variation in chlorophyll and carotenoid levels. With notable varietal differences, high phenolic content in Lamiaceae microgreens is confirmed, and alternative phenolics-rich microgreens from the Apiaceae are discovered. Based on retention times, elution orders, UV-vis and high-resolution mass spectra, an internal database. Sun *et al.* (2013) identified a total of 164 polyphenols, including 105 flavanol glycosides and 29 phenolic acid derivatives.

Popular Microgreen Vegetables in India

Most popular microgreens vegetables in India are mentioned from the following families:

- **Brassicaceae:** cauliflower, cabbage, watercress, radish, and arugula.
- **Asteraceae:** lettuce, endive and chicory,
- **Apiaceae:** Dill, carrot, fennel and celery
- **Amaryllidaceae:** garlic, Onion and leek
- **Amaranthaceae:** Amaranth, quinoa, swiss chard, beet and spinach
- **Cucurbitaceae:** Melon, cucumber and squash

Table 1: Non-edible species – Tomato, Brinjal, Pepper (because they have some anti-nutrients)

Crop	Soaking/ pre-soaking	Germination days	Harvesting days	Colour	Flavour	Nutrients	Health benefits
Broccoli	No	2-3 days	7-14 days	Green	Strong broccoli type	Vit A, C, Ca, P, Fe	Prevent lung & colon cancer, improve bone health and digestion
Cabbage	No	1-3 days	8-10 days	Green	Strong brassica type	Vit C, E, K, β-carotene, Fe, K	Reduce risk of cardiovascular diseases, cancer, and lower cholesterol level
Cauliflower	No	2-3 days	8-12 days	Green	Strong brassica type	Vit C, E, K, β-carotene, Fe, K	Reduce risk of cancer and heart diseases.
Brussel sprout	No	2-3 days	8-14 days	White/Pink/ Purple	Like broccoli/ Cabbage	Vit B, C, K, Folic acid, fiber	Protect against caners of lung, kidney, breast and lower cholesterol
Kale	No	2-3 days	7-12 days	Pale green	Broccoli like	Rich in Vit C, K, protein, Fe, Cu	Good for skin, anticarcinogenic properties present
Amaranthus	No	2-3 days	8-12 days	Red/Pink	Mild/sweet	Vit C, E, K, Ca, Fe, β-carotene	Prevent of cancer and heart related diseases.
Beet	4-10 hr	2-3 days	7 to 12 days	Red stem & Green leaves	Sweet & earthy	Vit A, B, C, E, K, Mg, K, Fe, Zn	Reduce inflammation., boost digestion, Promote healthy skin
Arugula	No	1-2 days	6-8 days	Green	Nutty, Peppery, Pungent	Vit A, C, β-carotene, Ca, Fe, P	Prevent various cancers, improve bone development
Basil		2-3 days	10-13 days	Green or purple leaves	Slight-sweet, spicy	Vit A, C, K, Polyphenols, minerals	Anti-inflammatory effect, regulate body functions, anti-aging, prevent cancers.
Coriander	8-10 hr	2-3 days	21-28 days	Green	Coriander type	Vit A, c, Ca, Fe, P	Promote health for bone, vision, nervous system and more

Crop	Soaking/ pre-soaking	Germination days	Harvesting days	Colour	Flavour	Nutrients	Health benefits
Lettuce	No	3-4 days	14-16 days	Green	Lettuce like	Vit B, C, K, folic acid & fiber	Prevent inflammation. Improve eye- health, skin-health
Mustard	No	2-3 days	8-12 days	Green	Mustard like	Vit A, c, E, K. Fiber	Good for vision, heart health, cancer prevention and more.
Pea	4-12 hr	2-3 days	12-16 days	Green	Crunchy, Sweet	Vit A, C, Folic acid, fiber	Boost immune system, eye health, control blood sugar, anti-cancer effect
Radish	4-6 hr	1-2 days	6-12 days	Red stem & Green top	Spicy & pungent	Vit A, B c, E, K, folic acid, niacin	Promote eye health, reduce the risk of heart related diseases, prevent some cancers and more.
Turnip	No	2-3 days	8-12 days	Deep green	Fresh turnip like	Vit C, E, K, β-carotene, Fe	Prevent cancers, improve skin & hair health
Swiss-chard	4-10 hr	3-4 days	11-21 days	Dark red	Sweet & earthy	Vit A, c, E, K, Ca, Mg, Fe, Zn	Prevent heart diseases 7 ling cancers
Fenugreek	No	2-3 days	10-14 days	Green	Bitter, mild, spicy, nutty	Vit c, Protein, Fibre, antioxidants, K, Fe	Controls diabetes, relives constipation, lower cholesterol levels
Cress		3-4 days	8-12 days	Green leaves, yellowish white stem	Peppery, tangy	Vit A, C, K, isothiocyanates	Improve eye health, prevent cancer, good for teeth, and maintaining heart health.
Sorrel		3-4 days	10-12 days	Vibrant green leaves and stems.	Lemony, tangy	Vit E, C, K, Oxalic acids, minerals	Improve skin health, heart health, lower blood pressure, boost immune system and regulate body functions

Radish (Red) Mung Amaranthus

Radish (white) Cabbage Sona mung

Strategies to Improve Nutritional Quality of Microgreens

Although the quantity and composition of plant secondary metabolites (PSM) in microgreens vary depending on genetic make up (i.e. species), many other factors, such as cultural practices, cultivation circumstances, and environmental factors, are also implicated in regulating PSM (Kyriacou *et al.*, 2016; Kyriacou *et al.*, 2017; Kyriacou *et al.*, 2019). The pre-harvest interventions can improve nutritional quality, boost yield, get rid of pathogens, and reduce safety risks. To improve the nutritional content of microgreens, a number of preharvest intervention measures have been used.

Light-emitting Diodes Technology

All higher plants' photo-physiological reactions are influenced by light, one of the most important environmental elements (Ma *et al.*, 2014). Photoreceptors in plants allow them to sense a very wide range of the light spectrum, from ultraviolet to far red (280-750 nm). According to Galvo and Fankhauser (2015), Casal *et al.* (2014), Pierik, de Wit (2014), and others, photoreceptors respond to red-far-red light (phytochromes), blue-UVA light (cryptochromes, phototropin, and "Zeitlupes"), and UVB (UVR8) light via the photoreception mechanism. The morphogenetic and photosynthetic responses of plants are significantly influenced by light quality (wavelength), light intensity (irradiation), and photoperiod (duration of day/night) as well (Björkman *et al.*, 2011).

In order to assure plant development in a greenhouse, which is an artificial growing environment, additional light sources must be used. Despite having improved in terms of intensity, spectral range, and energy efficiency, light-emitting diodes (LEDs) as the primary or supplemental lighting source can be used to manipulate metabolic responses with the goal of increasing plant productivity and quality (Olle and Viril, 2013; Darko *et al.*, 2014; Carvalho and Folta, 2014; Duchovskis *et al.*, 2015).

The phytonutrient contents of red pak choi, mustard, tatsoi, and basil microgreens, including phenols, anthocyanins, and total ascorbic acid (in basil), as well as antiradical activity, could be increased by supplementing HPS lights with light of a particular wavelength or by pulsing the LEDs at a specific frequency, (Vastakaite *et al.*, 2017, 2018). Alrifai, Hao, Marcone, and Tsao (2019) explain that red, blue, and combined red plus blue light is more effective than white light and other wavelengths for enhancing photosynthesis and regulating plant metabolism. Samuoliene *et al.* (2011) found that different ˙ supplemental LED wavelengths in addition to the basal components of blue (455 nm), red (638 nm), deep red (669 nm), and far red (731 nm) had different effects on the antioxidant compounds in the sprouted seed. Supplementation with green light (510 nm) improves mineral content in beet microgreens (Brazaityte *et al.*, 2018). Red and blue lights are the primary LEDs involved in stimulating the growth and accumulation of vitamins, antioxidants, and minerals in microgreens. Lobiuc *et al.* (2017) found that while blue light enhanced growth, cotyledon area, fresh mass, chlorophyll a, and anthocyanin pigment content of both red and green microgreens, phenolic content and free radical scavenging activity were improved by the application of mostly red light in the green cultivar and mostly blue light in the red cultivar. Supplementation with amber light (595 nm) enhanced antioxidant properties of radish sprouts (Samuoliene *et al.*, 2011). Mlinarc *et al.* compared the nutrients contents of chia microgreens under a dark room or constant light (100 μmol photons/m^2) and noted a significantly higher carotenoid, chlorophyll, and ascorbic acid level in the light treatment group. Short-term red LED lighting (638 nm) increased total ananthocyanins and total ascorbic acid and decreased nitrate content in purple mint Perilla frutescens microgreens (Brazaityte, Jankauskien, and Novickovas, 2013) and also increased P, K, Ca, Mg, S, and Mn, ˇ but reduced Na, Fe, Zn, Cu, and B in beet microgreens (Brazaityte˙ *et al.*, 2018). Total phenolic content (TPC) increased in all microgreens except for amaranth; total ascorbic acid content increased in amaranth, kale, broccoli, mustard, and pea and declined in basil and borage microgreens; total anthocyanins increased in amaranth, kale, broccoli, tatsoi, and pea and declined in mustard, borage, beet ,and parsley microgreens (Samuoliene *et al.*, 2012). Samuoliene *et al.* also noticed higher concentrations of carotenoids in beet, mustard, and parsley, after 10 to 14 days of continuous LEDs exposure with 16% to 33% blue light. Blue lights also increase the contents of macrominerals, trace minerals and glucosinolates in microgreens. Total phenolic contents and total flavonoid contents were enhanced under blue and ultraviolet-A (UV-A) lights. The combination of red and blue LEDs may increase the concentration of chlorophyll a, chlorophyll b, total phenolic contents, and anthocyanin content in various microgreens as compared to sole red and sole blue treatments. The supplementation of red LED light + HPS light increased the phenolic content, α-tocopherol, lutein, and β-carotene by 36.7%, 18.6%, 48.8%, and 47.9% in basil, respectively, as compared to those solely under HPS light. Glucosinolates contents are also increased by different combinations of light in Brassica vegetables.

Others

Other pre-harvest therapies have been researched in addition to the application of various illumination treatments. Nutritional fortification, salinity stress, and natural substrates were discovered to be excellent methods for boosting the nutrient and antioxidant contents in microgreens. Using various NaCl concentrations (0, 12.5, 25, 50, and 100 nmol/L), Islam *et al.* treated wheat microgreens. They found that the treatment at 12.5 mmol/L increased the contents of total chlorophyll, beta-carotene, phenolic acid, flavonoids, VC, and sodium by 1.48, 4.65, 1.06, 1.22, 1.17, and 2.5 times, respectively. Islam *et al.* treated white winter wheat with Se biofortification for 10 days in addition to the salt stress, and it was discovered that this increased the levels of Se, phenolic compounds, flavonoids, VC, and anthocyanin in the wheat microgreens. According to Kyriacou *et al.*, microgreens grown on natural peat moss media contained noticeably more phosphorus, potassium, calcium, and sulfur than microgreens grown in synthetic substrates (i.e., capillary mat and cellulose sponge). Therefore, a number of pre-harvest treatments have shown an improvement in the micronutrients, minerals, and phytochemicals of microgreens, raising the likelihood that they may be useful as functional meals.

Post-harvest Interventions

Since microgreens are delicate and extremely perishable items, post-harvest interventions, such as chlorine wash, ozone wash, coating, and modified environment packaging, are typically used to ensure the safety and lengthen the shelf life of such products. Some of these post-harvest procedures also had an impact on the microgreens' nutritional value. For instance, the composition and concentration of nutrients in plants are influenced by the technique of packaging, storage temperature, and sunlight. Microgreens' nutrient levels have been demonstrated to be effectively maintained and/or improved by storage at low temperatures, specifically 4°C. According to Xiao *et al.*, radish microgreens had higher levels of ascorbic acid and antioxidant capacity after 16 days of exposure to light than samples kept in the dark. Supapvanich *et al.* observed considerably greater phenols, flavonoids, and ascorbic acid when they soaked sunflower sprouts, daikon sprouts, and red amaranth microgreens with low dose (0.1 mol/L) of cyanocobalamin (vitamin B12).

Limitations of Growing Microgreens

Rapid quality degradation after harvest is one of the main factors limiting the development of the microgreen industry. Due to their high surface area to volume ratio, high respiration rate, and delicate leaves that easily wilt, microgreens are challenging to store (Berba & Uchanski, 2012; Chandra, Kim, & Kim, 2012; Kou *et al.*, 2013). They also exhibit rapid postharvest decay transpiration, leakage of nutrient-rich exudates, tissue damage, and early senescence.

Future Perspective

Microgreens are great for urban gardening in areas with limited resources since they can be cultivated with few resources in practically any setting. Urban microgreen gardening offers home gardeners the chance to generate income from a small space in their homes with little initial investment. Most microgreen analyses and studies are conducted at a comparatively low level, are restricted to a small number of researchers, and have constrained geographic scopes. More of the large area needs to be studied. Few species have undergone research and analysis, but the majority have not seen commercialization. Networks' food security can be improved by growing microgreens, which also promotes household independence. Microgreens may be produced quickly, and they are currently generating growing attention from customers who, more than ever, are looking for nearby, sustaining hotspots for food.

References

Brazaityte, A., Vaštakaite, V., Viršile, A., Jankauskien e, J., Samuolien e, G., Sakalauskien E,S. and Duchovskis, P. (2018). Changes in mineral element content of microgreens cultivated under different lighting conditions in a greenhouse. Acta Horticulturae 1227: 507–16.

Choe, U., Yu, L. L. and Wang, T.T. (2018). The science behind microgreens as an exciting new food for the 21st century. Journal of Agricultural and Food Chemistry 66(44): 11519-11530.

Kyriacou, M. C., De Pascale, S., Kyratzis, A. and Rouphael, Y. (2017). Microgreens as a component of space life support systems: a cornucopia of functional food. Front. Plant Sci. 8: 1587. doi: 10.3389/fpls.2017.01587

Kyriacou, M. C., El-Nakhel, C., Graziani, G., Pannico, A., Soteriou, G.A. and Giordano, M. (2019a). Functional quality in novel food sources: Genotypic variation in the nutritive and phytochemical composition of thirteen microgreens species. Food Chem. 277: 107–18. doi: 10.1016/j. foodchem.2018.10.098

Kyriacou, M. C., El-Nakhel, C., Pannico, A., Graziani, G., Soteriou, G. A., Giordano, M. and Rouphael, Y. (2019). Genotype-specific modulatory effects of select spectral bandwidths on the nutritive and phytochemical composition of microgreens. Frontiers in Plant Science 10: 1501.

Kyriacou, M. C., Rouphael, Y., Di Gioia, F., Kyratzis, A., Serio, F. and Renna, M., *et al.* (2016). Micro-scale vegetable production and the rise of microgreens. Trends Food Sci. Technol. 57: 103–115. doi: 10.1016/j.tifs.2016.09.005

Nagothu, U. S. (2014). Food security in India: the need for sustainable food systems. In Food Security and Development, 143-162.

Paradiso, V. M., Castellino, M., Renna, M., Gattullo, C. E., Calasso, M., Terzano, R. and Santamaria, P. (2018). Nutritional characterization and shelf-life of packaged microgreens. Food & Function, 9(11): 5629-40.

Pinto, E., Almeida, A. A., Aguiar, A. A. and Ferreira, I. M. (2015). Comparison between the mineral profile and nitrate content of microgreens and mature lettuces. Journal of Food Composition and Analysis, 37:38–43. https://doi.org/10.1016/j.jfca.2014.06.018.

Renna, M., and Paradiso, V. M. (2020). Ongoing research on microgreens: Nutritional properties, shelf-life, sustainable production, innovative growing and processing approaches. Foods 9(6): 826.

Samuoliene, G., Urbonavi ˙ ciˇ ut¯ e, A., Brazaityt ˙ e, A., ˙ Sabajevien ˇ eG., Sakalauskas ˙ eJ., and Bukovskis, P. (2011). The impact of LED illumination on antioxidant properties of sprouted seeds. Central European Journal of Biology 6: 68–74. https://doi.org/10.2478/s11535-010-0094-1

Samuoliene, G., Brazaityt ˙ e, A., Sirtautas, R., Sakalauskien ˙ e, S., Jankauskien ˙ e, J., Novi ˙ ckovas, ˇ A., and Duchovskis, P. (2012). The impact of supplementary short-term red LED lighting on the antioxidant properties of microgreens. Acta Horticulturae 956: 649–655. https://doi.org/10.17660/ActaHortic.2012.956.78

Sharma, S., Dhingra, P. and Koranne, S. (2020). Microgreens: Exciting new food for 21st Century. Ecology, Environment and Conservation 26: 2020.

Sreenivasa RJ, Sri TSD, Shivudu G and Kumari V (2019) Micro greens: A New Initiative and Harmonizing Approach for Promoting Livelihood and Nutritional Security-India. August, 2019.

Turner, E. R., Luo, Y. and Buchanan, R. L. (2020). Microgreen nutrition, food safety, and shelf life: A review. Journal of Food Science 85(4): 870-882.

Vastakaite, V., Virsile, A., Brazaityt e, A., Samuolien e, G., Jankauskien e, J., Novi ckovas, A., and Duchovskis, P. (2017). Pulsed light-emitting diodes for a higher phytochemical level in microgreens. Journal of Agricultural and Food Chemistry 65: 6529–34.

Vaˇstakaite, V., Vir ˙sile, A., Brazaityt ˙ e, A., Samuolien ˙ e, G., Miliauskien ˙e, J., Jankauskien ˙ e, J. and Duchovskis, P. (2018). Pulsed light increases the phytochemical level of basil microgreens. Acta Horticulturae 1227: 579–84.

Xiao, Z., Lester, G. E., Luo, Y. and Wang, Q. (2012). Assessment of vitamin and carotenoid concentrations of emerging food products: edible microgreens. Journal of agricultural and Food Chemistry 60(31): 7644-51.

www.kenresearch.com/blog/2022/12/global-microgreens-market-analysis-future-outlook-and-growth-ken-research/

14

Hydroponics: Production Technology for High-Value Vegetables

J. Suresh Kumar, Sandra Jose and A.U. Akash

Vegetable Science, ICAR- Central Tuber Crops Research Institute Thiruvananthapuram 695 017, Kerala, India

Hydroponics is a technique of growing plants using mineral nutrient solutions with or without the use of an inert medium such as gravel, soil, vermiculite, rock wool, peat moss, saw dust, coir dust and coconut fibre-provide mechanical support. The term Hydroponics was derived from the Greek words 'hydro' means water and 'ponos' means labour and means water work. The word hydroponics was coined by Professor William Gericke in the early 1930s; describe the growing of plants with their roots suspended in water containing mineral nutrients. Researchers at Purdue University developed the nutriculture system in 1940. During 1960s and 70s, commercial hydroponics farms were developed in Arizona, Abu Dhabi, Belgium, California, Denmark, German, Holland, Iran, Italy, Japan, Russian Federation and other countries.

Due to rapid urbanization and industrialization, not only the cultivable land is decreasing but also conventional agricultural practices causing a wide range of negative impacts on the environment. Modification in growth medium is an alternative for sustainable production and conserve fast depleting land and available water resources. In the present scenario, soil less cultivation might be commenced successfully and considered as an alternative option for growing healthy food plants, crops or vegetables. Agriculture without soil includes hydro agriculture (Hydroponics), aqua agriculture (aquaponics) and aerobic agriculture (aeroponics) as well as substrate culture. Among these hydroponics techniques is gaining popularity because of its efficient management of resources and food production. Various commercial and specialty crops can be grown using hydroponics including leafy vegetables, fruit vegetables like tomato, cucumber, pepper, and other high-value fruit crops, cut flowers and medicinal plants. This chapter covers different types of hydroponics, its benefits and limitations, various components and aspects of hydroponics, different environmental factors be considered, different crops under hydroponic system, global hydroponic market and pest and disease management aspects in hydroponics and soil less production systems.

Hydroponic Systems

Hydroponic system is customized and modified according-recycling, reuse of nutrient solution and supporting media. Commonly used systems are wick, drip, ebb-flow, deep water culture, nutrient film technique (NFT) and aeroponics systems which are described below.

Wick System

This is simplest hydroponic system requiring no electricity, pump and aerators. Plants are placed in an absorbent medium like coco coir, vermiculite, perlite with a nylon/cotton wick running from plant roots into a reservoir of nutrient solution. Water or nutrient solution supplied-plants through capillary action. This system works well for small plants, herbs and spice and doesn't work effectively that needs a lot of water.

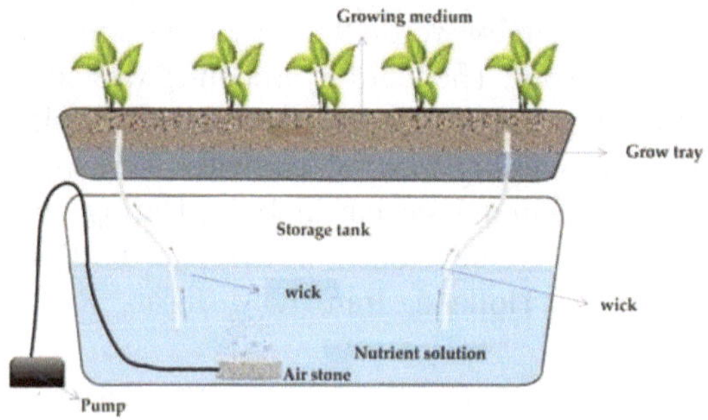

Fig 1a: Wick system with aerator for more number of plants (line diagram)

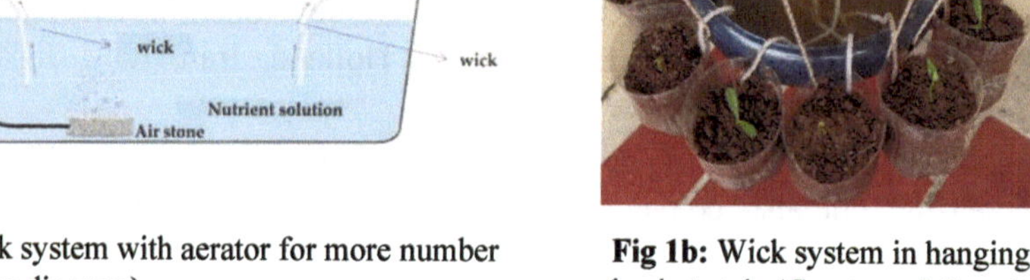

Fig 1b: Wick system in hanging bucket style (Courtesy: J Suresh Kumar)

Deep Water Culture System

In deep water cultures, roots of plants are suspended in nutrient rich water and air is provided directly-the roots by an air stone. Hydroponics buckets system is classical example of this system. Plants are placed in net pots and roots are suspended in nutrient solution where they grow quickly in a large mass. It is mandatory-monitor the oxygen and nutrient concentrations, salinity and pH as algae and moulds can grow rapidly in the reservoir.

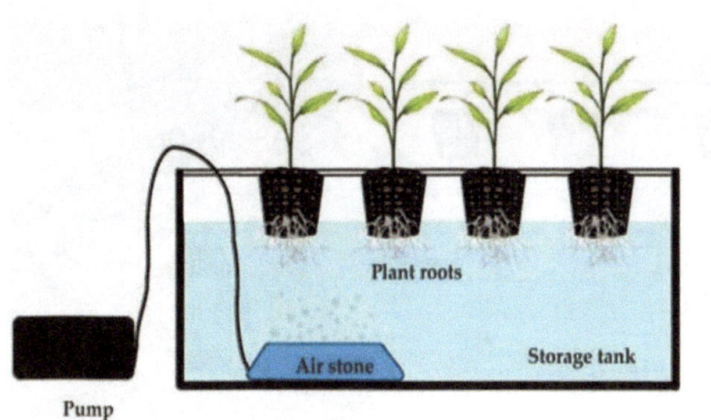

Fig. 2a: Deep water culture (line diagram)

Fig. 2b: Deep water culture of lettuce in net pots fixed in thermocol sheet floating on nutrient solution (Courtesy: J Suresh Kumar)

Ebb and Flow System

This is first commercial hydroponic system which works on the principle of flood and drain. Nutrient solution and water from reservoir flooded through a water pump-grow bed until it reaches a certain level and stay there for certain period of time so that it provides nutrients and moisture-plants. Besides, it is possible-grow different kinds of crops but the problem of root rot, algae and mould is very common therefore, some modified system with filtration unit is required.

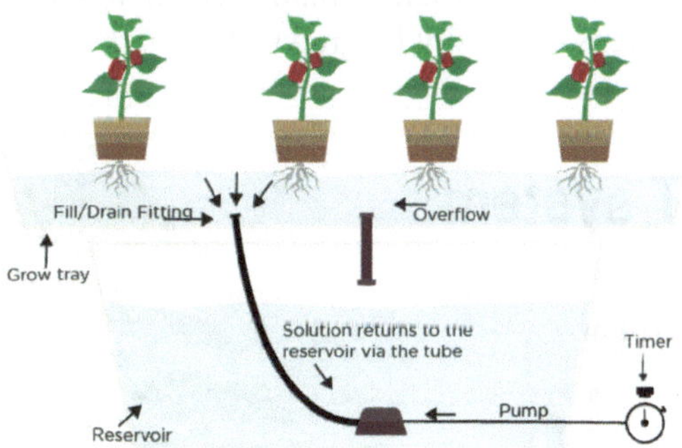

Fig. 3a: Ebb and flow system line diagram

Fig 3b: Ebb and flow model available in market

Drip System

The drip hydroponic system is widely used method among both home and commercial growers. Water or nutrient solution from the reservoir is provided-individual plant roots in appropriate proportion with the help of pump. Plants are usually placed in moderately absorbent growing medium so that the nutrient solution drips slowly. Various crops can be grown systematically with more conservation of water.

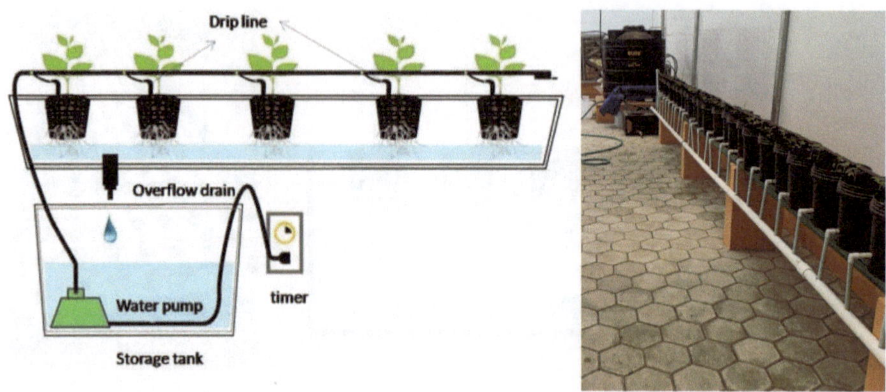

Fig. 4a: Drip system of hydroponics line diagram

Fig. 4b: Drip system of tomato cultivation in hydroponics (Courtesy: J Suresh Kumar)

Nutrient Film Technique (NFT) System

NFT was developed in the mid-1960s in England by Dr. Alen Cooper-overcome the shortcomings of ebb and flow system. In this system, water or a nutrient solution circulates throughout the entire system; and enters the growth t ray via a water pump without a time control. The system is slightly slanted so that nutrient solution runs through roots and down back into a reservoir. Plants are placed in channel or tube with roots dangling in a hydroponic solution. Although, roots are susceptible-fungal infection because they are constantly immersed in water or nutrient. In this system, many leafy greens can easily be grown. This is the most widely and commercially used for production of lettuce.

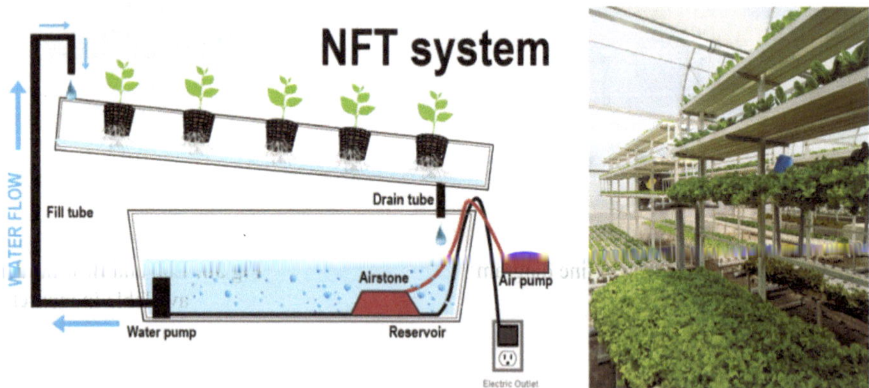

Fig. 5a: Nutrient film technique line diagram

Fig. 5b: Nutrient film technique - cultivation of leafy veggies (Photo courtesy: J Suresh Kumar)

Aeroponics

The basic principle of aeroponic growing is-grow plants suspended in a closed or semi-closed environment by spraying the plant›s dangling roots and lower stem with an atomized or sprayed, nutrient-rich water solution. The leaves and canopy extend above. The roots of the plant are separated by the plant support structure like cell foam is compressed around the lower stem and inserted into an opening in the aeroponic chamber.

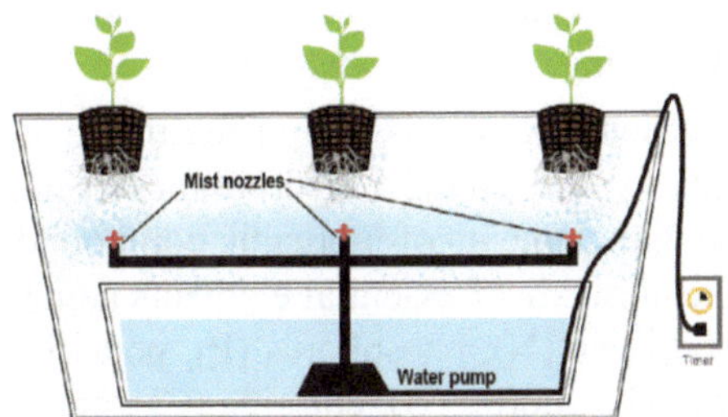

Fig. 6a: Aeroponics line diagram

Fig. 6b: Aeroponics seed potato tuber production (Source: https://www.patrika.com)

Benefits and Limitationsof Hydroponics

Hydroponic technique is becoming popular because this is clean, relatively easy method and there is no chance of soil-borne disease, insect pest infection-the crops thereby reducingor eliminating use of pesticides and their resultingtoxicity. Besides, plants require less growing time as compared-crop grown in field and growth of plant is faster as there is no mechanical hindrance-the roots and the entire nutrient are readilyavailable for plants. This technique is very useful for the area where environmental stress (cold, heat, dessert) is a major problem. Crops in hydroponic system are not influenced by climate change therefore, can be cultivated year-round. Further, commercial hydroponic systems are automatically operated and expected-eliminate the traditional agricultural practices like weeding, spraying, watering and tilling. Hydroponics saves large amount of water. The problem of pest and disease can be controlled easily. Higher yields can be obtained since the number of plants per unit is higher compared-conventional agriculture.

Although soil less cultivation is an advantageous technique but some limitations are significant. Technical knowledge and higher initial cost are fundamental requirement for commercial scale cultivation. Plant in a hydroponics system is sharing the exact same nutrient, sometimes contamination of nutrient solution may easily spreadfrom one plant-another. Hot weather and limited oxygenation in the nutrient solution may limit production and can result in loss of crops. Maintenance of pH, EC and proper concentration of the nutrient solutions are of prime importance. Finally, light and energy supply are required-run the system under protected structures.

Different Components of Hydroponics

Medium or substrate required: One of the most important factors for hydroponics system is-choose the right substrate or medium. Different media or substrate can be used for different growing techniques. The substrate which can be used for hydroponics farming is an inert material and it doesn't provide any nutrients-the plant, it only acts as base-grow the better plant roots and holds moisture, it shouldn't hold any poisonous materials. The most common materials used as growing media are coarse sand, gravel, perlite, vermiculite, shredded coconut fibers, expanded clay pellets and Rockwool either alone or mixing of these substrate.

Plant nutrients used in hydroponics: Nutrient solution is one of the most vital factors which influence crop growth, quality and yield. A hydroponics system contains mainly aqueous solution of essential elements of inorganic compounds, some organic compounds such as iron chelates. All 17 essential elements namely carbon (C), hydrogen (H), oxygen (O), nitrogen (N), phosphorus (P), potassium (K), sulfur (S), calcium (Ca), magnesium (Mg), boron (B), chlorine (Cl), copper (Cu), iron (Fe), manganese (Mn), molybdenum (Mo), nickel (Ni), and zinc (Zn) are required for plant growth are supplied using different chemical combinations. *Hoagland's* solution is used as most common nutrient solutions for hydroponic systems. *Cooper's* 1988 and *Imai's* 1987 nutrient solutions were also used for growing different leafy vegetables and fruit vegetables.

pH and electrical conductivity(EC): Proper pH and EC of the nutrient solution is very essential and should be maintained properly for optimum plant growth and yield. Optimum range of EC and pH values for different hydroponic crops mentioned in Table 1.

Total amount of ions of dissolved salts in nutrient solutions determine the plant growth, development and production of plants which exert a force called osmotic pressure. It is fully dependent on the measure of dissolved solutes (Landowne, 2006). The ions associated with EC are Ca^{2+}, Mg^{2+}, K^{+}, Na^{+}, H^{+}, NO^{3-}, SO_4^{2-}, Cl^{-}, HCO_3^{-} and OH^{-}. The supply of micro nutriments, namely Fe, Cu, Zn, Mn, B, Mo, and Ni, are very small in ratio-the others elements (macro nutrients), so it has no a significant effect on EC (Sonneveld and Voogt, 2009). Ideal EC range for hydroponics for most of the crops is between 1.5 and 2.5 dS m^{-1} or μS cm^{-1}. Higher EC will prevent nutrient absorption due-osmotic pressure and lower level severely affects plant health and yield. So, appropriate management of EC in hydroponics technique can give effective tool for improving yield and quality.

Table 1: Optimum range of EC and pH values for hydro-ponic crops (Sharma *et al.*, 2018)

Crops	EC (dSm^{-1})	pH
Asparagus	1.4-1.8	6.0-6.8
African Violet	1.2-1.5	6.0-7.0
Basil	1.0-1.6	5.5-6.0
Bean	2.0-4.0	6.0
Banana	1.8-2.2	5.5-6.5
Broccoli	2.8-3.5	6.0-6.8
Cabbage	2.5-3.0	6.5-7.0
Celery	1.8-2.4	6.5
Carnation	2.0-3.5	6.0
Courgettes	1.8-2.4	6.0
Cucumber	1.7-2.0	5.0-5.5
Egg plant	2.5-3.5	6.0
Ficus	1.6-2.4	5.5-6.0
Leek	1.4-1.8	6.5-7.0
Lettuce	1.2-1.8	6.0-7.0
Pak Choi	1.5-2.0	7.0
Peppers	0.8-1.8	5.5-6.0
Parsley	1.8-2.2	6.0-6.5
Rhubarb	1.6-2.0	5.5-6.0
Rose	1.5-2.5	5.5-6.0
Spinach	1.8-2.3	6.0-7.0
Strawberry	1.8-2.2	6.0
Sage	1.0-1.6	5.5-6.5
Tomato	2.0-4.0	6.0-6.5

The pH is a parameter that measures the acidity or alkalinity of a solution. This value indicates the relationship between the concentration of free ions H^+ and OH^- present in a solution and ranges between 0 and 14. In a nutrient solution, pH determines the availability of essential plant elements (Fig. 7). Optimum pH range of nutrient solution for development of plants is 5.5-6.5 (Trejo-Tellez and Gomez, 2012) for most species but some can differ from this range. pH values below 6.0 causing the solubility of phosphoric acid, calcium and magnesium-drop and above 7.5 causes iron, manganese, copper, zinc and boron ions-be less available-plants. Once the plants grow, it will change the composition of nutrient solution by depleting specific nutrients more rapidly than others, removing water from the solution and altering the pHby excretion of either acidity or alkalinity. Hence, care is required for maintaining optimum pH, EC and nutrient level in hydroponic solution. Crops such as vegetables, spices, flower and ornamentals, medicinal plants, fruits like straw berry, fodders, and up-some extent cereals like maize can be raised through soil less hydroponic technique (Table 1).

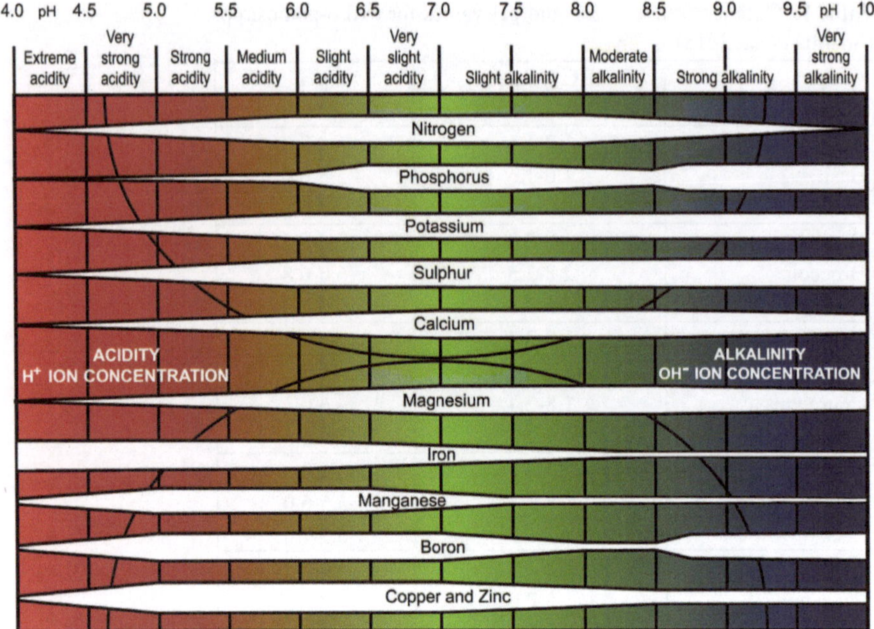

Fig. 7: Troug diagram of nutrient availability. Each nutrient is represented with a band; the thickness is proportional-the availability.

The following steps should be performed for EC and pH management:

- Set the desired EC and pH value of the solution.
- Calibrate the meters probe with buffer solution. Make sure the nutrient solution is stirred properly and allow the reading-stabilize, which may take a couple of minutes.
- If the pH reading is high, add phosphoric acid, citric acid, vinegar or pH down products into the solution and stir it well. Repeat the process until the pH reaches the optimum range.
- If the pH value is low, add potassium hydroxide, potassium carbonate or a pH up product slowly. Repeat until the pH reaches the optimum range.
- If the EC reading is higher than the required level, then dilute the nutrient solution by adding more water, repeat the process until it reaches the required level.
- If the reading is below the optimum level, add stock nutrient solution until the required level is attained.
- Clean the probe using distilled water.

Water Quality: Water quantity is a key factor in hydroponic system used in recirculating nutrient solution. In general, water quality is a complex concept.

However, water quality of hydroponics can be limited-the quantity of specific ions and growth promoting elements relevant for plant nutrition and also with the presence of organisms and substance that clog the irrigation systems.

First step for hydroponics is-have the water analyzed, pH, electrical conductivity (EC) and alkalinity of the water has-be tested. Nutrient toxicity or deficiency problems may occur in the initial or later stages of production if poor quality water is used. Normally water contains salts like chlorides and sulfates of sodium, calcium, magnesium, carbonates and bicarbonates. These salts should not be above an acceptable level (Table 2) because it has an effect on EC and pH of the nutrient solution.

Table 2: Acceptable values for common salts found in water (Singh and Bruce, 2016)

Salt	Acceptable value (ppm)
Sodium	<50
Calcium	<150
Magnesium bicarbonate	<50
Chloride	<140
Sulfate	<100

Dissolved Oxygen

Dissolved oxygen indicates the amount of oxygen available in solution. Concentration of oxygen required in nutrient solution varies with crop demand, it changes with the photosynthetic rate. The consumption of O_2 increases with increase in temperature of nutrient solution. If the root aeration is not enough, concentration of carbon dioxide in the root vicinity surge due-root respiration. First symptom of oxygen deficiency is inhibited root growth, followed by brown colored roots. Usually the oxygen level is 8 ppm, but it can be maintained within the range of 7-10. In case of small units, it is better-aerate the system using an air pump and aquarium air stone.

Nutrient Management in Hydroponics

Nutrient management programs should begin with an understanding of the nutrient solution concentrations in parts per million (ppm) for the various nutrients required by different crops. By managing the concentrations of individual nutrients, growers can control the growth and yield of the crop.

Nutrients in Hydroponics Solution

Steiner created the concept of ionic mutual ratio which is based on the anions (NO_3^- , $H_2PO_4^-$ and SO_4^{2-}) and the mutual ratio of cations (K^+, $Ca^{2+,}$ Mg^{2+}). Such a relationship is not just about the total amount of each ion in the solution, but in the quantitative relationship that keep the ions together; if improper relationship between them take place, plant performance can be negatively affected.

Nutrient Interaction in Hydroponics Solution

Nutrients in the solution may interact in various ways that may gain either positive or negative effects on crop production, depends on crop growth stages, amounts, combinations, and balance (Sheng-Xiu *et. al.,* 2009). Inadequate or excessive concentrations of elements or an imbalanced ion composition in the nutrient solution may inhibit plant development, resulting in toxicity or nutrient-induced deficiencies (Chrysargyris *et. al.,* 2021). Nutrient interactions can be either synergistic or antagonistic and also possible-have no interactions. Interaction between nutrients occurs when the supply of one nutrient affects the absorption and utilization of other. Upon the addition of two nutrients, an increase in crop yield that is more than adding only one, the interaction is positive or synergistic. Similarly, if addition of nutrients produced less yield as compared-individual ones, the interaction is said-be negative or antagonistic (Fageria, 2021).

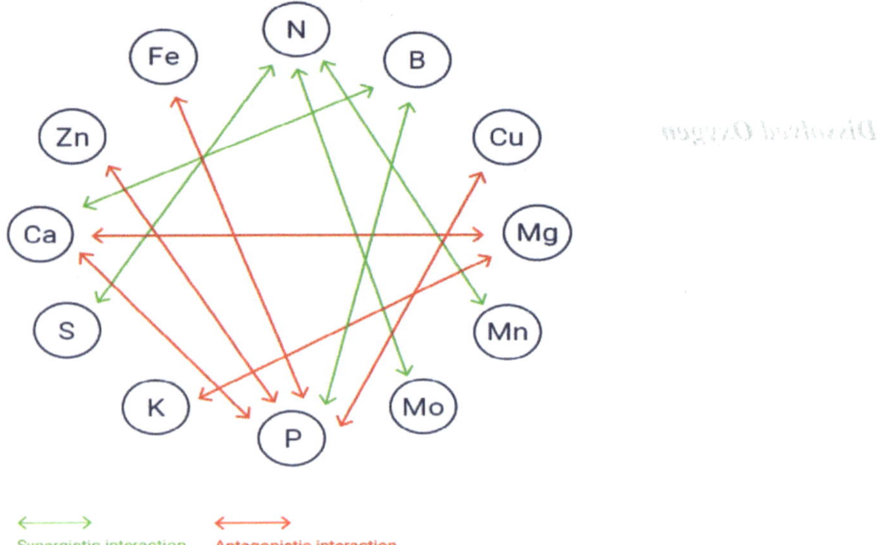

Fig. 8: Synergistic and antagonistic interaction
(*Source:* ttps://images.app.goo.gl/sioAUj19U AkqcWVT)

Monitoring the ratio between Ca^{2+} Mg^{2+} and K^+ in the solutions is very important-avoid K^+ /Ca^{2+} induced Mg^{2+} deficiency. Similarly, very high rates of Mg^{2+} fertilizers will depress K+ absorption by plants, but this antagonism is not nearly as strong as the inverse relation of K^+ on Mg^{2+}. Calcium, magnesium, and potassium compete with each other and the addition of any one of them will reduce the uptake rate of the other two. The uptake of nitrogen, sulfur, and iron is not entirely depending on its availability in the hydroponic solution but also on presence of other elements (Marschner, 2012).

Nutrient solution management: An optimized and well-balanced supply of nutrients is a prerequisite for efficient use of the resources by hydroponically grown vegetables and short duration high value crops, not only-ensure a high yield but also-guarantee the quality of the edible plant parts. In hydroponics, frequency and volume of the nutrient solution applied depends on the type of substrate, the crop and growth stage, the size of the container, the irrigation systems used, and the prevailing climatic conditions. Depending on the stage of plant development, some elements in the nutrient solution will be depleted more quickly than others and as water evaporates from the nutrient solution, the fertilizer becomes more concentrated and can burn plant roots. In hydroponics, nutrient management is very important and must be done as highly efficient as possible-improve productivity without harming the environment.

Nutrient management included application of the right fertilizers source (e.g., ammonium or nitrate as nitrogen source), balanced nutrient solution according-plants needs and according-plant growth stages and climatic conditions. Strong difference between the ion ratios presented in the nutrient solution and those absorbed by plants led-the accumulation of certain ions in the nutrient solution, which caused an imbalance of mineral elements in the nutrient solution and created more energy-absorb the suitable ions. Recycling exhausted solutions may also represent an efficient strategy-prevent groundwater and environmental pollution.

Fertilizer formulation: There are basically two methods-supply the fertilizer nutrients-the crop, premixed products and grower formulated solutions. The two methods differ in the approach-formulating the fertilizer and the resulting nutrient use efficiency

Pre-mix method: There are several commercial pre-mixed fertilizer formulations in the market. The pre-mixed materials contain nutrients at a fixed proportion and quantity making it difficult-achieve the desired level.

Grower Formulated Solution

The grower formulated solution contain individual ingredients formulated based on grower's preference. Additional acidification might be required for some water and this can be accomplished with phosphoric acid. This formulation also uses potassium chloride-provide K. There is no problem with using potassium chloride as a partial source of K. It provides K in the same form as potassium nitrate and potassium chloride is less expensive.

Environmental Factors

There are many environmental factors affecting the growth and development of plants under protected condition, especially temperature, light intensity and humidity have a strong effect on the growth, yield and nutritional quality of crops grown under hydroponic condition.

Temperature

Optimum temperature is one of the important factors for nutrient uptake, proper growth, and yield. Chilling and high temperatures affect growth and crop performance. Temperature can influence the amount of oxygen consumption of plants and uptake capacity of roots. Different plant species have different optimum temperature range for better growth, yield. Nutrient solution temperature also had an effect on the photosynthetic efficiency. In general terms, the effective quantum yield and the fraction of PSII reaction centers were higher in plants grown at cold solution (Calatayud *et al.,* 2008). The higher the temperature of the solution the lesser the oxygen solubility in water (table 3)

Table 3: Solubility of oxygen in water pure at various temperatures (Trejo-Tellez and Gomez, 2012).

Temperature (°C)	Oxygen solubility, mg L^{-1} of pure water
10	11.29
15	10.08
20	9.09
25	8.26
30	7.56
35	6.95
40	6.41
45	5.93

Light: One of the most basic pre-requisites of plant growth and survival is light. It helps the leaves-produce energy required for growth and development. Three principal characteristics of light affect plant growth: quantity, quality and duration. Light quantity refers-the intensity, or concentration, of sunlight. Up-a point, the more sunlight a plant receives, the greater its capacity for producing food via photosynthesis, and the plant growth is promoted under a certain range of light intensity.

Light quality refers-the wavelength of light. Blue and red light, which plants absorb, have the greatest effect on plant growth (Kong and Nemali, 2021). Duration, or photoperiod, refers-the amount of time a plant is exposed-light. Photoperiod controls flowering in plants. In order-attain maximum yield with a high quality from crops grown in a closed production system, optimum light intensity is very important.

Humidity: Growing plants contain about 90 percent water. Water vapor present in the surrounding has a major role in plant growth. The amount of moisture in the air is generally expressed as relative humidity (RH). Humidity inside a polyhouse depends on the temperature; Warm air has a higher moisture holding capacity than cool air, therefore, as the temperature of air increases relative humidity decreases.

Proper watering and adequate plant spacing, having well-drained floors, warming plants, moving air and venting moisture are ways-reduce humidity in greenhouses.

Watering early in the day time allow plant surfaces-dry before evening. The highest relative humidity in a greenhouse is generally found inside plant canopies, where moisture is generated from transpiration and trapped due-insufficient air movement. Adequate plant spacing and mesh benches will help-improve air circulation at the plant level.

Crops under Hydroponic System

A large number of plants and crops can grow by hydroponics system. Quality, taste and nutritive value of end products are generally higher than the natural soil-based cultivation. Various experimental findings outline that leafy greens (lettuce, spinach, parsley, celery and atriplex, etc) can be successfully grown in hydroponic systems. Lettuce and spinach are most promising species-grow in hydroponics because of its higher growth and nutrient uptake capacity. However, there are some other high value crops which can be grown under hydroponics (table 4).

Table 4: Various species of plants grown under soil less hydroponic system

Crops	Crops
Cereals	Rice, Maize
Fruits	Strawberry
Vegetables	Tomato, Chilli, Brinjal, Green bean, Beet, Winged bean, Bell pepper, Cucumbers,Melons, green Onion
Leafy vegetables	Lettuce, Spinach, Celery, Swiss chard, Atriplex, Mint
Condiments	Coriander leaves, Methi, Parsley, Mint, Sweet basil, Oregano
Flower/Ornamental crops	Marigold, Roses, Carnations, Chrysanthemum
Medicinal crops	Indian Aloe, Coleus, Indian Ginseng (Ashwagndha), Basil

Economics of Hydroponics Market

Globally hydroponics market has crossed USD 2.1 billion in 2020. It is expected-expand at a compound annual growth rate (CAGR) of 20.7% from 2021-2028 (Anonymous, 2021). Global hydroponics market includes tomato, cucumber, pepper, lettuce, herbs, and other leafy vegetables. Tomato forms the largest market segment and it accounts for 30.4% share of the global market, during 2018. Leading countries in hydroponic technology are Netherland, Australia, France, England, Israel, Canada and USA. Income generated through hydroponics crop production is expected-be more in tomatoes, spinach and other leafy vegetables. However there are some other short duration high value crops suitable for hydroponics.

Pest and Disease Management

In a properly managed soil less cropping system, the threat posed by pests and diseases is minimal compared-farming in the soil in an open environment. However, negligence and improper hygienic maintenance in soil less cropping systems might result in pest and disease occurrences, which may even result in entire crop damage. Soil less cropping system involves a continuous supply of water and nutrients, in the open or closed circulation. When compared-systems with a closed circulation, systems with open circulation have a lower frequency of pests and diseases. Although, open circulation cannot be opted over closed circulation systems due-economic concerns regarding wastage of inputs and the environmental harm brought on by excessive nutrient solution runoff (Ehret *et al.*, 2001; Schnitzler, 2003). This makes proper management of pests and diseases crucial for the success of the soil less cropping system.

The same pests and pathogens that affect plants traditionally grown in soil may also affect plants grown in the soil less cropping system but will differ in frequency and severity. Major pests found in the soil less cropping system include insect pests like aphids, whiteflies, spider mites, thrips, leaf miners, caterpillars, and pinworms. Among the phytopathogens in the hydroponic system, phytopathogenic fungi are most destructive followed by viruses, bacteria, and nematodes. Due-their propensity for humid/aquatic environment conditions, soil born fungi like *Fusarium* spp., *Phytophthora* spp., *Pythium* spp., and *Olpidium* spp. are the most problematic among the variety of phytopathogens that occur in soil less cropping system (Stouvenakers *et al.*, 2019). *Olpidium* spp. of fungi are not a very serious threat themselves but acts as a vector for virus infestations such as Lettuce Big Vein Virus (LBVV) on lettuce and Tobacco Necrosis Virus (TNV) on pepper, lettuce, cucumber and tomato.

To manage the pest and diseases occurring in the protected cultivation in general and soil less cropping systems in particular, integrated pest disease management strategies have been developed by various researchers (Gullino et al., 2020). Once the pest or disease enters the system it will spread rapidly throughout the system due-the recirculation of nutrient solution. Therefore, non chemical prophylactic measures are highly efficient for pest and disease prevention in soil less cropping systems (Folorunso *et al.*, 2021). Preventive measures-be applied in hydroponics include using healthy planting material, ensuring that nutrient solution and substrate are free of pest and disease inoculums, optimizing the growing conditions (temperature, RH, and pH), and maintaining good hygiene in cropping systems. Chances of pest and disease occurrence are more likely in closed systems with re-circulated nutrient solution therefore proper disinfection of nutrient solution should be ensured. Disinfection of water can be achieved by physical methods such as heat treatment, UV radiation, and filtration (membrane filter or slow sand filter) or by use of chemicals (ozone, hydrogen peroxide, chlorine, iodine, surfactants) (Folorunso *et al.*, 2021). The substrate used in the soil less cropping system also needs-be free of contaminants. Contamination of substrate may occur during processing, handling, transportation, or storage and care should be taken-minimize contamination at all levels. Reuse of substrates is not advisable and should be completely avoided in the case of seedling producers. Other costly alternatives like steam disinfection of substrate which can tolerate heat can be done, but are not generally considered by farmers.

Even after the application of preventive measures pests and diseases cannot be entirely managed, this is where IPDM comes in. Pest and disease monitoring is crucial in IPDM as it helps in identifying whether the pest has crossed the action threshold (economic injury level) thereby determining the time-apply management practices. Pest scouting, monitoring using traps (sticky, pheromone, and light traps), frequent inspection for disease symptoms and presence of vectors organisms, etc. should be done-ascertain the presence of pests and diseases in the system. Once sighted the pest should be correctly identified and quantified before selecting the suitable management practice. For the identification of insects various available guides can be used while pathogens can be detected by direct macroscopic and microscopic observation of the pathogen, isolation of the pathogen, or by use of serological and molecular methods.

IPDM in the soil less cropping system involves a combination of cultural, physical, mechanical, biological, and chemical control measures. The primary objective is-prevent the entry of pests or pathogens into the system while the secondary objective is-limit the infestation, development, and spread during the growing period (Stouvenakers *et al.*, 2019). Cultural methods are a no cost, efficient, and best first approach for controlling pests and diseases in the soil less cropping system. These include the use of resistant plant varieties, certified seeds, optimum plant spacing, sanitation, environmental conditions management, avoidance of plant abiotic stresses, use of trap crops, tools disinfection, crop rotation, fallow period. Physical and mechanical measures involve the use of physical agents for disinfection, physical barriers, and traps against insect pests and vectors, hand removal of large larva, removal of infected plants, maintaining sanitation. Biological control is a more sustainable and effective strategy for controlling pests and diseases in the soil less cropping system. Antagonistic microorganisms control pest populations by competing for nutrients and niches, parasitism, antibiosis, or by induction of disease resistance in plants (Narayanasamy, 2013).

Natural enemies of insect pests can be promoted in the system for its control. Also, entomopathogenic microorganisms are effective in controlling insect pest population in the system which includes majorly entomopathogenic bacteria (*Bacillus thuringiensis*) and entomopathogenic viruses (NPV, BCV) followed by entomopathogenic fungi and nematodes (Osborne *et al.*, 2004; Khan *et al.*, 2012).-control phytopathogens bacterial (*Bacillus* spp., *Pseudomonas* spp.) fungal (*Trichoderma* spp., non-pathogenic *Fusarium* spp., Vascular arbuscular mycorrhiza), and actinomycetes (*Streptomyces* spp.) bio-control agents are used (Stouvenakers *et al.*, 2019; Folorunso *et al.*, 2021). Chemical control methods are often avoided and used only as a last resort when no other management practice could save the crop owing-their detrimental effects on non-target organisms and persistence (Stouvenakers *et al.*, 2019; Folorunso *et al.*, 2021). Fungicides, insecticides, nematicides and acaricides used for the management of pests in soil cultivation can be employed in the soil less cropping system. Less effective but more sustainable alternatives like essential oils and botanicals can also be used (Folorunso *et al*., 2021). Proper implementation of IPDM can ensure the creation of a pest and disease free soil less cropping system, which will be profitable and healthy.

Summary

Hydroponics and soil less cultivation is possibly the most intensive method of crop production in today's agricultural industry. In combination with greenhouses or protective covers, it is high technology and makes it possible-grow crops year-round in limited spaces with less labour requirement. But needs higher initial investments and needs qualified technician-operate and manage the soilless systems. The hydroponic industry is expected-grow exponentially in near future.-encourage commercial hydroponic farm, it is important-develop low cost hydroponic technologies and thereby overall startup and operational cost could be reduced. Hydroponic technology is an efficient mean for food production under extreme environmental ecosystems and in highly populated areas, it can provide locally grown high-value crops with better quality. Proper hygiene throughout the production cycle is very important in soil less production systems under closed structures else which lead-complete collapse of the production due-its quick spread of pest and diseases.

References

Chrysargyris A, Petropoulos S.A, Prvulovic D, Tzortzakis N. 2021. Performance of hydroponically cultivate[d] geranium and common verbena under salinity and high electrical conductivity levels. Agronomy 11:1237. DOI: 10.3390/ agronomy11061237

Ehret D, Alsanius B, Wohanka W, Menzies J and Utkhede R. 2001. Disinfestation of recirculating nutrient solutions in greenhouse horticulture. Agronomy 21(4): 323-339. https://dx.doi.org/10.1051/agro:2001127.

Fageria V D. 2021. Nutrient interactions in crop plants. Journal of Plant Nutrition 24(8): 1269-1290. DOI: 10.1081/ PLN-10010698

Folorunso E A, Roy K, Gebauer R, Bohata A and Mraz J. 2021. Integrated pest and disease management in aquaponics: A metadata-based review. Reviews in Aquaculture 13(2): 971-995. https://doi.org/10.1111/raq.12508.

Gullino M L, Albajes R and Nicot P C. 2020. Integrated pest and disease management in greenhouse crops, 2nd edn, Springer Nature, Swizerland. 681p.

Khan S, Guo L, Maimaiti Y, Mijit M and Qiu, D. 2012. Entomopathogenic fungi as microbial biocontrol agent. Molecular Plant Breeding 3(7): 63-79. doi: 10.5376/mpb.2012.03.0007

Kong Y and Nemali K. 2021. Blue and Far-Red Light Affect Area and Number of Individual Leaves-Influence Vegetative Growth and Pigment Synthesis in Lettuce. Frontiers ion plant science 12: 667407. https://doi.org/10.3389/fpls.2021.667407

Marschner P. 2012. Rhizosphere biology. In: Mineral Nutrition of Higher Plants, 3rd edn, Marschner P (Ed.). Academic Press, Amsterdam, pp 369-388.

Narayanasamy P. 2013. Biological management of diseases of crops. Springer, Dordrecht, Netherlands, pp. 295-429.

Osborne L S, Bolckmans K, Landa Z and Pena J. 2004. Kinds of natural enemies. In: Biocontrol in protected culture. Heinz K M, van Driesche R.G, Parrella PM. (Ed.). Ball Publishing, Batavia, IL, pp 95-127.

Schnitzler W H. 2003. Pest and disease management of soilless culture. In: South Pacific Soilless Culture Conference – SPSCC, held at Palmerston North, New Zealand, during 10-13th February 2003, pp 191-203. http://doi.org/10.17660/ActaHortic.2004.648.23.

Sharma N, Acharya S, Kumar K, Singh N and Chaurasia O P. 2018. Hydroponics as an advanced technique for vegetable production: An overview. Journal of Soil and Water Conservation 17(4): 364-371. doi: 10.5958/2455-7145.2018.00056.5

ShengXiu L, Wang H, Malhi S S, Li S Q, Ya-Jun G, Xiao-Ho ng T. 2009. Nutrient and water man agement effects on crop production and nutrient and water use efficiency in dry land areas of China. In: Advances in Agronomy. Sparks (Ed.). Academic Press, Burlington, 102: 223-265.

Singh H and Bruce D. 2016. Electrical conductivity and pH guide for hydroponics. http://osufacts.okstate.edu.

Stouvenakers G, Dapprich P, Massart S and Jijakli M H. 2019. Plant pathogens and control strategies in aquaponics. In: Aquaponics Food Production Systems. Goddek S, Joyce A, Kotzen B and Burnell G M. (Eds.). Springer, Cham. https://doi.org/10.1007/978-3-030-15943-6_14.

Trejo-Tellez L I and Gomez F C. 2012. Nutrient solution for hydroponics system. In: Hydroponics - A Standard Methodology for Plant Biological Research. Asao T (Ed.). 1: 1-22. DOI: 10.5772/37578.

www.grandviewresearch.com.. 2021. Hydroponics Market Size, Share & Trends Analysis Report. Accessed on 05th October 2022.

Section 03: Floriculture and Landscaping

15

Recent Research and Development in Floriculture and Landscaping

S.S. Sindhu and Namita

Division of Floriculture and Landscaping
ICAR- Indian Agricultural, Research Institute, New Delhi-110012

Floriculture is most diversified and potential component of horticultural industry in India. It is one of the fastest growing sectors in domestic and international trade markets and holds high business potential and prospects for trade. In ancient era, the flowers were used mainly for religious offering but now floriculture sector has emerged as a commercial venture. There is a dynamic shift from sustenance production to commercial production in this era and thereby significant increase in trade of flowers at domestic level because of rapid urbanization, increase in income level, *etc*. Flower business is considered as a most profitable business as commercial floriculture has higher potential. The cut and loose flowers have opened various avenues for establishment of flower business.

The institutions/universities like ICAR- Indian Agricultural Research Institute, New Delhi; ICAR- Directorate of Floricultural Research; ICAR- Indian Institute of Horticultural Research, Bengaluru; CSIR-National Botanical Research Institute, Lucknow; CSIR- Central Institute of Medicinal and Aromatic Plants, Lucknow; CSIR- Institute of Himalayan Bio-resource Technology, Palampur, Tamilnadu Agricultural University, Coimbatore; Punjab Agricultural University, Ludhiana etc. are pioneer in scientific research on floricultural crops. Many amateurs, nurserymen are also involved in developing flower varieties and technologies. In recent times a number of varieties was evolved from the division of Floriculture and Landscaping, ICAR-IARI for commercial production of flowers.

Varietal Development

Rose

Rose is universally known as "Queen of Flowers". It belongs to the family Rosaceae. Traditionally, roses (*Rosa* × *hybrida*) have been considered to be one of the most valued flowers of the world because of their high ornamental value.

The importance of beautiful flowers has grown over the years with the emergence of a global cut flower market (Namita *et al.*, 2015). It represents a major product for floriculture market as well as oil industry (Panwar *et al.,* 2015). Recently developed varieties of rose in India are given hereunder.

***Pusa Shatabdi*:** It is hybrid of varieties Jadis × Century Two. The variety produces very attractive light pink coloured flowers. The petals (numbering 35-40) are fleshy and are pink in colour. Each plant produces 20-30 flowers in winter and 35-40 flowers in spring season. This variety is moderately tolerant to powdery mildew and Leaf spot diseases. The blooms are mildly fragrant and are suitable for cut flower and exhibition purposes.

***Pusa Ajay*:** It is hybrid of varieties Pink Parfait × Queen Elizabeth. Foliage is pigmented and glossy. It has dark pink coloured flowers. The petals (numbering 35-40) are fleshy and are dark pink in colour and recurrent blooming. Each plant produces 15-20 flowers in winter and 35-40 flowers in spring season. This variety is moderately tolerant to powdery mildew and black spot diseases. The blooms are mildly fragrant and are suitable for cut flower and exhibition purpose.

***Pusa Mohit*:** It is hybrid of varieties Suchitra × Christian Dior. Variety is thorn less and has red petals with lighter shade on the reverse side of petals, tolerant to black spot. Each plant produces 20 flowers in winter and 45 flowers in spring season.

***Pusa Arun*:** Pusa Arun is a hybrid between an exotic variety Queen Elizabeth and indigenous variety Jantar Mantar. It is a variety of Hybrid tea category, produces medium to tall bushes often reaching a height of about one meter with broad and moderately glossy leaves. The shinning foliage contains higher pigmentation in early stages. The flower buds are pointed and are dark red in colour which opens to dazzling dark red, large sized, double blooms on strong and long shoots. This variety is tolerant to red scale, which is one of the major insect pests attacking roses in open cultivation. It takes 45-50 days to bloom after pruning in second fortnight of October. It produces 20-30 flowers during winter and 35-40 flowers during spring season. The blooms remain fresh for 8-10 days on the plant as well as in the vase when harvested at bud stage.

***Pusa Komal*:** Pusa Komal is a hybrid between an exotic variety Pink Parfait and an indigenous variety Suchitra. It is a highly floriferous bushy variety, which belongs to floribunda group. The short to medium stature bushes are completely thornless. The variety possesses strong stems with multiple flower buds arranged in different heights, which open in a sequential manner. The variety produces as many as 60 flowers, which are mostly borne in clusters. The pink coloured buds are pointed in shape and unfurl in to double flowers. The variety produces very attractive pink coloured flowers, which last almost 6-7 days in vase.

***Pusa Mahak*:** It is a Hybrid Tea variety of rose. The plants are tall and vigorous with a height of 100 - 120 cm. The Flowers are dark pinkish in colour and have outstanding fragrance. The flowering starts in 40-45 days after pruning. Flowers are large and semi-double with 22-23 petals. It is a recurrent flowering and floriferous variety and each plant produces on an average 50-60 flowering shoots in a season. The variety is ideal for garden display and the fragrant flowers can be used for floral arrangements.

***Pusa Alpana*:** The variety is a selection from the open pollinated population of cv. Rose Sherbet. It is a floribunda type variety of rose. The Flowers are compact, light pinkish in colour and have fragrance. Flowers are medium and semi-double with 35-43 petals. It is a recurrent flowering and floriferous variety. This variety is directly propagated by semi hard wood cuttings. The variety is ideal for loose flower production and garden display purpose.

Pusa Shatabdi Pusa Ajay Pusa Mohit

Pusa Arun Pusa Komal Pusa Mahak

Pusa Alpana

Recently Released Rose Varieties from ICAR-IARI

Marigold

Marigold, a versatile crop with golden harvest is grown commercially in different parts of the country for loose as well as for cut flower production. It has beautiful flowers with wide adaptability to various soils and climate conditions (Kumar *et al.* 2017). It has long duration of flowering which makes it one of the most popular loose flower crops with considerable acreage. The dwarf type French marigold is most ideal for landscaping purpose such as bedding, sunken garden, rockery, edging, hanging, window boxes, etc. Marigolds are well known for their medicinal (Abad *et al.*, 1999), insecticidal (Perich *et al.*, 1995) and nematicidal properties (Govindaiah *et al.*, 1993). Recently developed varieties of marigold in India are given hereunder.

Pusa Arpita: It is a selection from heterozygous population of French marigold. It produces medium sized, light orange flowers during mid December to mid February in Northern plains of India. The average fresh flower yield of the variety is 18-20 tons /ha.

Pusa Bahar: It is an open pollinated variety belonging to African marigold group which flowers in 90-100 days after sowing. Its plants are vigorous having height of 75-85 cm. It produces compact, flattened, attractive and large size (8-9 cm) flowers of yellow colour (RHS Yellow Group: 9A). The variety is very floriferous producing on average 50-60 large sized flowers per plant. In northern plains, ideal time for seed sowing is mid October to mid November. The main flowering time is mid January to March. It is suitable for bedding in gardens as well as other floral decorations.

Pusa Deep: It is an early flowering variety of French marigold which flowers in 85-95 days after sowing. The variety produces medium statured spreading plants having 55-65 cm plant height and 50-55 cm plant spread. It produces compact and medium sized flowers of dark red colour (red group 46 A-RHS colour chart). The variety is very floriferous and produces on average 80-90 flowers per plant resulting in high flower yield (18-20 ton. /ha). In Northern plains, it flowers during October-November. It is suitable for loose flower production and flowers during festive season.

Pusa Arpita

Pusa Bahar

Pusa Deep

Recently Released Marigold Varieties from ICAR-IARI

Chrysanthemum

Chrysanthemum is one of the most important ornamental crops grown worldwide as cut flower and pot mums (Kumar *et al.,* 2015). It holds a prominent position in both the domestic and international trade and ranks second after rose at the

Dutch auctions (Steen, 2010) indicating its global significance. In modern times, no single cultivar can sustain in market for a longer period, hence arises the need of development of novel cultivars with quality traits and therefore a large number of cultivars are being bred every year with desired characteristics and aesthetic values (Kumar *et al.,* 2014). Therefore, the recently developed cultivars by different institutes are explained hereunder.

***Pusa Anmol*:** This variety is a Gamma ray induced mutant of cv. Ajay. This is a highly floriferous bushy variety with yellowish pink flowers. The variety is a thermo and photo insensitive variety and produces three flower flushes in a year (October-November, February-March and June-July) as against one in a majority of the cultivars. The variety flowers in 85-100 days after transplanting. Ideal for loose flowers and whole plant cut flower. Blooms remain fresh for 20-22 days in field conditions.

***Pusa Centenary*:** This variety is a Gamma ray induced mutant of cv. Thai Chen Queen. This is a vigorous growing variety producing very big yellow flowers. The variety blooms in 100-110 days after transplanting. This variety is Ideal for cut flower production. The blooms remain fresh for 20-22 days in field conditions.

***Pusa Arunodaya*:** It is a gamma ray induced pink coloured mutant of cv. Thai Chen Queen, which is orange in colour. Pink coloured ray florets (65D) appeared as chimers which were used for in vitro regeneration to establish as a new variety. The plants are medium in height (50-55 cm) with good spread (60-65 cm). The semi double flowers are big (7-8 cm diameter) with incurving ray florets. The variety is suitable as a cut flower and pot culture for exhibition purposes.

***Pusa Kesari*:** It is a gamma ray induced red coloured mutant of cv. Thai Chen Queen, which is orange in colour. Red coloured ray florets (171A) appeared as chimers which were used for in vitro regeneration to establish as a new variety. The plants are tall in height (65-70 cm) with good spread (60-65 cm). The semi double flowers are big (9-10 cm diameter) with incurving ray florets. The variety is suitable as a cut flower and pot culture for exhibition purposes.

***Pusa Aditya*:** It is an open pollinated seedling of cv. Jaya. The Plants are bushy (30-35 branches) medium in height (55-60cm) with moderate spread (45-50 cm). The variety is a spray type that produces star shaped semi-double flowers that resemble that of gazania flowers, which is unique in chrysanthemum. The flowers are yellow in colour (5 A) at the periphery with orange red colour (45A) in the center. The florets are spatulate with distinct keel. This attractive variety is suitable as a cut flower and as a potted plant.

***Pusa Chitraksha*:** It is an open pollinated seedling of cv. Lal Pari. The Plants are bushy (24-30 branches), tall in height (60-65cm) with excellent spread (60-65 cm). The variety is a spray type that produces single flowers that are deep magenta in colour, which is unique in chrysanthemum. The florets are spatulate in shape with magenta colour (59A) at the periphery with silvery white ray floret tube.

The disc florets are yellow (12A) that provide a very good contrast. Owing to its floriferous nature the variety is suitable as a potted plant and for garden display.

***Pusa Sona*:** It is an open pollinated seedling of cv. Sadbhawana. The Plants are bushy (20-25 branches), extremely dwarf in height (25-30 cm) with excellent spread (50-55 cm). The variety is a spray type that produces single flowers that are yellow in colour (8A). The disc florets are also yellow (12C). The variety is early flowering by at least 20 days in advance when compared to other varieties. The variety is ideally suited for pot mums.

***Pusa Guldasta*:** It is an open pollinated seedling of cv. Lalpari. Plant of this variety attains a height of 58 cm with a good spread of 50 cm. The plant is of upright growth habit, very sturdy and branches do not droop down. It bears semi double medium sized flowers (3.8 cm) with orange red ray florets and yellow disc. The inflorescence is corymb and flowers are borne at almost same height. The flowers stay for longer duration (48 days) under field conditions. This does not require pinching and staking. It is suitable for spray and pot culture purposes.

***Pusa Shwet*:** It is an open pollinated seedling of cv. Lalpari. Plant attains a height of 41 cm with a good spread of 48 cm. The plant is of Semi-spreading growth habit. It bears semi double medium sized flowers (6.0 cm) with white ray florets and yellow disc. The flowers stay for longer duration (40 days) under field conditions. When in full bloom its magnificent white flowers give beautiful picturesque effect of snow. This variety is suitable for pot culture and garden display purposes.

***Pusa Sunderi*:** It is developed from a cultivar Himanshu by gamma ray irradiation and selection as an EDV. Plant attains height of 38 cm with a good spread of 67 cm. It bears medium size flowers (4.44 cm) with yellow ray florets and yellow disc. The flowers stay for longer duration (40 days) under field conditions. When in full bloom its magnificent yellow flowers give beautiful picturesque effect of golden colour. This variety is suitable for pot mum and garden display purposes.

Pusa Anmol

Pusa Centenary

Pusa Arunodaya

Pusa Kesari Pusa Aditya Pusa Chitraksha

Pusa Sona Pusa Guldasta Pusa Shwet

Pusa Sunderi

Recently Released Chrysanthemum Varieties from ICAR-IARI

Gladiolus

Gladiolus (*Gladiolus grandiflorus* L.) popularly known as the 'Queen of the bulbous ornamentals' or Sword lily is an ornamental bulbous plant native to South Africa, and belongs to the family Iridaceae. Gladiolus is a major source of income for the farmers due to high demand for their cut flowers and corms in domestic as well as export market (Swaroop *et al.,* 2017). The major emphasis on gladiolus improvement has been on the development of varieties having attractive colour, large size of florets with long spikes. (Mehra *et al.,* 2016) Hence, recently developed varieties of gladiolus are given hereby.

***Pusa Kiran*:** A selection among the open pollinated population of the variety 'Ave'. The selection produces white colour florets (16 – 18 in number) on a long and sturdy spike. It was developed through clonal selection (Selection among open pollinated population). It produces very good spikes well suited for vase decoration, bouquet preparation and for floral arrangements which lasts for more than 10 days.The florets are white (155 B) in colour with ray like red colour (53 C) markings on throat. It has good spike length (>95.00 cm), rachis length (>55.00 cm) and number of florets per spike (>16). Further it is a very good multiplier producing more than two corms (on an average 2.53) and 20 cormels from each mother corm. It is an early variety flowering in about 75 days.

***Pusa Shubham*:** A selection among the progeny obtained from the cross of Lucky Shamrock and Green Lilac. It was developed through hybridization followed by selection among the segregants. The variety produces very good spikes well suited for vase decoration, bouquet preparation, floral arrangements, etc which lasts for more than 10 days. An early variety flowering in about 72 days The inter-floret length is less which results in compact arrangement of florets on one side of the spike. About 6-7 florets remain open at a time which makes it excellent for vase decoration or for making floral bouquets or arrangements. Further it is a very good multiplier producing two corms and more than 23 cormels from each mother corm.

Pusa Unnati: Florets are purple red/pink in colour (72B) with white stripe on inner two tepals. Spikes are very long about 126 cm in length with 16-20 florets. It produces 2-3 shoots from each mother corm.

***Pusa Manmohak*:** A selection among the progeny obtained from the cross of Mayur x Hunting Song. The plants are healthy, medium green in colour and straight. It is mid season variety takes 100-105 days to bloom. The florets are saffron red (40B) with thin whitish stripes on the throat of two oppositely placed lower tepals. Spikes are more than 93.00 cm in length with good rachis length (55.00 cm) and about 19-21 number of florets per spike. The inter-floret length is less which results in compact arrangement of florets on all sides of the spike. About 5-6 florets remain open at a time which makes it excellent for vase decoration, bouquet preparation, floral arrangements and for garden display.

***Pusa Red Valentine*:** A selection among the open pollinated population of the variety 'Regency'. Plants are healthy, green in colour and straight reaching a height of 125.00 cm. Each corm produces on an average two shoots with 7-8 leaves. A mid season variety bloom in about 95 days. Spikes are straight and long with good rachis length (50-55 cm) and close arrangement of 18-19 florets on spike. Florets are brick or bloodred (33B) in colour with sun ray like small lines on the lower tepals which makes them more attractive and very good for vase decoration as well as for bouquet preparation during valentine day. This variety produces on an average 2.33 corms and more than 28 cormels from each mother corms.

***Pusa Vidushi*:** A selection among the progeny obtained from the cross of Melody x Berlew. The plants are healthy, light green in colour and straight. An early and mid-maturing variety, first floret opens 80-85 days after planting. Spikes are straight with good rachis length and about 15-16 number of florets per spike. Florets are purplish white in colour with grayed purple spots on the base of the throat. Compact arrangement of florets on one side of the spike coupled with 5-6 florets remaining open at a time makes the spikes excellent for vase decoration as well as for bouquet preparation and floral arrangements.

***Pusa Sindoori*:** It is a selection from the open pollinated seedlings of the variety 'Little Fawn". The florets base colour is bright red (44c as per R.H.S. colour chart). Two yellowish spots on base of inner tepals with red coloured rainbow type stripe on throat add novelty in colour and make it more attractive. This is mid season variety (105.22 days) with robust and compact spikes. The variety produces long spike having more number of florets per spike. In addition, it is a very good multiplier on an average producing 2.88 corms and 47.77 cormels per corm which makes it more suitable for commercialization.

***Pusa Shanti*:** It is a selection among the progeny obtained from the cross of Yellow Stone x Melody, Yellow Stone produces 15-17 florets in yellow-white group 158 C and inner two tepals are in yellow-orange group 14C, leaves are green group 138 C; whereas the Melody produces 12-15 florets in red group 41A. The number of florets remain open at a time are very less in both the parents (4-5).

Pusa Kiran

Pusa

Shubham

Pusa Red Valentine

Pusa Vidushi

Pusa Srijana

Pusa Manmohak

Pusa Sindoori

Pusa Shanti

Recently released gladiolus varieties

Production Technology

The efforts have been made by the division to develop various technologies *viz.* dwarfing in bougainvillea using growth retardants like. Paclobutrazol (PBZ), daminozide, cycocel (CCC), ancymidol and uniconazole were used with different concentrations to induce dwarfing plant architecture in bougainvillea. On the basis of study, it was concluded that in order to obtain potted bougainvillea for var. Parthasarthy for rigorous and recurrent bloom Paclobutrazol (PBZ) at 20 ppm is effective. As there is a gradual reduction in vegetative growth which cause dwarf growth and increasing flower production and these components are most suitable for pot production of bougainvillea mostly in major metro-cities particularly for high rise building where different small potted plants of coloured bougainvillea can be placed to decorate balcony. The cleft grafting technology in bougainvillea has also been standardized and this technique is being used for grafting to develop high value product of multiple colourful bougainvillea varieties on a single plant.

Different turf species were evaluated for different seasons and soil type and paspalum and centipede grass found good to tolerate salinity and Tift dwarf 419 performed comparatively better in lawn than other others under shaded conditions.

Vitamins colour shade net and growing media were used in LA hybrid lily. A red colour 50% shade net was found more suitable for vegetative growth, quality flowers and bulb production in the media containing cocopeat+soil (1:1) for production of flower, bulb and bulblet production. NPK (19:19:19) spray of 0.5 % with GA_3 @ 550 ppm performed better in most of the vegetative growth parameter, better quality flower, bulb & bulblet production.

Value-addition Through Dry Flowers

Most demanding and valuable segment of floriculture is floral value addition. Value- added products such as dry flowers, pot pourris, essential oil extraction, aromatherapy, pharmaceutical and nutraceutical, dyes, petal embedded handmade paper, etc. is an important sector which provides a great scope for the Indian entrepreneurs. Division has standardized different drying techniques for roses, chrysanthemum, gerbera, annual flowers such as annual chrysanthemum, marigold, larkspur, sweet sultan, pansy, etc. for preparation of value-added products to empower the women and signed MOU with M/s Floral Images, owned by Mrs. Poonam Quamra. The flower drying techniques are very easy, simple which require little knowledge and experience. The technology is low cost as the drying techniques involved are air drying, press drying, solar drying etc. Once the person is trained, one can make very good quality products by using both cultivated and wild flora. Use of value added products made from dry flowers are eco-friendly as these are easily biodegradable and does not release any toxic compounds like VOC in the atmosphere.

Landscaping

Another integral segment of Floriculture industry is Landscaping which in very high demand now a days. It include arboriculture which is defined as systematic studies of trees and shrubs, their health is deteriorating due to old age plantations and developmental activities resulting to high pollution exposure due to vehicular emission and various kind of industrial smokes. All the plantation of Lutyen Delhi is 100 years old and it needs systematic replacement with the selection of native trees having functional values towards pollution mitigation along with their aesthetic look that is flowering and its architectural canopy. Earlier tree species were selected based on native origin and natural habitat. Tree mortality is high because of higher air pollution level, construction activities by means of physical damage, metro line, street-scaping and underground development activities of MTNL, Electric departments PWD, MCD, DDA and other developmental agencies causes' damage in root zone and trunk of trees. Recently, NGT has imposed the penalty and directives on all such developmental bodies for ten time compensatory plantations and third party tree audit has become mandatory.

The division of Floriculture and Landscaping is coordinating with local departmental bodies such as, DMRC, EDMC, SDMC, DJB and other local department of horticulture for tree auditing and vertical gardening especially metro pillars, central verge and green belt development. Technical expertise with IIT (Delhi) for green campus development using minimum energy, best use of green waste and water harvesting devices is the another initiative in this area. The air qualities across Delhi-NCR are some time very 'severe' during festival season like Dushera and Diwali, the AQI goes very higher than prescribed level up to 465 and even beyond which is hazardous to health. There is an urgent need of systematic information on effect of pollution, role of tree plantations and green cover in Delhi. The scientific studies are needed to classify the trees species for their functional qualities to mitigate the air pollution for ecological balance for healthy and sustainable environment. The Air Pollution Tolerant Index (APTI) of various tree species suited under Delhi conditions are very much needed to study for the selection of right kind tree to be planted and this type of experiments are in progress at IARI.

Teaching and practicing CAD is another thrust area under landscape gardening for the development of gardens and tree plantation in urban area. In many developing countries like Dubai, Singapore which was once upon a time just considered as port and business hub but today they are known for its world class urban landscape with systematic tree plantations. The students and faculty in NARS would be needs to be trained and guided for next generation to save our eco-system. Science led tree plantation, will reduce the amount of storm, water runoff, reduce erosion and sequestrations of carbon reducing overall impact of greenhouse gases and will provide sustainable environment for clean and green campus comforts.

Services

Service Sector and input supply is very prime area where, IARI is role model and provide all kinds of technical knowhow and inputs like seeds, planting material and ornamental plants like roses, chrysanthemum, bougainvillea, bulbs of SAUs, KVKs, ICAR Institutes and more particular to public in Delhi and NCR. All the varieties developed by this division are in the seed chain and quality breeder and certified (TL) seeds of marigold are being produced at institute HQ and its regional station, Karnal to meet the farmers demand.

References

Abad, M.J., P. Bermejo, S. Sanchez-Palomino, X. Chiriboga and L. Carrasco (1999). Antiviral activity of some South American medicinal plants. Phytotherapy Res. 13: 142-46.

Babita Singh, S. S. Sindhu, Ajay Arora & Harendra Yadav (2019). Evaluation of physiological and growing behavior of warm season turfgrass species against salinity stress Indian Journal of Agricultural Sciences 89 (3): 482–8

Govindaiah, S.B. Dandin, T. Philip and R.K. Datta (1993). Effect of marigold (*Tagetes patula*) intercropping against *Meloidogyne incognita* infecting mulberry. Indian J. Nematol. 21: 96-99.

Kumar, G., Singh, Kanwar P., Prasad, K. V., Rana, M. K., Namita and Panwar, S. (2014). Genetic diversity analysis of chrysanthemum (*Chrysanthemum gradiflorum*) cultivars using RAPD markers. Indian Journal of Agricultural Sciences. 84 (11) 1323-28.

Kumar, G., Singh, Kanwar P., Prasad, K. V., Verma, N., Namita and Panwar, S. (2015). Characterization of chrysanthemum (*Chrysanthemum grandiflorum*) varieties using ISSR markers. Indian Journal of Agricultural Sciences, 85(4): 566-70.

Mehra, T.S., Kalkame C.H. Momin, Tomar, K.S., Nilay Kumar and Pandey A.K. (2016). Performance of gladiolus (Gladiolus grandiflorus L.) cultivars under Pasighat conditions of Arunachal Pradesh. Journal of Ornamental Horticulture. 19 (1&2): 19-22.

Namita, Raju, D.V.S., Prasad, K. V., Singh, K. P. and Kumar, S. (2015). Standardization of protocol for *in vitro* multiplication of rose (*Rosa* × *hybrida*) cv. Happiness. Indian Journal of Agricultural Sciences. 85(11): 1513–7.

Panwar, S., Singh, K. P., Sonah, H., Deshmukh, R. Namita, Prasad, K.V. and Sharma, T. R. (2015). Molecular fingerprinting and assessment of genetic diversity in rose (*Rosa x hybrida*). Indian Journal of Biotechnology 14(4): 518-24.

Perich, M.J., C. Wells, W. Bertsch and K.E. Tredway (1995). Isolation of the insecticidal components of *Tagetes minuta* (Compositae) against mosquito larvae and adults. J. American Mosquito Control Assoc., 11: 307-10.

R. Rohith, S. S. Sindhu, Kishan Swaroop, Babita Singh (2021. Standardization of growth retardants for production of potted bougainvillea. Chem. Sci. Rev. Lett. 10 (38):214-20.

Swaroop, K., Singh, Kanwar Pal, Panwar, S., Namita and Sunita Dhakar. (2017). Advances in integrated nutrient management of bulbous flower crops — A review. Journal of Ornamental Horticulture. 20 (1&2): 1-20.

16

Entrepreneurial Floriculture A Profitable Venture

S.L. Chawla[1], Saryu Trivedi[1], Roshni Agnihotri[2] and Mallika Sindha[1]

ASPEE College of Horticulture, Navsari Agricultural University Navsari-396450.

[1]Dr. Rajendra Prasad Central Agricultural University, Pusa Samastipur, Bihar- 848 125

Pandemic brought in tough times, especially for different business sectors. However, the floriculture business was one such sector that people embraced, especially in India. In the current scenario, the floriculture industry in India has emerged as a huge trend setter and a survivor, albeit with a drop in earnings and the convenience of working from home. Flower Market report summarizes top key players overview as Syngenta, Floret Flowers, among others. The Indian economy has been starved of sustainable ideas, especially when the outbreak of COVID-19 saw the loss of jobs and the reverse migration of rural people. The floriculture business provided a vast opportunity that involves specific technical know-how and provided flower and allied business as a medium to earn income and generate opportunity for others. It performed as a most subtle sector in Indian agriculture. This business allows individuals to thrive in a competitive environment and earn immense profits and has lots of venues of novel innovative ideas. Floriculture business can be started anywhere in India with basic knowledge of flower farming.

Floriculture in India

Flower business in India or Floriculture is not only growing ornamental plants, cut flowers, foliage plants, potted plants but also turf grasses and their value addition in market. In today's age, flowering or ornamental plants are used for decoration or gifting purposes on a variety of occasions. They are also utilised as essential raw materials in oil extraction, pharmaceutical and perfume industries. Thus, the use of the agri-business in its marketing and production must be understood for starting a Floriculture business in India. Although flowers have been a vital component of Indian society and were cultivated for several purposes ranging from aesthetic to social as well as religious purposes, the commercial floriculture market has been of recent origin. Floriculture has risen as one of the most significant trades in Indian agriculture due to a rapid increase in the demand for cut and loose flowers. The metros and the major urban cities in India currently represent a major chunk of significant consumers of flowers in the boundaries of India. As a result of elevating urbanized life style and the impact of western cultures, people are buying flowers to draw the attention of one and all on several occasions such as festivals, anniversaries, Valentine's Day, birthdays, farewell parties, marriages, religious ceremonies, etc. The consumption of flowers is likely to increase further as trends of urbanization as well as the influence of modern culture are anticipated to get the much-needed boost in the forthcoming period. The IMARC Group expects the Floriculture market in India to showcase accelerating growth during 2021-26. While exports remain a vital motivator for cultivators, the demand for flowers within India's geographical boundaries is also increasing rapidly, especially in the metros and larger cities. Besides aesthetic and decorative purposes, a significant amount of flower consumption is bound to happen in industrial applications. It includes fragrances, flavours, natural colour, medicines, etc. The consumption of these products would witness an upward trend during the forecast period leaving a positive impact on the floriculture industry and lure more floriculture business enthusiasts to come forward and commence the Floriculture business in India.

Floriculture Includes

Cacti and succulents, Perennial, annual, or biennial ornamental plants, Shrubs, Trees, Orchids, Bamboos, Bromeliads, Bamboo grass, ornamental grass and lawn grasses, Bulbous plants, Foliage, House plants, Climbers, Bedding plants, Cycads, Palms, Loose flowers, cut flowers and seeds, Ornamental plants, Ferns, Fillers, Dried plant and flower parts, Plants used in food decoration, edible pigments, oil extraction, essential oils, etc. Among which major aspects are described as below:

Nursery

A lucrative business where individuals have a wide range of nurseries based on wholesale or retail, indoor/shade loving plants, tissue culture plants, commercial flowering plants, shrubs, climbers and tree seedling for landscaping, annual plants, bulbous flowers, etc. are flourishing well throughout the country. Besides, there is a high demand of lawn species for landscaping as well as plant rental services in the nurseries.

The nursery business is highly competitive, and it is important to maintain a high level of customer service, offer quality products, and give the consumer a wide variety of choices when shopping for plants and trees. Our approach will be to offer a diverse selection of plants, trees, and garden supplies. Exceptional customer service will be important in meeting the needs of our target markets. Healthy plants will be a top priority, and we will display the plants artistically.

Plants are a commodity that will always have a market as people consider them to be very important to have around their residences and we will have to provide top quality service to be succeeded.It consists of many aspects *viz.,* flower nursery, vegetable nursery, fruit nursery, medicinal nursery.

Wholesale Nurseries

Setting up wholesale nurseries consists of several decision-making strategies, which must be made before things like what are you going to grow, who the customers are going to be, and your plans on delivering the flower plants to the customers. For instance, let's suppose that your largest customers are going to be landscapers, in this case, it more than likely means that you will need to grow plenty field grown plants on the larger scale and the plants and flowers for the summer will need to be plenty of container grown for transplanting. Some of the major nursery raising pockets is Pune, Mumbai (Maharashtra), Kadiyam (Andhra Pradesh), Bangalore (Karnataka), Agra (Uttar Pradesh), Thrissur (Kerala) *etc.*

Retail Nurseries

It is not hard to guess who retail nurseries sell too, of course; they sell their plants to their retail customers. There are a few retails garden centres who will grow many of their plants themselves, however, most of them will purchase the plants from a wholesale nursery too resell to their customers. It is important that retail nurseries stay focused on their customer's needs, this will let them know just what it is they should be selling. The retail nurseries high end or higher profit items includes plants in the larger sizes, and accessories, such as window boxes, garden tools, and birdbaths, among many other accessories. The retail nurseries can make some of their profit from creating custom planters and replanting using decorative containers.

Mail-order Nurseries

Selling their products through the mail are called as mail-order nurseries. Many are still growing their own plants, while others are purchasing from the wholesalers to resell. More than any of the other nursery businesses, when in the mail-order nursery business, it is significant to be able to define your customer base. The nurseries that specialize in growing plants that are unique, and plants that are a specialty, which they can ship anywhere they want will have the highest profit margin, this is because the mail-order nurseries selling their plants to be picked up locally at a retail nursery are going to be less popular.

Online Nursery/One Stop Store

There are many websites which sell plants/seeds/pots and other flower products online like nurserylive, ferns and petals, Ugaoo, Flower aura where they sell all flower indoor décor products and seeds online. These markets are markets of future and really good avenue of investment now a days.

Value-addition

Value addition in flowers is a good source of income by self-employment which includes many components described as under:

Floristry: It includes fresh flower products from cut flower arrangements like bouquets, baskets, bunch, boutonniere, corsage, etc. and secondly loose flower products like garland, floral strings, pomanders, wreaths, floral jewellery. Floristry is an art of designing and arranging flowers aesthetically in a vase or bowl retaining their freshness for a longer time. It consists of floral arrangements in various occasions like weddings, birthdays, parties and any other social events by making bouquet, decorating stage, banquet hall, car, etc. with different styles of floral crafts and arrangements. Floristry business has an immense potential for self employment generation on small as well as large scale to obtain maximum returns in a short period.

Dry flowers: Dry flower value added products includes dried flower arrangements, products of press dried flowers and pot pourries, etc. Dry flowers contribute around 70 per cent revenue of total floriculture export. Unfortunately, very little importance has been given to this sector, which has tremendous export potential. Our country has 10 per cent share of the total global dry flower market. USA, UK, the Netherlands, Germany, Italy, *etc*. are major export destinations for India. Tuticorin (Tamil Nadu), Calcutta (West Bengal) and Mumbai (Maharashtra) are major centres of dry flower industry in India.

Processed products: These include essential oils, absolutes, concrete, petal jam (rose, rhododendron),gulkand, jelly, ready to serve beverages, wine, floral tea, rose hip juice, poultry feed, insect repellent, floral dyes, petal embedded handmade paper, cosmetics like calendula cream, rose water, rose cream, etc. Natural dyes, colours and pot pourries are also value-added products of flowers. Natural dyes make from marigold, hibiscus, Bixa, *Butea monosperma, Calendula, Chrysanthemum, Bougainvillea* etc. are substitute for the synthetic colours which are harmful for human health.

Protected Cultivation

The protected cultivation is technique of crop cultivation when the micro climate of the plant body is controlled partially or fully as per requirement of plant growth. The different designs of greenhouses developed for crop production varied from Hi-tech to shade net. In floriculture, India is emerging country but floriculture crops have high tendency for export. The demand of flowers has been increased as now day's people like to celebrate their occasions with flowers. The export of cut flowers is emerging as the commercial flower productions under the protected conditions have emerged in India. The floriculture crops have high potential for per unit area than the other field crops so it is more profitable business. The floricultural commodities for export consist cut flowers, pot plants, cut foliage, seeds bulbs, tubers, rooted cuttings and dried flowers or leaves. The demand of cut flowers is increasing in domestic and international market. The important cut flower in the international trade is rose, carnation, chrysanthemum, gerbera, gypsophila, orchids, anthurium and lilies. To meet the demand floriculture crops like rose, gerbera, carnation, lilium, *etc.* are grown in green house for quality flower production. Flowers other than the major cut flowers are referred to as specialty flowers or high value flowers *i.e.,* Asiatic ginger lily, protea, heliconia, orchids, bird of paradise, alpinia, etc. are high value flower crops cultivated in comparatively smaller area but these fetch higher returns per unit area owing to high value. Intensive cultivation of specialty flower crops under suitable agro climatic conditions have potential to increase the income of farmers. Western coastal area is highly suitable for protected cultivation due to high humidity and other favourable climatic conditions.

Cut greens or cut foliages are used as fillers along with cut flowers in floral arrangements to increase aesthetic value are in great demand. Asparagus, ferns, dracaena, cycads, caladium, *Murraya exotica,* thuja, *etc.* cut foliages are in great demand due to their attractive form, colour and freshness.

Seed Production

Seed production of annuals is an important component in landscape design and indoor gardening. Since it has short duration and colourful flowers, it is utilized more in landscaping to change the colour pattern of the gardens according to the season and need. Most of the annuals are propagated only through seeds. Due to the demand in flowering annuals in the world floricultural trade, the seed market of the same is too become a vital sector. Hybrid seeds of these plants have high vigorous and superior quality. The production of hybrid seeds for export market is also in essential. Ludhiana (Punjab), Bangalore (Karnataka), *etc.* are major contributing states in seed production.

In India, summer, rainy and winter season flowering annuals are available. Summer and rainy seasonal annuals in India are available in limited number whereas winter annuals arerich in kinds. Thus, for commercial seed production, winter annuals need more importance. Ideal climate condition for seed production is long duration of cool and dry season which helps in good seed setting of bold size. While excessive hot and dry season hampers seed setting of summer annuals in north Indian plains. Excessive rain at the time of flowering washes away pollen grains resulting in poor seed set. However, flower seed production is labour intensive. According to climatic requirements, the production of flower seed is divided in following climatic zones. 1. Mild climate (Kashmir Valley, Kullu Valley etc.): Delphinium, Giant Pansy, Zinnia, etc. 2. Sub-Topical area: Antirrhinum, Anchusa, Ageratum, Calendula, Brachycome, Lineria, Californian poppy, Candytuft, Carnation, Dianthus, Daisy, Dimorphotheca, Nasturtium, Petunia, Portulaca, etc. 3. Tropical: Tagetes, Salvia, Ipomea, etc.

Initial work on hybrid seed production in ornamentals was started by M/S Indo American Hybrid Seeds (India) Pvt. Ltd., Bangalore. The company started producing F_1 hybrid seeds of Petunia for 100% export during mid-sixties. Production of seeds of open pollinated flower crops was started by M/S Beauscape Farms, Sangrur, Punjab who started flower seed production involving farmers on large scale. Now many companies have started producing seed on large scale for export to Holland, UK, USA, France, Germany, and Japan etc.

Essential Oil and Perfumery

The scented oil obtained from natural sources is called essential oil. An essential oil may be defined as a volatile perfumery material derived from a single source of plant or animal origin, which has been separated from that source by a physical process. There are about 2000 plants in India which are known to be aromatic plant. Out of which 65 plant species have demand in the world market. Essential oils from aromatic plants are low volume and of high value. They have a longer shelf life at room temperature than horticultural plants. Some of them can be grown in marginal lands through contract farming. The world production and consumption of essential oils and perfumes are increasing very fast. Essential oils are used in a wide variety of consumer goods such as detergents, soaps, refreshment products, cosmetics, pharmaceuticals, perfumes, confectionery food products, soft drinks, distilled alcoholic beverages (hard drinks) and insecticides. Rose (bourbon rose and damask rose), jasmine (chameli, mogra, juhi), tuberose, carnation and champaka are widely used flower crops for making essential oils.

Manufacturers use essential oils to create a range of products. The cosmetic and makeup industry use essential oils to create perfumes, add fragrance to creams and body washes, and even as sources of natural antioxidants in some beauty care products.Many natural medicine practitioners, such as aromatherapists, use essential oils. Aromatherapy involves diffusing these essential oils into the air.

Turfgrass Production

Sprawling, manicured lawns in neighbourhood parks, golf clubs, hospitals, and the airport seem like oasis of greenery. Most landscape clients request green lawns as they are influenced by gardens they have seen in Europe or USA. There are food and non-food produce demanded by the city, from farms in its peripheries. Of the non-food demand, ornamental plants for landscaping actually help enterprising farmers with resources to earn a livelihood. Lawn grass has major demand from Bengaluru and other south Indian cities. It requires capital, machinery, hired labour, water and chemicals.

Lawns can be established by planting grass slips. But this takes time to establish and spread a green cover uniformly over the soil. There are different grass species requiring different levels of care and inputs to maintain uniform green cover throughout the year. Urban impatience, affluence ensures demand for quick green carpeting. Thus, transplanting readymade mats of grass grown elsewhere is the norm followed in most lawns we see."

The average size of lawn farms range between five and 10 acres. Turfgrass farms require an initial investment of around ` 1.5-2 lakh an acre, including land lease charges if any, setting up irrigation facility, buying equipment, etc. On the whole, the cost of cultivating a square foot of lawn grass (including soil, inputs and labour) is about five rupees. From Andhra Pradesh, Telangana, Maharashtra and Uttar Pradesh, turf pieces harvested with soil, for Rs. 15-20 per square foot. Buyers from the neighboring states visit lawn farms and do on-the-spot purchases in large volumes. A farmer receives nearly Rs. 1.5-2.0 lakhs an acre as net income, after every harvest, once in four months for lawn grass cultivation. Labour from surrounding villages is employed for turfgrass cultivation. Activities in these farms continue round-the-year, in cycles of 3-4 months duration. Women are, in fact, employed in large numbers and lawn grass cultivation is their most lucrative and stable job as compared to other jobs.

The turf industry has recently got a boost due to new specialized parks, golf courses and stadiums for sports events in different locations of the country. The corporate houses are evincing their commitment to mitigate the pollution by resorting to landscape the corporate premises and their factories. The turfgrass industry consists of many diverse beneficiaries including millions of homeowners, athletic field managers, lawn care operators, golf course superintendents, architects, developers and owners, landscape designers and contractors, seed and sod producers, parks and grounds superintendents, roadside and vegetation managers and cemetery managers.

Urban Landscape Designing and Maintenance Contracting

Landscape contracting is a profession that involves the art and technology of landscape and garden project planning, construction and landscape management, and maintenance and gardening; for garden aesthetics, human enjoyment and safety, and ecosystem-plant community sustainability. A landscape designer designs attractive and functional outdoor spaces for various spaces. The work consists of installing and maintaining landscaped and natural spaces. This too is an upcoming remunerative non-traditional business venture for youngsters. Vertical garden and Green roof making are one of the upcoming specialized facets of this business.

Bonsai

Bonsai are ordinary trees and shrubs that have been trained in pots to grow into naturally beautiful shapes. As a result of urban growth, and more people moving to the city where their yards are small, the bonsai tree business has boomed. Today the art of bonsai is a well-known and that is practiced worldwide. Many ornamental plants can be converted into a miniature form with specific designs and sold at high prices according to its age. This trend is blooming in big cities to beautify the houses. There is a great appeal in owning tiny bonsai trees, which might be only 18" high instead of 60' high. Business owners can be very successful since both local sales and Internet sales are becoming extremely popular. The cheapest bonsai plants at the nursery cost Rs 800, the most expensive vertical penjing plant is priced at ` 2.5 lakh

Terrarium and Aquascaping

Terrariums developed as a miniature landscape are an interior-scaping masterpiece, made in just about any glass container. It is planted to look like a miniature garden or forest enclosed in its own little world depending solely on style or type. For people who live in cities and have minimal or no garden space, their thirst for a small touch of nature has been quenched by terrarium. As researched there are not many people involved in this business but a few community of hobbyist turned into entrepreneurs but the demand of this seemed very high especially in times of pandemic when people are stuck indoors for more than 1.5 year worldwide. If seized properly, terrarium making would prove to be good export commodity along with highly profitable venture.

A riparium is an aquarium with marginal plants planted in the background. Marginal plants are the plants that grow on the banks of a waterbody. In riparium, it is possible to keep plants with a wider range of fishes which were otherwise not compatible with plants, Aqua Terrariums are more theatrical in creations. Focal point could be anything from winding driftwood to a textured rock or a beautiful focal plant. A sufficient bioload of fishes might take care of the nutrient needs of some plants, but for others may need to be fertilized, especially if they are showing deficiency. This too is a wonderful business opportunity which will cater the living rooms of metro cities and small cities alike.

Consultancy Services

Consultancy service business too is sprawling in the big cities as the horticultural growth is sky rocketing and there is dearth of subject matter experts. Now-a-days, consultant visit service can also be made available online. Consultancy regarding landscaping and protected floriculture can be a good money-spinner business.

Planter/Pot Renting Services

These services are gaining lots of popularity across the corporate world as they wish to be tension free about maintenance and nursery raising but plants added to their indoors adds grace to the offices. Although not very popular in India yet, this business has very good possibility in near future especially in metro cities.

Pot plants either foliage or flowering plants, are considerable commercial trade for instantaneous gardening for both outdoor as well as indoor decoration. Poinsettia, phalaenopsis orchid, aglaonema, dracaena, begonias, cyclamens, money plant, fuchsia, geranium, kalanchoe, ficus, coleus, caladium, dracaena, dieffenbachia, pedilanthus, palms, cycads, ferns, crotons, cactus, etc. are some of the important pot plants. Large scale production of seedlings of many annual flower crops are in demand for the corporate and public landscaping for beautification. It has also opened up a newer avenue of plant rentals for interior decoration in corporate houses.

Flower Waste Management

Huge loads of floral-waste are collected from religious places, mainly in Uttar Pradesh. The waste is then sorted, dried and pulverised to make flower powder, which is then mixed with raw materials to make 'raw agarbattis'. By using pioneering flower recycling technology and establishing a first of a kind Circular Economic model in the country, HelpUsGreen are the leaders in starting the movement to clean the rivers and the places of worship and have opened a new venture in this field.

Pharmaceutical and Nutraceutical Compounds

Ornamental plants are rich in natural pigments that include anthocyanins, carotenoids and betalains which are pharmacologically valuable compounds that are known as nutraceutical compounds, which have a wide range of applications in healthcare as medicine and dietary supplements. Marigold flowers are rich in carotinoids, lutein and zeaxanthin. Lutein and zeaxanthin are essential nutrition for healthy eyes and vision. For the preparation of dietary supplement products, lutein is isolated from marigold flower petals, and includes about 5% of zeaxanthin. Lutein is also effective in preventing or controlling free radical generation and oxidation damage associated with cancer, coronary heart disease, cataract and age-related macular degeneration. Carotenoids pigments are used for intensification of yellow colour of egg yolks and broiler flesh. Rose hips are rich source of Vitamin C, riboflavin, pectin, nicotinic acid, and malic acid which has been used for the prevention and cure of certain cancers. Rose tea prepared from rose petals contain vitamins and other compounds like astringent and tannin which can help in reduction of stomach problems and can also be used to control bleeding. Besides rose, marigold and calendula flowers are rich source of lutein which prevents blindness.

Anthocyanins, flavonoids, carotenoids and xanthophyll pigments are also used in food, beverage and confectionery industries and act as anti-oxidants. Besides pharmaceutical use of pigments natural dyes can prepared from some flowers such as marigold, hibiscus, Bixa, *Butea monosperma*, etc. are substitute for the synthetic colours which are harmful for human health. These dyes have great demand in textile industry.

Allied-Industries

Allied-industries associated with floriculture like green house construction and installation, shade net manufacturing, tissue culture laboratories, grading - packaging machines, packaging materials for post harvest handling of flowers, etc. have come up and flourishing well due to high demand with the expanding area under high tech floriculture

Women Empowerment

Women empowerment is not a sufficient condition, it is still a necessary condition in order to stabilize and in turn to have sustainability of the development process. Women empowerment as a matter ofkey concern in national and international policy making and activities of social life. Nations cannot achieve their development goals if their women are discriminated. Floriculture is a labour-intensive activity, wage payment forming roughly one-third of the costs of production. It indicates, that even a modest floriculture programme can generate millions of jobs, predominately for young men and women, quite apart from significantly, contributing to national income. By and large, women are engaged in cultivation, harvesting and post harvesting activities including packaging, while men perform activities linked to pre-cultivation, maintenance of nurseries, irrigation and fumigation since these involve hard work and health and safety considerations.

References

Patel, N.L., Chawla, S.L. and Ahlawat, T.R. 2015. Commercial Floriculture. Published by NIPA, New Delhi. 561 p.

Patil, A.A., Mantur S.M., Mannikeri, I.M. and Biradar, M.S. 2009. Protected Cultivation. Winter School Compendium, UAS, Dharwad.

Reddy, B.S.S., Janakiram T., Kulkarni B.S. and Misra R.L. 2007 High–Tech Floriculture. Published by Indian Society of Ornamental Horticulture, New Delhi.

Sumangala, H.P., Kumar, P., Sindhu, S.S. and Jamakiram, T. 2015 Commercial Floriculture and Landscape Gardening for Urban and Peri-Urban Horticulture. Confederation of Horticulture Associations of India (CHAI).

Section 04: Tuber and Plantation Crops

17

Breeding Strategies for Development of Bio-fortified Tuber Crop Varieties

Kalidas Pati[1] Visalakshi Chandra[2], K.M. Senthil Kumar[2]

[1]Regional Centre, ICAR- Central Tuber Crops Research Institute Bhubaneswar-751019, Odisha, India
[2]ICAR- Central Tuber Crops Research Institute, Trivandrum-695 017 Kerala, India

An emerging, promising, affordable, and sustainable method of providing micronutrients to a large population with limited access to varied meals and other micronutrient therapies is referred to as biofortification. The biofortified tuber crops, particularly sweet potato, cassava, yam, and taro, are giving the targeted people adequate amount of nutrients. When genetic diversity is present in a gene pool of targeted crop, biofortification through conventional breeding approaches has been most successful. The breeding strategies used to elevate micronutrients include pre-breeding, large-scale germplasm screening, molecular marker development and understanding genetic control and Genotype x Environment interaction (GxE). Transgenic approach is one of the appropriate techniques to develop biofortified tuber crops with required nutritional and agronomic characteristics which do not naturally contain the target nutrient in the gene bank. Additionally, biofortification utilising traditional breeding methods was not subject to rules, but when using contemporary biotechnological methods, the proper regulatory was necessary.

Biofortification is a process of increasing the amount of vitamins and minerals in a crop through plant breeding, transgenic methods or agronomic practice. Regular use of bio fortified staple crops will result in measurable benefits in human nutrition and health (Bouis and Saltzman, 2017). One or more of the staple crops represent the main source of food for a sizeable section of the population in developing countries, and these crops are the focus of biofortification efforts using both traditional breeding and contemporary biotechnology techniques. The prevalence of micronutrient deficiencies will significantly decline with increased production of foods high in micronutrients and improved dietary diversity. Consuming biofortified crops in the near future can assist in addressing micronutrient deficiencies by boosting daily adequacy of micronutrient intakes among people throughout the lifecycle (Bouis *et al.*, 2011). The breeding programme for development of biofortified varieties includes germplasm screening, creation of micronutrient-rich germplasm, pre-breeding parental genotypes, genetic research, applying rapid generation advance through "speed breeding," and developing and utilizing molecular markers and genomic selection strategies to lower costs, expedite the procedure, and boost rates of genetic gain. Once promising lines or hybrids have been created, they are evaluated for genotype x environment interaction (GxE), or how the growing environment affects the levels of micronutrients, agronomic performance, and end-use attributes of the studied varieties or hybrids. (Bouis and Saltzman, 2017). A number of orange sweet potato varieties with a n high vitamin A content have been produced and distributed by HarvestPlus and the International Potato Centre (CIP). Six varieties (Ejumula, Kakamega, Vita, Kabode, Naspot 120, and Naspot 130) and three in Zambia have been made available (Twatasha, Kokota, and Chiwoko). (Garg *et al.*, 2018). ICAR-Central Tuber Crops Research Institute (ICAR-CTCRI), research and development works resulted in the development of high yielding carotene rich orange fleshed sweet potato variety like Bhu Sona, Bhu Kanti and Bhu Ja. Carotene content of such genotypes is found to vary in the range of 4.8-14.0mg/100g (Pati *et al.*, 2021). Anthocyanin is abundant in purple fleshed sweet potatoes variety Bhu-Krishna (Bansode *et al.*, 2020). Orange flesh sweet potato variety (Bhu Sona, Bhu Kanti and Bhu Ja) are suitable for cultivation in the coastal region of Odisha and West Bengal due to salt tolerance nature (Mukherjee *et al.*, 2015).

Conventional Breeding

Without sacrificing yield or farmer-preferred agronomic features, plant breeding can improve nutrient levels in staple crops to target levels required for enhancing human nutrition. In order to develop new crops, it is necessary to screen germplasm for genetic diversity, develop and test micronutrient-dense germplasm, perform genetic research, and create molecular markers to speed up and reduce the cost of breeding (Bouis and Saltzman., 2017). Crop improvement initiatives for biofortification initially focused on evaluating diversity in the largely unimproved germplasm in core collections in gene banks, including wild relative species and unimproved stocks like landraces, as well as in the genetic diversity for iron, zinc, and carotenoids in the germplasm in active breeding programmes. Agronomic and end-use characteristics are described concurrently, in field evaluation or during future screening, as varieties must provide farmers both strong crop yields and marketing alternatives. There are no legal barriers and traditional breeding is widely recognized. The laws governing transgenic crops vary between nations. Instead of using a transgenic strategy, conventional breeding has been used to create biofortified cultivars for the Harvest Plus programme (Bouis and Saltzman, 2017).

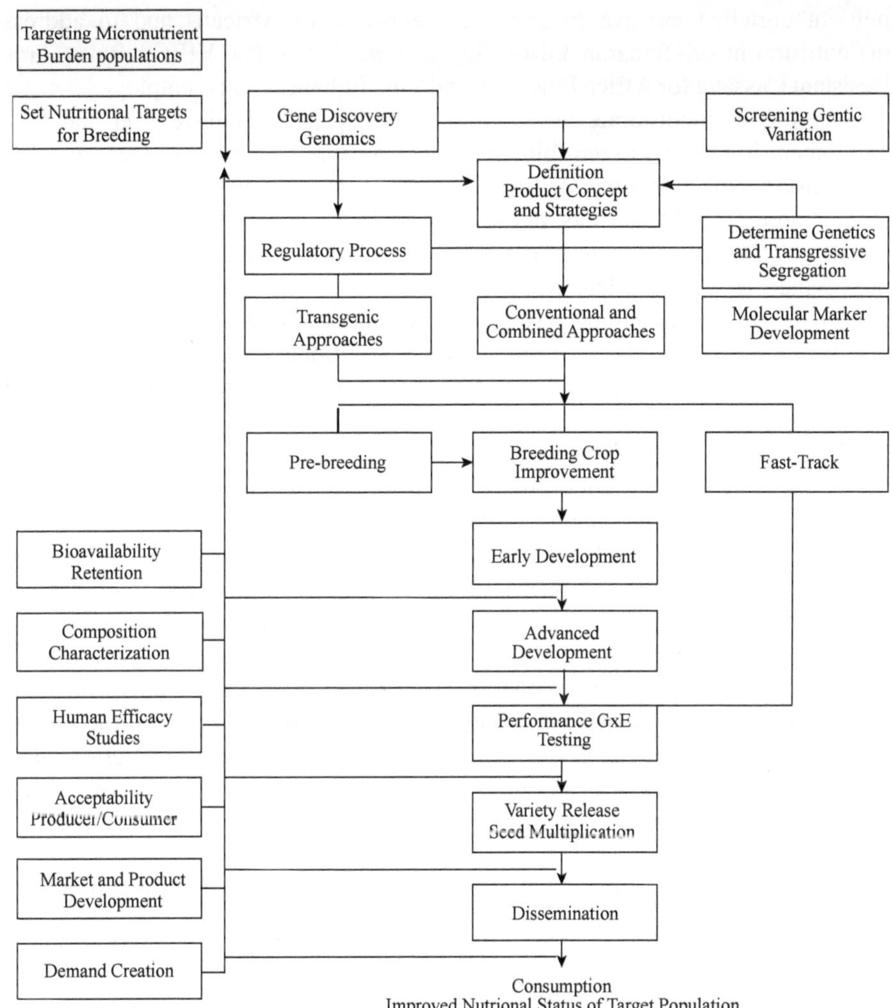

Fig. 1: Crop Development Framework (Bouis and Saltzman 2017).

Cassava Biofortification

Cassava is an excellent source of calories (1800 calories) but low in iron and zinc content providing less than the minimum daily requirement (MDR) and only 5–8% and 13–14% of the Estimated Average Requirement (EAR), respectively, for children 2–5 years old (Chavez *et al.*, 2000; Abah *et al.*, 2015; Gegios *et al.*, 2010, Stephenson *et al.*, 2010). There is insufficient variability for iron (6–10 µg/g DW) and zinc (7–10 µg/g DW) in cassava and therefore fortification by conventional breeding efforts is improbable (Chavez *et al.*, 2000).The BioCassava Plus program, 2005 of Bill and Melinda Gates Foundation aimed at developing nutrient enriched cassava to improve the health of Africans and to address malnutrition in sub-Saharan Africa (Sayre *et al*., 2011). The VIRCA Plus (Virus Resistant Cassava for Africa Plus Iron And Zinc Enhancements) employed genetic engineering to incorporate genes from Arabidopsis to achieve iron and zinc enrichment in cassava to ten-fold and higher bioavailability than other varieties. The VIRCA plus also aims to incorporate this trait via conventional breeding in varieties preferred by farmers and consumers in Nigeria and other West African countries (Nutrition Enhancement - Cassava Plus). HarvestPlus also aims at improving the micronutrient content of cassava via exploration of diversity for micronutrient content among germplasm, exploitation of heterosis, understanding genotype x environment (GxE) and release of nutritionally enhanced cassava after multi-location testing over multiple seasons (Bouis and Saltzman, 2017).

Vitamin A rich cassava

Vitamin A is a micronutrient essential for normal development and functioning of the human body and the vitamin A deficiency (VAD) affects 33.3% of preschool-age children across the globe and 13% of the women of childbearing age in Africa Among the other carotenoids such as α-carotene, β-carotene, and β-cryptoxanthin which are precursors of vitamin A, cassava is rich in β-carotene occurring in trans and cis forms accounting to 54-77% of the total carotenoid (Carvalho *et al*., 2016). The yellow flesh colour of cassava roots is attributed to the total carotenoids content and a strong positive correlation has been reported between the yellow flesh colour and carotenoids (Akinwale *et al*., 2010; Sánchez *et al*., 2014). The average carotene content of the roots ranges from 0.81 to 0.89. The carotenoid content of deep coloured roots is higher than white or dull coloured roots of cassava (Chavez *et al*., 2000).

Natural variation for high carotenoid accumulation was observed in landraces of cassava from the center of origin and Brazil and indigenous clones of cassava grown by natives of amazon region (Nassar *et al*., 2005). The indigenous cultivar Amarelinha do Amapá showed a very high content of α-carotene, β-carotene, and both cis- and trans-β-carotene, equivalent to 27 mg/100 g greater than that of human daily nutritional needs. Also, this variety has 40 % higher protein content than the white cassava storage root and therefore can be used in cassava breeding to address to the nutritional challenges in cassava since cassava has low levels of fat, protein and micronutrients that are essential for normal growth, eyesight and cognitive development in spite of being a rich source of energy (Carvalho *et al*., 2004; Black *et al*., 2013).

Genetic diversity present within the germplasm have been used to generate biofortified cassava and maize using conventional breeding approaches by various initiatives, including Harvest plus. 100% of daily Vitamin A needs for women and children are provided by the Vitamin A cassava developed by Harvest Plus (Pfeiffer and McClafferty 2007). Biofortification efforts in cassava has progressed significantly increasing the original concentration of carotenoids in cassava roots and provide up to 40% of the daily recommended vitamin A intake for children (Ceballos *et al.*, 2012, De Moura *et al.*, 2015). La Frano *et al.*, 2013 reported that consumption of biofortified cassava roots increased the concentration of β-carotene and retinyl palmitate triacylglycerol-rich lipoprotein plasma in healthy well-nourished adult women. For a successful breeding project to increase carotenoids content and realise significant genetic gains in cassava, it is essential to phenotype the large number of segregating progenies through spectrophotometer and/or HPLC, by selecting the best parents and making new crosses through recurrent selection breeding scheme (Sánchez *et al.*, 2014).

Carotene Rich Sweet Potato

Sweet potato is one of the major tropical tuber crops rich in various phytochemicals, anthocyanins, vitamin C, carbohydrates, potassium, and dietary fibre (Teow *et al.*, 2007). Huang *et al.*, 1999 analysed the β-carotene of eighteen cultivars of sweet potato grown in Hawaii and reported 0.1 to 0.6 mg per 100 grams fresh weight. Vimala *et al.*, 2011 reported higher ratio of β carotene to total carotenoids (78.56-94.23%). Orange fleshed sweet potato (OFSP) is the most successful example of biofortification of a staple crop (Jongstra *et al.*, 2020) and has improved vitamin A intake and serum retinol concentrations in young children in rural Mozambique (Low *et al., 2007*). HarvestPlus and International Potato Centre (CIP) have developed and released several varieties (Ejumula, Kakamega, Vita, Kabode, Naspot 12O, and Naspot 13O, Twatasha, Kokota, and Chiwoko) of orange sweet potato with high vitamin A. Zambia Agriculture Research Institute has developed 15 new varieties of vitamin A fortified sweet potatoes. The significant nutrient and health improvement achieved in Sub Saharan Africa (SSA) via HarvestPlus orange sweet potato consumption was duly recognised and honoured with World Food Prize-2016.

Sweet potato has potential genetic variation for β-carotene (BC) with low genotype × environment (G × E) interactions and significant selection response (Grüneberg *et al.*, 2015). Andrade et al.2017 reported that out of the 64 high-yielding OFSP breeding clones selected and evaluated using accelerated breeding scheme (ABS), the genotype × environment interaction was highly significant for storage root and vine yields, dry matter (DM) and BC content. BC content, iron and zinc ranged from 5·9 to 38·4, 1·6 to 2·1 and 1·1 to 1·5 mg/100 g dry weight, respectively.

The International Potato Center (CIP), in conjunction with the national sweet potato breeding programme in Mozambique, employed Accelerated Breeding Scheme (ABS) to develop OFSP cultivars rich in BC content in the shortest period compared to conventional methods which require 7–9 years (Grüneberg *et al.*, 2005). Significant heterosis was achieved applying reciprocal recurrent selection (RRS), a cyclic breeding procedure in sweet potato since 2010 at the International Potato Center (CIP) in Peru (Grüneberg *et al.*, 2009; David *et al.*, 2018).

Biofortification in Yams

Yams are potential food source with high levels of carbohydrates including fiber, starch and sugar, proteins, lipids, vitamins, minerals, bioactive compounds such as flavonoid, diosgenin and dioscorin, tannin, saponin and total phenols (Kumar *et al*., 2017). The carotene content in Yams ranges from traces to low concentrations (96.3–326 µg/ 100 g DW) and is lower than sweet potato and cassava. Zeaxanthin is only found in all yams except *D. Pentaphylla* (Champagne *et al*., 2010). *D. Cayennensis,* the yellow variety is reported to have higher carotenoid content (Bhattacharjee *et al*., 2011). Although variability for carotenoids are present in yams, biofortification for provitamin A content of *Dioscorea* in early stages behind that of cassava and sweet potato (Price *et al*., 2018). The complex genetic background, lack of genomic information and genetic resources of yams hinders genetic improvement of yams via breeding.

Biofortification in Taro

Lack of information is a serious constraint to biofortification efforts through breeding in taro. Variability for lack of acridity, high starch and low soluble sugar, cellulose, fiber contents, secondary metabolites was reported by Lebot *et al*., 2004 and Champagne *et al*., 2009. Wide diversity for 4 anthocyanins, phenolic compounds, 20 flavonols, 9 flavanols and 2 phenolic acids was observed in the corms (Champagne *et al*., 2011). One anthocyanin (Terasawa *et al*., 2007), 11 hydroxycinnamic acids and 30 *O*- and *C*-glycosylated flavones (Ferreres *et al*., 2012) were reported in taro leaves. The variation reported for major components are genetically controlled and hence presents immense potential for genetic improvement by crossing parents with suitable chemotypes. Champagne *et al.*2013 reported that carotenoid contents can be rapidly improved by selecting plants of good agronomic performance and corm shape with increased density of yellow and orange colours. According to Lebot *et al.* (2011) biofortification of sugars, proteins and minerals can be achieved and the genetic improvement for higher starch content and dry matter contents can be jointly improved as the two are positively correlated. Biofortification of taro through breeding for increased carotenoids and anthocyanins contents was reported by Champagne *et al.*2013. Explorative studies on diversity of secondary metabolites and other nutrients, selection of potential parents, efficient screening procedures of hybrids in taro can accelerate development of biofortified hybrids thereby resulting in better corm quality.

Transgenic Approaches

In cassava, two transgenes, phytoene-synthase (crtB) gene from the bacteria *Erwinia herbicolor* and 1-deoxyxylulose-5-phosphate synthase (DXS) gene from *Arabidopsis* were used in transgenic programs for development of the cassava transgenic lines. crtB gene drives the β carotene synthesis with the geranylgeranyldiphosphate (GGDP) localized in the plastid, whereas, DXS is gene enhances the concentration of GGDP carotene synthesis. The transgenic lines developed using these genes expressed 40 μg/g dry weight (DW) of beta carotene in the storage root (Fregene *et al.*, 2010). Similarly, transgenic expression of iron-specific metal transporter (FEA1) gene from the single cell green algae *Chlamydomonas reinhardtii* increased the iron content in the transgenic lines of cassava storage roots (Fregene *et al.*, 2010). Narayanan *et al.*, 2015 achieved iron enrichment (60 μg/g vs 9 μg/g DW) by overexpression of the Arabidopsis thaliana vacuolar iron transporter gene VIT1 in cassava with successful field. Consumption of IRT1+FER1 biofortified cassava was found to beneficially impact the health of consumers (Narayanan *et al.*, 2019).

Gene	Gene description	Gene Source	Trans host	Trait	Reference
crtB	phytoene-synthase	*Erwinia herbicolor*	Cassava	β carotene	Fregene *et al.*, 2010
DXS	1-deoxyxylulose-5-phosphate synthase	*Arabidopsis*	Cassava	β carotene	Fregene *et al.*, 2010
FEA1	iron-specific metal transporter	*Chlamydomonas reinhardtii*	Cassava	Iron	Fregene *et al.*, 2010
IbMYB1	Transcription factor	sweet potato	Sweet potato	Anthocyanins	Park *et al.*, 2015a
IbOr	Sweet potato orange (IbOr) protein	sweet potato	Sweet potato	Carotenoids	Park *et al.*, 2015b

In sweet potato, expression of *IbMYB1* gene increased the anthocyanin pigments in the sweet potato transgenic lines (Park *et al.*, 2015a). *IbMYB1* is a R2R3-MYB transcription factors (TFs), that regulates the anthocyanin biosynthesis in sweet potato (Park *et al.*, 2015a). Similarly, expression of the *IbOr* gene, a protein increased the caretonoid pigments in the transgenic lines (Park *et al.*, 2015b).

Molecular Breeding

Morphological and molecular marker based information are used to study the diversity of the germplasms/breeding lines. Molecular markers such as RAPD, AFLP, ISSR, SSRs, etc., were used to study the genetic diversity of the tuber crops such as cassava, sweet potato and yams etc. The genes controlling the carotenoid biosynthesis pathway are well known (Giuliano 2014; Nisar *et al.*, 2015). The gene phytoene synthase (PSY2) belonging to a family with three copies in the cassava reference genome has been reported to be associated with carotenoid accumulation in cassava and is found to be in chromosome 1 and 2 (Welsch *et al.*, 2010; Esuma *et al.*, 2016). Single nucleotide polymorphisms (SNPs) in candidate carotenoid genes namely lcyE, PSY2, lcyB, and crtRB in maize and cassava was reported by Welsch *et al.*, 2010; Fu *et al.*, 2013, Udoh *et al.*, 2017. cDNA libraries have been constructed for varieties selected for carotene content and composition

(Udoh *et al.*, 2017). The contents of carotene, lutein, and total carotenoids was increased by overexpressing orange *IbOr-Ins* gene in white fleshed sweet potato (Kim *et al.*, 2013). β-Carotene content in sweetpotato is associated with the *phytoene synthase* gene and it was reported that β-carotene and starch content are negatively correlated (Gemenet *et al.*, 2020). Overexpression of the golden SNP-carrying orange gene enhanced the carotenoid accumulation along with heat stress tolerance in sweet potato plants (Kim *et al.*, 2021). So far, the markers tightly linked to the beta carotene, anthocyanin, iron and Zinc content are yet to be identified for the different tuber crops. Thus, development of the mapping population and mapping of the QTLs linked to these traits will be helpful in improvement of these crops through marker assisted breeding.

Conclusion

Research and development of nutrient-enriched biofortified tuber crops should be undertaken to address this issue because people in underdeveloped countries cannot afford to eat diversified and supplemented diets. There are many strategies for development of biofortified crops, but conventional breeding approaches are most common and easiest to implement. Therefore, a modern definition of food security is access to a sufficient supply of nutrient-rich foods. The ICAR-CTCRI, Regional Center, Bhubaneswar, has developed biofortified OFSP varieties such as Bhu Sona, Bhu Kanti, and Bhu Ja through conventional breeding to enhance the health and nutrition of the people at risk.

References

Abah R O, Okolo S N, John C and Ochoga M O. 2015. Prevalence of zinc deficiency among school children in a rural setting in North-Central Nigeria. International Journal of Public Health Research 3: 214-17.

Akinwale M G, Aladesanwa R D, Akinyele B O, Dixon A and Odiyi A. 2010. Inheritance of β-carotene in cassava (*Manihot esculenta crantza*). International Journal of Genetics and Molecular Biology 2:198–201.

Andrade M I, Ricardo J, Naico A, Alvaro A, Makunde G S, Low J, Ortiz R and Gruneberg W J. 2017. Release of orange-fleshed sweetpotato (*Ipomoea batatas* [L.] Lam.) cultivars in Mozambique through an accelerated breeding schem. Journal of Agricultural Science 155: 919–29.

Bansode V, Chauhan V B S, Pati K, Nedunchezhiyan M, Giri N, Krishnakumar T and Mahanand S. 2020. Development and storage study of Anthocyanin rich jelly from purple fleshed sweet potato (Variety Bhu-Krishna). Journal of Pharmacognosy and Phytochemistry 9 (2): 388-91.

Bhattacharjee R, Gedil M, Sartie A, Otoo E, Dumet D, Kikuno H, Kumar PL and Asiedu R. 2011. Wild Crop Relatives: Genomic and Breeding Resources. Springer, Berlin, Heidelberg, pp:71-96.

Black R E, Victoria C G, Walker S P, Bhutta Z A, Christian P and de Onis M. 2013. Maternal and child undernutrition and overweight in low-income and middle-income countries. Lancet 382: 427–51.

Bouis H E and Saltzman A. 2017. Improving nutrition through biofortification: a review of evidence from Harvest Plus, 2003 through 2016. Global Food Security 12: 49-58.

Bouis H E, Hotz , McClafferty B, Meenakshi J V and Wolfgang H P. 2011. Biofortification: a new tool to reduce micronutrient malnutrition. Food Nutr Bulletin 32: 31-40.

Carvalho L J, Agostini M A V and Junior C B.2004. Chromoplast associated proteins from color storage root diversity in cassava (Manihot esculenta Crantz). In 6th International Scientific Meeting of Cassava Biotechnology Network. CIAT, Cali, pp. 26-28.

Carvalho L J, Agustini M A, Anderson J V, Vieira E A, de Souza C R, Chen S, Schaal B A and Silva J P.2016. Natural variation in expression of genes associated with carotenoid biosynthesis and accumulation in cassava (*Manihot esculenta* Crantz) storage root. BMC Plant Biology 16(1):133.

Ceballos H. Luna J, Escobar A F, Ortiz D, Pérez J C, Sánchez T, Pachon H and Dufour D. 2012. Spatial distribution of dry matter in yellow fleshed cassava roots and its influenceon carotenoid retention upon boiling. Food Research International 45: 52–59.

Champagne A, Legendre L, Malapa R and Lebot V. 2009. Chemotype profiling to guide breeders andexplore traditional selection of root crops in Vanuatu, South Pacific. Journal of Agricultural and Food Chemistry 57:10363-10370.

Champagne A, Bernillon S, Moing A, Rolin D, Legendre L and Lebot V. 2010. Carotenoid profiling of tropical root crop chemotypes from Vanuatu, South Pacific. Journal of Food Composition and Analysis 23(8): 763–71.

Champagne A, Hilbert G, Legendre L and Lebot V. 2011. Diversity of anthocyanins and other phenolic compounds among tropical root crops from Vanuatu, South Pacific. Journal of Food Composition Analysis 24: 315-25.

Champagne A, Legendre L and Lebot V. 2013. Biofortifcation of taro (*Colocasia esculenta*) through breeding for increased contents in carotenoids and anthocyanins. Euphytica 194:125–36.

Chavez A L, Bedoya J M, Sánchez T, Iglesias C, Ceballos H and Roca W. 2000. Iron, Carotene, and Ascorbic Acid in Cassava Roots and Leaves. Food and Nutrition Bulletin 21(4): 410–13.

Chavez A L, Bedoya J M, Sánchez T, Iglesias C, Ceballos H, Roca W. 2000. Iron, Carotene, and Ascorbic Acid in Cassava Roots and Leaves. Food and Nutrition Bulletin 21(4): 410–13.

David M C, Diaz F C, Mwanga R O, Tumwegamire S, Mansilla R C and Grüneberg, W. 2018. Gene Pool Subdivision of East African Sweetpotato Parental Material. Crop Science. 58 (6): 2302-14.

De Moura F F, Moursi M, Lubowa A, Ha B, Boy E, Oguntona B, Sanusi R A and Maziya-Dixon B. 2015. Cassava Intake and Vitamin A Status among Women and Preschool Children in Akwa-Ibom, Nigeria. PLoS One 17:10(6):e0129436.

Esuma W, Herselman L and Labuschagne M T (2016) Genomewide association mapping of provitamin A carotenoid content in cassava. Euphytica 212:97–110.

Ferreres F, Lopes G, Gil-Izquierdo A, Andrade P B, Sousa C, Mouga T and Valentão P. 2012. Phlorotannin extracts from fucales characterized by HPLC-DAD-ESI-MSn: approaches to hyaluronidase inhibitory capacity and antioxidant properties. Marine Drugs 10(12):2766-81.

Fregene M, Sayre R, Fauquet C, Anderson P and Taylor N. 2010. Opportunities for biofortification of cassava for sub-Saharan Africa: the biocassava plus program. Promoting Health by Linking Agriculture. Food and Nutrition. NABC Report 22:81–90

Fu Z, Chai Y, Zhou Y, Yang X, Warburton M L, Xu S, Cai Y, Zhang D, Li J and Yan J. 2013. Natural variation in the sequence of PSY1 and frequency of favorable polymorphisms among tropical and temperate maize germplasm. Theoritical and Applied Genetics 126(4):923-35.

Garg M, Sharma N, Sharma S, Kapoor P, Kumar A, Chunduri V and Arora P. 2018 Biofortified Crops Generated by Breeding, Agronomy, and Transgenic Approaches Are improving Lives of Millions of People around the world Frontiers in Nutrition. 5:12.

Gegios A, Amthor R, Maziya-Dixon B, Egesi C, Mallowa S, Nungo R, Gichuki S, Mbanaso A and Manary M J. 2010. Children consuming cassava as a staple food are at risk for inadequate zinc, iron, and vitamin A intake. Plant Foods for Human Nutrition. 65(1):64-70.

Gemenet D C, da Silva Pereira G, De Boeck B, Wood J C, Mollinari M, Olukolu B A, Diaz F, Mosquera V, Ssali R T, David M, Kitavi M N, Burgos G, Felde T Z, Ghislain M, Carey E, Swanckaert J, Coin L J M, Fei Z, Hamilton J P, Yada B, Yencho G C, Zeng Z B, Mwanga R O M, Khan A, Gruneberg W J and Buell C R. 2020. Quantitative trait loci and differential gene expression analyses reveal the genetic basis for negatively associated β-carotene and starch content in hexaploid sweetpotato [Ipomoea batatas (L.) Lam.]. Theoritical and Applied Genetics 133(1):23-36.

Grüneberg W J, Ma D. Mwanga, R O M, Careyeds E E, Huamani K and Diaz F, 2015. "Advances in sweetpotato breeding from 1992 to 2012," in Potato and Sweetpotato in Africa-Transforming the Value Chains for Food and Nutrition Security, eds J. Low, M. Nyongesa, S. Quinn, and M. Parker (Boston, MA: CAB International). 3:68.

Gruneberg W J, Manrique K, Dapeng Z and Hermann M, 2005. Genotype x environment interactions for a diverse set of sweet potato clones evaluated across varying ecogeographic conditions in Peru. Crop Sciences 45: 2160–71.

Grüneberg W J, Mwanga R, Andrade M, and Espinoza J. 2009. Selection methods: part 5. breeding clonally propagated crops in Plant Breeding and Farmer Participation, eds S. Ceccarelli, E. P. Guimarães, and E. Weltzien (Rome: FAO). pp: 275–322.

Huang A S, L Tanudjaja and Lum D.1999. Content of Alpha-, Beta-Carotene, and Dietary Fiber in 18 Sweetpotato Varieties Grown in Hawaii, Journal of Food Composition and Analysis 12(2): 147-51.

Jongstra R, Mwangi M N, Burgos G, Zeder C, Low J W, Mzembe G, Liria R, Penny M, Andrade M I, Fairweather-Tait S, Zum Felde T, Campos H, Phiri K S, Zimmermann M B, Wegmüller R. 2020. Iron Absorption from Iron-Biofortified Sweetpotato Is Higher Than Regular Sweetpotato in Malawian Women while Iron Absorption from Regular and Iron-Biofortified Potatoes Is High in Peruvian Women. Journal of Nutrition 150(12): 3094-102.

Kim S E, Lee C J, Park S U, Lim Y H, Park W S, Kim H J, Ahn M J, Kwak S S, Kim H S. 2021. Overexpression of the Golden SNP-Carrying Orange Gene Enhances Carotenoid Accumulation and Heat Stress Tolerance in Sweetpotato Plants. Antioxidant (Basel) 10(1): 51

Kim S H, Ahn Y O, Ahn M J, Jeong J C, Lee H S, Kwak S S. 2013. Cloning and characterization of an Orange gene that increases carotenoid accumulation and salt stress tolerance in transgenic sweet potato cultures. Plant Physiol Biochem 70: 445-54.

Kumar S, Das G, Shin H S, Patra J K. 2017. *Dioscorea* spp. (A wild edible tuber)" A study on its ethanopharmacologicak potential and traditional use by the local people of simlipal biosphere researve, India. Frontiers in Pharmacology 8: 52.

La Frano M R, Woodhouse L R, Burnett D J, Burri B J.2013. Biofortified cassava increases β-caroteneand vitamin A concentrations in the TAG-rich plasma layer of American women. British Journal of Nutrition 110:310–20.

Lebot V, Malapa R, Bourrieau M .2011. Rapid estimation of taro quality by near infrared spectroscopy. Journal of Agricultural and Food Chemistry 59:9327–34.

Lebot, V, Prana M S, Kreike N, Van H H, Pardales J, Okpul T, Gendua T, Thongjiem M, Hue H, Viet N, Yap TC. 2004. Characterisation of taro (*Colocasia esculenta* (L.) Schott) genetic resources in Southeast Asia and Oceania. Genetic Resources and Crop Evolution 51: 381–92.

Low J W, Arimond M, Osman N, Cunguara B, Zano F and Tschirley D. 2007. A food-based approach introducing orange-fleshed sweet potatoes increased vitamin A intake and serum retinol concentrations in young children in rural Mozambique. Journal of Nutrition 137(5):1320-7.

Mukherjee A, Naskar S K, Ray R C, Pati K and Mukherjee A. 2015. Sweet Potato and Taro resilient to stresses: Sustainable livelihood in fragile zones vulnerable to climate changes. Journal of Environment and Sociobiology 12 (1): 53-64.

Narayanan N, Getu B, Raj D C, Eliana G S, Michael A G, Nigel T and Paul A. 2015. Overexpression of Arabidopsis VIT1 increases accumulation of iron in cassava roots and stems. Plant Sci 240: 170–81.

Narayanan N, Getu B, Raj D C, Eliana G S, Jackson G, Paula B, Dimuth S, Ihuoma O, Arthur W, Dulce M J A, Erick B, Michael A. G, Paul A and Nigel J T. 2019. Biofortification of field-grown cassava by engineering expression of an iron transporter and ferritin. Nature Biotechnology 37:144-51.

Nassar N M A, Vizzoto C S, Silva H L, Schwartz C A. 2005. Potentiality of cassava cultivars as a source of carotenoids. Gene Conservation 4: 257-83.

Nisar N, Li L, Lu S, Khin N C, Pogson B J. 2015. Carotenoid metabolism in plants. Molecular Plant 8:68–82.

Park S C, Kim S H, Park S, Lee H U, Lee J S, Park W S, Ahn M, Kim Y H, Jeong J C, Lee H S and Kwak S S. 2015b. Enhanced accumulation of carotenoids in sweet potato plants overexpressing IbOr-Ins gene in purple-fleshed sweet potato cultivar. Plant Physiology and Biochemistry 86: 82-90.

Park, S C, Kim Y H, Kim S H, Jeong Y J, Kim C Y, Lee J S., Bae J Y, Ahn M J, Jeong J C, Lee H S and Kwak S S. 2015a. Overexpression of the IbMYB1 gene in an orange-fleshed sweet potato cultivar produces a dual-pigmented transgenic sweet potato with improved antioxidant activity. Physiologia plantarum 153(4):525-37.

Pati K, Chauhan VBS, Bansode VV, Nedunchezhiyan, M. 2021. Biofortification in sweet potato for health and nutrition security. In: More SJ, Giri NA, Suresh KJ, Visalakshi CC, Tadigiri S (eds.), Recent Advances in Root and Tuber Crops, Brillion Publishing House, New Delhi, India. pp: 21-30.

Pfeiffer W H and Mc Clafferty B. 2007. Harvest Plus: breeding crops for better nutrition. Crop Science 47:88.

Price E, Bhattacharjee R, Lopez-Montes A and Fraser P. 2018. Carotenoid profiling of yams: Clarity, comparisons and diversity. Food Chemistry 259: 130-38

Sánchez T, Ceballos H, Dufour D, Ortiz D, Morante N, Calle F, Zum Felde T, Domínguez M and Davrieux F. 2014. Prediction of carotenoids, cyanide and dry matter contents in fresh cassava root using NIRS and Hunter color techniques. Food Chemistry 151:444-51.

Sayre R, Beeching J R, Cahoon E B, Egesi C, Fauquet C, Fellman J, Fregene M, Gruissem W, Mallowa S, Manary M, Maziya-Dixon B, Mbanaso A, Schachtman D P, Siritunga D, Taylor N, Vanderschuren H and Zhang P. 2011. The BioCassava plus program: biofortification of cassava for sub-Saharan Africa. Annual Reviews in Plant Biology 62: 251-72.

Stephenson K, Amthor R, Mallowa S, Nungo R, Maziya-Dixon B, Gichuki S, Mbanaso A and Manary M. 2010. Consuming cassava as a staple food places children 2-5 years old at risk for inadequate protein intake, an observational study in Kenya and Nigeria. Nutrition Journal 9: 9.

Teow C C, Truong V D, McFeeters R F, Thompson R L, Pecota K V and Yencho G C. 2007. Antioxidant activities, phenolic and b-carotene contents of sweet potato genotypes with varying flesh colors. Food Chemistry 103: 829–838.

Terasawa N, Saotome A, Tachimura Y, Mochizuki A, Ono H, Takenaka M and Murata M. 2007. Identification and some properties of anthocyanin isolated from Zuiki, stalk of Colocasia esculenta. The Journal of Agricultural and Food Chemistry 55: 4154-59.

Udoh L I, Gedil M, Parkes E Y, Kulakow P, Adesoye A, Nwuba C. 2017. Candidate gene sequencing and validation of SNP markers linked to carotenoid content in cassava (*Manihot esculenta* Crantz). Molecular Breeding 37(10):123.

Vimala B, Nambisan B and Hariprakash B. 2011. Retention of carotenoids in orange-fleshed sweet potato during processing. Journal of Food science and Technology 48: 520-24.

Welsch R, Arango J, Bär C, Salazar B, Al-Babili S, Beltrán J, Chavarriaga P, Ceballos H, Tohme J and Beyer P. 2010. Provitamin A accumulation in cassava (*Manihot esculenta*) roots driven by a single nucleotide polymorphism in a phytoene synthase gene. Plant Cell 22:3348–56

18

Scope and Potential of Vertical Farming in Root and Tuber Crops Production

J. Suresh Kumar[1*], M. Nedunchezhiyan[2], S. Sunitha[1] and T. Janakiram[3]

[1]ICAR-Central Tuber Crops Research Institute, Thiruvananthapuram Kerala-695 017

[2]Regional Centre ICAR-Central Tuber Crops Research Institute Bhubaneswar, Odisha-751019

[3]Vice Chancellor, Dr.YSRHU, Venkataramannagudem Andhra Pradesh-534101

Vertical farming is a system of growing crops in multiple layers in skyscrapers, to maximize the use of land by having a vertical design, whereby plants, animals, fungi and other life forms are cultivated for food, fuel and fibre by artificially stacking them vertically above each other (Despommier, 2009; Kalantari *et al.*, 2017). It is estimated that nine acres in a horizontal farm is equal to 1 acre in a vertical farm. The concept of vertical farming started during the mid 90s by Dr Dickson Despomier, professor of Public and Environmental Health at Columbia University. This method of farming got more momentum due to increasing trend of conversion of fertile agriculture land to urban areas and un-reducing trends of population growth and urbanization, age problem of agriculture labourers and reduced availability of younger population in traditional agriculture, desertification and natural calamities, increased pest and disease phenomenon, and increased food demands.

There are various advantages of vertical farming like year-round production, less input required (water and nutrients), high yield per unit area compared to field-grown crops, reduce shipment charges of produce due to production in cities for the urban markets, no or very less post-harvest loss, produce free from pesticide and herbicide residues, and farm fresh produce available to urban dwellers.

There are two types of vertical farming in practice coloumn cultivation and multilayer cultivation. It depends up on the crop, growing conditions of the production system. There are different types of production of vertical farming, i.e with and without supplemental lighting in controlled environments and ambient conditions. Most of the non-commercial tropical urban farming units cultivate the crops on terraces and open urban fields under polyhouse or net house with rain out shelters with minimal temperature control mechanism and with no artificial lighting.

Column cultivation under normal light and soilless production system in a polyhouse [*Source*: goodnewsnetwork.org)]

Vertical faming in multilayers in indoors with artificial lighting and soil less production system (*Source*: https://www.usda.gov/media/blog/2018/08/14/vertical-farming-future)

Scope of Vertical Farming in Tropical Tuber Crops

Tropical tuber crops like cassava, sweet potato, yams, elephant-foot yam, taro, tannia and Chinese potato are energy-dense starchy vegetables that ensures food and nutritional security besides sustenance and resilience under climate change. The edible energy production in terms of mega joules/ha/day is estimated to be higher in some tuber crops than cereal crops. It is 194 for sweet potato, 182 for yams, 121 for cassava, and they are quite comparable with major cereal food crops rice (151), maize (159) and wheat (135) (Horton and Fano, 1985; Edison, 2006). Of these, sweet potato, taro and Chinese potato have scope for vertical farming and soil less production systems like hydroponics and aeroponics due to their short stature, early growth and early harvestibility (3-6 months). These crops are nutritionally important as "health foods" due to the presence of anti-oxidants like β carotene, anthocyanins, minerals (K, P etc., high K:Na ratio), vitamins (sweet potato and taro), mucilage (taro), alkaloids like flavonoids (Chinese potato). These crops offer protection from diseases like cancer, liver malfunction; hyper tension, cardio-vascular disorders, and regular intake of these foods would help to ward off several life-style diseases like diabetes and helps to lead a healthy life. Hence, these are preferred for cultivation in kitchen gardens, home gardens and for terrace cultivation in urban and peri-urban areas and for vertical faming in controlled environment agriculture.

The production of sweet potato tubers in the hydroponic system started by a few workers due to its importance as a food, fodder, fuel, industrial purpose. In 1980s NASA (National Aeronautics and Space Administration) identified sweet potato as a future space mission crop for advanced space support programme due to its nutritional importance. As of now there is no commercial cultivation of this crop in vertical farms due to long duration, not known importance among the consumers. But recent past there were varieties released with rich in anthocyanin, β-carotene, and this crop is identified as a potential alternative to corn-based ethanol with high industrial value. The crop produces more edible energy than any other major food crop (194 MJ ha^{-1} day^{-1}; Woolfe, 1992). This promoted its research and production of sweet potato in soil less production system in controlled environment for higher yields and year round production.

Hydroponic sweet potato grown in a net house with rain out shelter at ICAR-CTCRI (*Courtesy*: J. Suresh Kumar, ICAR-CTCRI)

There is lack of research on vertical farming in tropical tuber crops at the global or national level, except a few reports on sweet potato from China (Yang, 2005), Japan (Kitaya *et al.*, 2008, 2009; Sakamoto and Suzuki, 2018, 2020) and USA (Morris *et al.*, 1989). Techniques for over ground production of sweet potato tubers was developed by Yang (2005). Kitaya *et al.* (2008) developed a new hydroponic method for producing tuberous roots and fresh edible leaves and stems of sweet potato for space farming. "Elegant summer" was found to be a promising variety to produce large amounts of food with high nutritional values in the above hydroponic system for space farming. Studies conducted by Kitaya *et al.* (2009) at Osaka, Japan, indicated that sweet potato performed successfully in the hydroponic system on the rooftop in summer. Sakamoto and Suzuki, 2020 able to grow sweet potato in hydroponics with a substrate culture and got good economical yield under controlled environments in poly house.

Presently, there is dearth of scientific information on vertical farming in other tropical tuber crops and evidence on impact of vertical farming on yield and quality, standardization of structures and techniques for vertical farming, as well as its socio-economic impact in tuber crops.

References

Edison S. 2006. Status of tropical tuber crops production, utilization and marketing in India. Book of Abstracts of the 14th Triennial Symposium of International Society for Tropical Root Crops, 20-26 November, 2006, ICAR. Tuber Crops Research Institute, Thiruvananthapuram, Kerala, India. p. 7.

Horton D and Fano H. 1985. Potato Atlas. CIP (International Potato Center), Lima, Peru.

Kalantari F, Tahir O M, Lahijani A M and Kalantari S. 2017. A review of vertical farming technology: A guide for implementation of building integrated agriculture in cities. Advanced Engineering Forum 24: 76-91.

Kitaya Y, Hirai H, Wei X, Islam A F M S and Yamamoto M. 2008. Growth of sweet potato cultured in the newly designed hydroponic system for space farming. Advances in Space Research 41(5): 730-35.

Kitaya Y, Yamamoto M, Hirai H and Shibuya T. 2009. Rooftop farming with sweet potato for reducing urban heat island effects and producing food and fuel materials. In: The Seventh International Conference on Urban Climate, 29 June - 3 July 2009, Yokohama, Japan.

Morris C E, Marinez E, Bonsi C K, Mortley D G, Hill W A, Ogbuehi C R, Loreton P A. 1989. Alabama A & M University, NASA-HBCU Space Science and Engineering Research Forum Proceedings. https://ntrs.nasa.gov/api/citations/19910018752/downloads/19910018752.pdf

Sakamoto M and Suzuki T. 2018. Effect of pot volume on the growth of sweet potato cultivated in the new hydroponic system. Sustainable Agriculture Research 7(1): 137-145. DOI:10.5539/sar.v7n1p137

Sakamoto M and Suzuki T. 2020. Effect of Nutrient Solution Concentration on the Growth of Hydroponic Sweet potato. Agronomy 10: 1708. DOI: 10.3390/agronomy10111708

Woolfe J A. 1992. Sweet potato: An Untapped Food Resource. Cambridge University Press, New York, USA.

19

In-vitro Propagation of Coconut (*Cocos nucifera*)

Neema M., Aparna Veluru and Sudha R.

ICAR – Central Plantation Crop Research Institute, Kasaragod Kerala-671124

Plant tissue culture is the process in which mass multiplication is possible from the somatic cells of the plant by utilizing the principle of totipotency. Each plant cell can grow into a plant by converting the somatic nature of the cells to embryogenic ones. In tissue culture, plantlets can be produced either directly through organogenesis or indirectly by somatic embryogenesis. Organogenesis is the process in which shoots develop from the explants directly and then rooting is induced. In indirect somatic embryogenesis, the explant cells pass through a callusing stage where the cells get multiplied. These cells then obtain embryogenic potential and develop somatic embryos. These somatic embryos later germinate and develop into plantlets. These plantlets are hardened *in vitro, ex vitro* and finally planted in the field. Palms are usually recalcitrant to tissue culture. In this chapter, we deal with the tissue culture of a specific palm namely coconut (*Cocos nucifera* L.). This is a mono-specific crop of the tropics. The palm is usually referred to as the 'Tree of life' as many of the small-scale holdings obtain their livelihood from this crop. Clonal propagation of coconut is gaining importance as most of the coconut plantations are old and senile, affected by pests and diseases or are destroyed by floods and fire.

In-vitro Propagation in Coconut

Research on *in vitro* propagation of coconut was initiated as embryo culture of the Makapuno coconut by Del Rosario and de Guzman during the 1970s. Makapuno coconuts are filled with jelly endosperm and fetch a good price in the market. But these nuts will not germinate naturally and embryo rescue is the only method of propagation. In India, ICAR- CPCRI had developed a protocol for embryo culture of coconut (Karun *et al.*, 1999) and this protocol was successfully used for embryo rescue of Ratnagiri, sweet kernel coconut (Karun *et al.*, 2017) which like makapuno were difficult for natural germination. Reports on the research published concerning coconut tissue culture way back in the 1970 s, when stem, leaf, and inflorescence were used as explants for callus induction. Eeuwens in 1976 standardized the media for callus induction (Y3). Later in 1995, an international collaborative project on tissue culture was initiated with researchers from France, Cote d'Ivoire, the U.K., Germany, the Philippines and Mexico which resulted in solving a large number of problems encountered in coconut tissue culture (Hocher *et al.*, 1998). Even though Y3 media was most commonly used for coconut tissue culture, del Rosario (1984) found no difference between Murashige and Skoog (1962) and Y3 media. To circumvent the problem of excessive browning in coconut tissue cultures Ebert *et al.*, 1993 used activated charcoal in the culture media. Activated charcoal could adsorb the phenols in the culture media thereby reducing the browning caused by phenolic interference. But it was considered a necessary evil as it also adsorbed the majority of the growth regulators like auxins and cytokinins supplied to the media.

Blake and Eeuwens (1978) reported initial success using inflorescence tissue for callus production. At ICAR Central Plantation Crops Research Institute, Raju *et al.*, (1984) successfully developed plantlets from immature leaf tissue explants of 1-2-year-old WCT seedlings. They also reported that NAA and IAA resulted in direct embryogenesis in leaf explants. However, the result was not reproducible in subsequent trials and their report that direct embryos and embryoids develop from vascular tissue was challenged by Blake (1989) who reported this to be unusual as vascular tissue area normally gives rise to root primordia. In the same year, Karunaratne and Periyaperuma reported that the embryogenic capacity of leaf explants was related to their physiological maturity in young palms of coconut. Subsequently, in 1991, they ascertained that only a particular leaf in a particular physiological state produces embryogenic cells and only a portion of this leaf yielded embryogenic explants (Karunaratne *et al.*, 1991). This might explain why Raju *et al.* (1984) were finding difficulty in reproducing the results.

Again root (Jones, 1983), and sub-apical leaf were of limited use as explants due to their limited embryogenic potential (Karunaratne *et al*, 1991). Calli initiated from embryos, leaves, leaf bases, and the apical meristem could not be regenerated. Likewise, callus induction from anthers and rachilla did not give a repeatable response (Neera Bhalla Sarin *et al*, 1986). Duffard Morel *et al*. (1992) successfully developed somatic embryos from leaf explants and the study was supported by detailed histological observations. Even though zygotic embryos, leaf base, apical meristem and endosperm were also tried (Verdeil and Buffard Morel, 1995), immature inflorescences and embryos were found to be promising. Blake and Eeuwens (1978) reported initial success using inflorescence tissue for callus production. Branton and Blake (1986) produced plantlets in 9 months from immature rachilla explants through somatic embryogenesis of nodular callus by reducing 2,4-D concentration in Y3 medium to 100µM 2,4-D, with 5µM each of 2ip and BAP, and 0.25% AC. Areza *et al* (1993) soaked the inflorescence tissue in antioxidants, viz., Citric acid (50mg/l) and Ascorbic acid (100mg/l), before slicing and culturing in Y3 medium supplemented with activated charcoal (AC), which resulted in reduced browning. Verdeil *et al.* (1994) reported successful embryo maturation via somatic embryogenesis from inflorescence explants, which further regenerated into plantlets. Immature inflorescences were most successful among the various explants tried, and plantlet regeneration was successful even though the transfer of plantlets to the nursery is yet to be achieved. The use of plumular tissues taken from germinating embryos was another source from where success was forthcoming, because of the juvenile nature of the tissue (Hornung, 1995). Chan *et al.* (1998) developed a protocol using plumules of zygotic embryos. A procedure for the regeneration of complete plantlets via organogenesis from the plumular tissues of coconut was reported by Rajesh *et al.* (2014). Meanwhile, the enhancement of callogenesis from plumular explants of coconut via exogenous supplementation of amino acids and casein hydrolysate was studied by Aparna *et al.* (2023). Perera *et al.* (2007) reported the induction of somatic embryogenesis and plantlet regeneration from callus cultures of unfertilized ovaries isolated from immature female flowers of coconut. When Griffis and Litz (1997) used anthers and filaments, unfertilized ovaries and immature leaf pieces as explants, callus initiation and direct formation of somatic proembryos could be stimulated with the addition of 2,4 -D to the culture medium. Fernando and Gamage (2000) induced nodular calli from 7-9 month-old immature zygotic embryos in BM72 medium supplemented with 24 μM 2,4-D. Samosir *et al.* (1999) indicated that the development and maturation of coconut somatic embryos could be improved by using ABA alone, or, with any of the osmotically active agents preferably Poly Ethylene Glycol (PEG). Samosir *et al.* (1998) used longitudinally sliced mature zygotic explants cultured on medium supplemented with 125-μM 2,4-D and 2.5 g/l activated charcoal. Adkins *et al.* (1998) used cotyledonary slices from embryos cultured on a medium with additives like amino ethoxy vinyl glycine (AVG) and silver thiosulphate (STS) which could reduce ethylene production, or used polyamines such as spermine, putrescine and spermidine. The use of zygotic embryo culture for germplasm collection, storage and retrieval were standardized in India (Karun & Sajini, 1994, Karun *et al.*, 1996). Koshy and Kumaran (1997) collected 15 accessions from the Indian Ocean Islands of Mauritius, Madagascar and Seychelles (Anon, 1998), and later, Parthasarathy (2001) used this technique to collect four accessions from Sri Lanka.

Zygotic plumular tissue can now be used to achieve clonal propagation via SE (Pe´rez-Nu´n˜ez *et al.* 2006). However, difficulties in this process are still preventing the establishment of an affordable and universal protocol for the production of plantlets on a large scale. Perera *et al.* (2008) have reported the production of doubled haploid plants via anther-derived embryogenesis. However, the *ex vitro* acclimatization of somatic embryo-derived plantlets has yet to be refined, with present rates of success of around 50 % so far (Fuentes *et al.*, 2005).

Further improvements may come from using a photoautotrophic culture system (Samosir &Adkins, 2014) and through the incorporation of fatty acids, notably lauric acid, into the plantlet maturation medium (López-Villalobos *et al*., 2011).

Bandupriya *et al* (2018) studied the genetic homogeneity of coconut plantlets derived from unfertilized ovaries through somatic embryogenesis. These plantlets showed no apparent differences among themselves and were comparable with the respective mother palm in the Simple Sequence Repeats analysis. The results suggest that no somaclonal variation or genetic instability is occurring in plantlets that are regenerated from ovary explants. It proved the suitability of regeneration protocol for large-scale micropropagation applications for coconut. Recent research in clonal propagation of coconut has shifted towards direct organogenesis when compared to indirect somatic embryogenesis. Direct multiple shoot induction is a reliable method for *in vitro* clonal propagation as the risk of genetic variability is less compared to other regeneration methods. Recent research publications on coconut organogenesis are by Shareefa *et al* (2019) and Wilms *et al.* (2021). Shareefa and co-workers have demonstrated the possibility of inducing direct *in vitro* plant regeneration under appropriate culture conditions using the right developmental stage of inflorescence. This study formed the basis for developing an *in vitro* regeneration protocol using rachillae bit explants, which will ultimately enable large-scale multiplication of disease-free and high-yielding coconut palms. Meanwhile, Wilms and co-workers could develop the first axillary *in vitro* shoot multiplication protocol for coconut palms using apical meristems of *in vitro* coconut seedlings. The apical meristems were cultured onto a Y3 medium containing 1 μM TDZ, which induced the apical meristem to proliferate through axillary shoots. These axillary shoots seen as white clumps of proliferating tissue could be multiplied at a large scale or regenerated into rooted *in vitro* plantlets. The quality and productivity of the seedlings produced by these methods are to be ascertained in future, but for the time now these protocols remain the repeatable and reliable ones as far as clonal propagation of coconut is considered.

Problems Associated with Coconut Tissue Culture

Varying Response of Explants

The varying response of the explants is a grave problem associated with coconut tissue culture. Since coconut is a cross-pollinated crop, the nuts obtained will be heterogeneous. Plumular explants exhibit variability at the initial stage itself. The response in an *in-vitro* culture of the explants taken from the same bunch of nuts varies. Hence finding the right explants responsive to tissue culture itself is a tedious process. The repeatable response of the once-responsive explants is also not certain in the case of coconut. The response varies between seasons, varieties as well as the maturity of explants. The inflorescence explants are the ones which give consistent results. But here, inflorescence of length 2-12 cm is usually used, which damages the shoot meristem during explants collection and finally results in the death of the plant. So this lethal sampling could be attempted only in extreme cases. Studies with unfertilized ovaries as the explants with the advantage of minimal damage to palms are also in progress. Even though scanty reports are present where somatic embryogenesis and plantlet regeneration is possible from mature and immature leaf explants, these were unrepeatable results. Again the leaf explants lose their embryogenic capacity within a short period

Inability to Induce Friability in Callus

Coconut callus is hard and till now friable callus is not obtained in coconut except in very few cases where fast-growing friable callus is reported (Perera *et al.*, 2021). The embryogenic callus of coconut is translucent and ear-shaped, from which the somatic embryos develop. So the cell lines with embryogenic potential are not yet identified. Hence suspension culture is also not yet possible in tissue culture. Hence mass multiplication of callus and subsequent production of embryogenic callus and plantlet regeneration in coconut is difficult. Again the callus loses its embryogenic potential very soon if not subcultured into the lower concentration of auxins. It becomes compact and hard and loses its regenerative potential.

Variation in Culture Medium Composition

Activated charcoal is an essential component of coconut tissue culture media. Charcoal is added to the media to reduce phenolic interference. The phenols extruded into the media results in the browning of the media as well as the explant. This charcoal adsorbs the hormones as well as nutrients in the media thus making an undefined culture condition to the media. Hence the optimum culture conditions cannot be standardized. Again the type of charcoal added to the media also determines the embryogenic potential of the callus.

Long Duration of in-vitro Culture

For a coconut plantlet to develop from the explant, it requires a minimum period of 1.5 to 2 years. In each stage of the *in vitro* culture, from callus induction to multiplication to somatic embryogenesis, germination of the somatic embryo, plantlet regeneration, *in vitro* and *ex vitro* acclimatization constant surveillance is required. In each stage, the cultures are under the threat of losing. Hence precise maintenance of the culture is very important and is a tedious process.

Loss of Cultures due to Contamination

Contamination due to fungus, bacteria and mites impose acute problem to *in-vitro* culture. Even though research has been carried out for reducing the contamination during initial inoculations (Neema *et al.*, 2022), rescuing the contaminated cultures is very difficult. The rescued cultures are more prone to recontamination easily. Sometimes the contaminants appear similar to the cell clumps of suspension culture (Neema *et al.*, 2023), detecting them at an earlier stage is not possible. Since coconut *in-vitro* culture is a very long-duration process, the chances of losing the cultures due to contamination are very high. Losing cultures to contamination is the main practical difficulty we face during coconut tissue culture. From the day of inoculation to the day of *ex-vitro* acclimatization the cultures are under the constant threat of losing to contamination.

Inability of Plant to Cope with Field Conditions

The survival rate of the in vitro regenerated coconut plantlets in the field is less. Plant death may happen in the in-vitro as well as ex-vitro acclimatization stages. The sudden death of the plantlets also happens in field conditions.

References

Aparna, V., Neema, M., Chandran, K.P., Muralikrishna, K.S. and Karun, A., 2023. Enhancement of callogenesis from plumular explants of coconut (*Cocos nucifera*) via exogenous supplementation of amino acids and casein hydrolysate. Current Horticulture 11(1): 66–68

Areza, M. B., Rillo, E. P., Paloma, M. B. F., Ebert, A. W., Cueto, C. A. and Orense, O. D. 1993. Prevention of browning of coconut (*Cocos nucifera* L) inflorescence and leaf tissues in *in vitro* cultures. Philippine J. Coconut Studies, 18: 20 - 23.

Bandupriya, H.D.D., Iroshini, W.W.M.A., Perera, S. A. C. N., Vidhanaarachchi, V.R.M, Fernando, S.C., Santha, E.S., Gunathilake, T.R. 2018. Genetic Fidelity Testing Using SSR Marker Assay Confirms Trueness to Type of Micropropagated Coconut (*Cocos nucifera* L.) Plantlets Derived from Unfertilized Ovaries. *The Open Plant Science Journal* 11: DOI: 10.2174/1874294701710010046

Buffard-Morel, J., Verdeil, J.L. and Pannetier, C. 1992. Embryogenese somatique du cocotier *(Cocos nucifera* L.) a partir d'explants folaires: etude histologique. *Can J Bot* **70**: 735-41.

Chan, J.L., Saenz, L., Talavera, C., Hornnng, R., Robert, M and Oropeza, C. 1998. Regeneration of coconut *(Cocos nucifera* L.) from plumule explants through somatic embryogenesis. Plant Cell Rep 17:515-21.

Del Rosario, A.G. and De Guzman, E.V. 1976. The growth of coconut makapuno embryos *in vitro* is affected by mineral composition and sugar level of the medium during liquid and solid culture. *Phil J Sci* **105**:215-22.

Del Rosario, E.J., Bergonia, H.A., Flavier, M.E., Samonte, J .L. and Mendoza, E.M.T. (1984). Chromatographic analysis of carbohydrates in coconut water. Trails. Nat. Acad. Science & Technnl. 6:127-151

Ebert, A., Taylor, F. and Blake, J., 1993. Changes of 6-benzyl amino purine and 2, 4-dichloro phenoxy acetic acid concentrations in plant tissue culture media in the presence of activated charcoal. Plant Cell, Tissue and Organ Culture, 33: 157-62.

Eeuwens, C. J. (1976). Effects of organic nutrients and hormones on growth and development of tissue explants from coconut (*Cocos nucifera* L.) and date (*Phoenix dactylifera*) palms cultured *in vitro*. Physiol. Plant 36: 23-28.

Fernando, S. C. and Gamage, C. K. A. 2000. Abscisic acid-induced somatic embryogenesis in immature embryo explants of coconut. Pl. Sci., 151: 193 -198.

Fuentes, G., Talavera, C., Desjardins, Y., Santamaria, J.M. 2005. High irradiance can minimize the neg Fuentes G, Talavera, C., Desjardins, Y., Santamaria, J.M. (2005a) High irradiance can minimize the negative effect of exogenous sucrose on the photosynthetic capacity of *in-vitro* grown coconut plantlets. *Biol Plant* **49**:7–15ative effect of exogenous sucrose on the photosynthetic capacity of *in vitro* grown coconut plantlets. Biol Plant 49:7–15.

Hocher, V., Verdeil, J. L., Grosemange, F., Huet, C., Bourdeix, R., N'Cho, Y., Sangare, A., Hornung, R., Jacobsen, H. J., Rillo, E., Oropeza, C. and Hamon, S. (1998). Collaboration internationale pour la maitrise de la multiplication vegetative in vitro du cocotier (*Cocos nucifera* L.). Cahiers Agril., 7: 499-505.

Hornung, R., Domas, R. and Lynch, P.T. (2001). Cryopreservation of plumular explants of coconut (*Cocos nucifera* L.) to support programmes for mass clonal propagation through somatic embryogenesis. CryoLetters 22: 211-220.

Karun, A., Sajini, K.K. and Shivashankar, S., 1999. Embryo culture of coconut: the CPCRI protocol. *Indian journal of horticulture* **56(4),** pp.348-353.

Karun, A., Sajini, K.K., Rajesh, M.K., Muralikrishna, K.S., Samsudeen, K., Kumar, P.A. and Nagwekar, D.D., 2017. In vitro retrieval via embryo rescue of 'MohachaoNarel', a sweet endosperm coconut from Maharashtra. International Journal of Innovative Horticulture, 6(2): 147-50.

Karunaratne, S. and Periyperuma, K. 1989. Culture of immature embryos of coconut (*Cocos nucifera* L.): Callus Proliferation and Somatic Embryogenesis. *Pl. Sci.*, 62: 247-253

Karunaratne, S., Gamage, S and Kovoor, A. 1991. Leaf maturity is a critical factor in embryogenesis. J. Pl. Physiol, 139: 27-31

Koshy, P. K. and Kumaran, P. M. 1997. 1962. Exotic germplasm of coconut collected through embryo culture techniques. *ICAR News*, **3**: 2.

Murashige T.; Skoog F. A revised medium for rapid growth and bioassays with tobacco tissue cultures. *Physiol. Plant.*, **15**: 473–497.

Neema, M., Aparna, V. and Chandran, K.P., 2022. Contrast analysis recommends flame sterilization for surface depuration in coconut (*Cocos nucifera*) meristem culture. Current Horticulture, 10(1):.41-44.

Neema, M., Gupta, A., Gopal, M. and Karun, A.,2023, Basidiomycete (*Pseudolagarobasidium acaciicola*) in coconut (*Cocos nucifera*) suspension culture—a report. Current Horticulture, 11(1): 66-68.

Parthasarathy, V.A., Geethalakshmi, P. and Niral, V. (2004) Analysis of coconut cultivars and hybrids using isozyme polymorphism. Acta Botanica Croatica, 63: 69-74.

Pe´rez-Nu´n˜ez MT, Souza R, Saenz L, Chan JL, Zuniga-Aguilar JJ, Oropeza C (2009) Detection of a SERK-like gene in coconut and analysis of its expression during the formation of embryogenic callus and somatic embryos. Plant Cell Rep., 28:11–19.

Perera, P. I .P., Hocher, V., Verdeil, J. L., Doulbeau, S., Yakandawala, D. M. D., Weerakoon, L. K. (2007). Unfertilized ovary: a novel explant for coconut (*Cocos nucifera* L.) somatic embryogenesis. Pl. Cell Rep., 26: 21-28.

Perera, P.I.P., Pathirana, R. and Vidhanaarchchi, V.R.M., 2021. Somatic embryogenesis in anther-derived fast-growing callus as a long-term source for doubled-haploid production of coconut (*Cocos nucifera* L.). Journal of the National Science Foundation of Sri Lanka, 49(1).

Perera, P.I.P., Perera, L., Hocher, V., Verdeil, J.L., Yakandawala, D.M., Weerakoon, L.K. (2008). Use of SSR markers to determine the anther-derived homozygous lines in coconut. Plant Cell Rep. 27:1697–1703.

Rajesh, M.K., Radha, E., Sajini, K.K. and Anitha, K., 2014. Polyamine-induced somatic embryogenesis and plantlet regeneration in vitro from plumular explants of dwarf cultivars of coconut (*Cocos nucifera*). Indian Journal of Agricultural Sciences, 84(4): 527-30.

Samosir, Y.M.S and Adkins, S.W. 2014. Improving acclimatization through the photoautotrophic culture of co Samosir YMS, Adkins SW 2014. Improving acclimatization through the photoautotrophic culture of coconut (*Cocos nucifera*) seedlings: an *in vitro* system for the efficient exchange of germplasm. *In Vitro* Cell Dev Plant, 50:493–501.

Samosir, YMS, ID Godwin and SW Adkins. 1999. A new technique for coconut (*Cocos nucifera* L.) Germplasm collection from remote sites: Culturability of embryos following low-temperature incubation. Australian Journal of Botany, 47:69-75.

Shareefa, M., Thomas, R.J., Sreelekshmi, J.S., Rajesh, M.K. and Karun, A., 2019. In vitro regeneration of coconut plantlets from the immature inflorescence. Current Science, 117(5): 813-20.

Wilms, H., De Bièvre, D., Longin, K., Swennen, R., Rhee, J. and Panis, B., 2021. Development of the first axillary in vitro shoot multiplication protocol for coconut palms. Scientific Reports, 11(1): 18367.

20

Conservation and Utilization of Genetic Resources of Arecanut

N. R. Nagaraja[1] and T. N. Ranjini[2]

[1]*ICAR-Central Plantation Crops Research Institute, Regional Station Vittal-574 243, Karnataka*
[2]*ICAR-Central Plantation Crops Research Institute Kasaragod-671 124 Karnataka*

Arecanut is an important traditional plantation crop of India and South-Eastern Asia. Also known as "betel nut", the kernel is obtained from its fruit. It is mostly used by the people as masticatory and is an essential requisite during several religious, social and cultural functions of India. Arecanut has been widely used in East Africa, South-East Asia, and the Pacific Ocean islands, providing economic security for millions of people and for many sole means of livelihood in the Indian sub-continent. Arecanut is commercially cultivated in Bangladesh, China, India, Indonesia, Malaysia, Myanmar, Philippines, Sri Lanka, Thailand and Vietnam. Its current world production is 17.96 lakh tonnes from 12.26 lakh ha (FAO STAT, 2020). India stands top in both area and production of arecanut in the world, with an area of 5.27 lakh ha and total production of 9.04 lakh tonnes.. In India, arecanut is predominantly cultivated in Karnataka, Kerala, Assam, Maharashtra, West Bengal, and Andaman and Nicobar group of islands etc.

History, Origin and Distribution

The exact centre of the origin is not known. No fossil remains of the genus *Areca* exist, but the fossil records of closely related genera indicate its presence during tertiary period. The maximum diversity with 24 species and other indicators recommend that original habitat in Philippines, Malaysia, and Indonesia (Raghavan, 1957; Bavappa, 1963). According to Watt (1889) it is a native of Cochin China and Malay Peninsula. Although it is not precisely known as to when arecanut found its way to Indian subcontinent, several evidences about its antiquity exist (Mohan Rao, 1982). Arecanut is mentioned in various ancient Sanskrit scriptures (650-1300 BC); and its medicinal properties were known to a famous Scholar Vagbatta (41 century A.D.). A well-known cave in central India, Ajanta caves (200 B.C.

to 900 A.D.) has one exquisitely painted arecanut palm providing a backdrop to Padmapani Budha. According to Furtado one of the earliest references to arecanut was in 1510 A.D. The abundant use of arecanut in chewing and religious functions was indicated even during the times of Aryans in India.

Botany

The arecanut (*Areca catechu*) belongs to the genus *Areca* L., sub-tribe Arecinae and tribe Arecae in the family Arecaceae, growing in hot and humid tropical regions of the world. *Areca catechu* L. is the only cultivated species under the genus *Areca* L. *Areca* species and their geographical distribution are (Table 1). It is an erect, unbranched palm with a prominent crown shaft, reaching a height of up to 18-20 m with a trunk 25-40 cm in diameter. The stem has scars of fallen leaves in regular annulated forms. It possesses an adventitious root system and more than 70 per cent of roots are distributed within a radius of about 50 cm. Almost 80 per cent of roots do not go beyond a radius of 85 cm. Depth wise, 0–50 cm contained about 70 percent of the roots and further from 51 to 100 cm about 18-24 percent. The slender stem had crown by spirally arranged large elegant 7-12 leaves which are pinnatisect and consist of a sheath, a rachis, and leaflets. Fronds are 1-1.5 m long with 30-50 pinnae (leaflets). Depending on the position of the leaf, leaflets length and breadth varies from 30.0 to 70.0 cm and 5.8 to 7.0 cm respectively.

Arecanut is a cross-pollinated, monoecious palm, bears male (=staminate) and female flowers (=pistillate) in the same spadix, which is surrounded by one or more bracts or spathes that become woody at maturity (Bavappa and Ramachander, 1967). The number of spadices produced depends upon the number of leaves. The mean number of spadices produced is 3 to 4, depending on the age of plant (Murthy and Bavappa, 1960). Spadix are crowded with much branched panicles. Each terminal branch has a few female flowers borne at the base and numerous male flowers extending from there out to the branch tip. Flowers of both sexes have six tepals, are stalkless (=sessile), creamy-white, fragrant. Male flowers are minute, deciduous, have six stamens, arrowhead-shaped anthers, rudimentary ovary. Female flowers are larger (1.2–2 cm long), with six small sterile stamens and a three-celled ovary bearing a triangular stigma with three points at the apex. Male flowers last 4–7 weeks, and female flowers, which are cream colored at flowering, turn green within a week. Flowers open between 2and 10 am, female flowers last up to 10 days. The stigma remains receptive up to a week (Murthy and Bavappa, 1960). Pollen grains are generally carried by wind.

The fruit is a monocular one-seeded berry that is orange-red to scarlet in color when ripe, encircled by a thick fibrous outer layer, akin to a husk. Each branch bears 100–250 fruits. Seed usually, ovoid, globose, or ellipsoidal, base sometimes flattened; endosperm ruminate, embryo conical, located at seed base. Fruit development takes place in three stages. Increase in size is rapid in the first phase, followed by increase in volume and dry matter accumulation in kernel in the second phase. During this period, the embryo becomes macroscopic and develops rapidly. In the third stage, the fruit swells, and gradually the green colour fades giving rise to a bright yellow colour. In all, it takes about 9 months to a full year for fruit to develop for harvesting (Bavappa *et al.*, 1982).

Table 1: Areca species and their geographical distribution

Country	Species
India	*Areca catechu, Areca triandra*
Andaman Islands (India)	*Areca catechu, Areca laxa*
Sumatra	*Areca catechu, Areca triandra, Areca latiloba*
Sri Lanka	*Areca catechu, Areca concinna*
Malaysia	*Areca catechu, Areca triandra, Areca latiloba, Areca montana, Areca ridleyana*
Borneo	*Areca catechu, Areca borneensis, Areca kinabaluensis, Areca arundinacea, Areca bongayensis, Areca amojahi Areca mullettii, Areca minuta, Areca furcate*
Java	*Areca catechu, Areca latiloba*
Celebes	*Areca celebica, Areca oxicarpa, Areca paniculata, Areca henrici*
Australia	*Areca catechu, Areca alicae*
Solomon Islands	*Areca niga-solu, Areca rechingeriana, Areca torulo, Areca guppyana*
New Guinea	*Areca congesta, Areca jobiensis, Areca ladermaniana Areca macrocalyx, Areca nannospadix Areca warburgiana*
Philippines	*Areca catechu, Areca hutchinsoniana, Areca vidaliana Areca costulata, Areca macrocarpa, Areca parens Areca caliso, Areca whitfordii, Areca camariensis, Areca ipot*
Moluccas	*Areca glandiformis*
Bismarck Islands	*Areca novo hibernica*
Laos	*Areca laosensis*
Cochin China	*Areca triandra*
Lingga Islands	*Areca hewittii*

Cytology

The commercially cultivated species, *Areca catechu* L., chromosome number was first reported by Venkatasubban (1945) as 2n=32. Later the chromosome number of its was confirmed by Sharma and Sarkar (1956), Raghavan and Baruah (1958), Abraham *et al.* (1961) and Bavappa and Raman (1965). Darlington and Janaki Ammal (1945) reported chromosome number of *A. triandra* as 2n=32. Meiotic chromosome number of *A. macrocalyx* Becc as n=16 was reported by Nair and Ratnambal (1978). In the somatic chromosome complement of *A. catechu*, three pairs of long chromosomes, six pairs of medium-sized chromosomes and seven pairs of short chromosomes were observed (Sharma and Sarkar, 1956). They also observed that the chromosomes of *A. triandra* were longer than those of *A. catechu*.

The chromosomes of *A. catechu* and *A. triandra* differ in size, total chromatin length, position of primary and secondary constrictions and number and position of satellites (Bavappa and Raman, 1965). Bavappa and Raman (1965) considered *A. catechu* as more advanced than *A. triandra*.

Genetic Resources

Its organized crop improvement programmes was started at ICAR-Central Plantation Crops Research Institute (CPCRI), Regional Station, Vittal, Karnataka, to improve is productivity. Comprehensive collections of germplasm from within the country and abroad were collected and screening was done under uniform conditions was undertaken since 1957 (Bavappa and Nair, 1982). A total of 178 accessions were collected so far and being maintained in the field gene bank at the station. Out of which, 155 indigenous ecotypes of arecanut collected from different parts of India as well as 23 exotic accessions introduced from other areca growing countries of the world especially South-East Asian countries such as Fiji, Mauritius, South China, Sri Lanka, Indonesia, Saigon, Singapore, British Solomon Islands and Australia which represents five species *viz.*, *Areca catechu* L., *A. triandra* Roxb., *Areca concinna* and related genera *Normanbya normanbyii*, *Actinorhytis calapparia* (Ananda, 2010, CPCRI, Kasaragod, 2017).

The indigenous arecanut germplasm are collected from Assam, Goa, Gujarat, Karnataka, Kerala, Maharashtra, Meghalaya, Tamil Nadu, West Bengal, Andaman and Nicobar group of Islands etc. Some of the germplasm collections are sub sampled and planted in ICAR-CPCRI, Research Centre, Mohitnagar, West Bengal, as alternative germplasm repository for arecanut. Described 75 accessions based on plant morphology, reproductive and fruit component traits and also arecanut germplasm database has been created for 43 indigenous and exotic accessions which include salient features, bunch, inflorescence and nuts of accessions and are illustrated through photographs of different species and genera (Ananda, 2007).

On the basis of comparative yield trials of indigenous and exotic accessions, promising tall and semi-tall cultivars were selected and released as varieties. Under the evaluation of exotic accessions and selection for high yield and its attributes varieties released were Mangala, Sumangala, Sreemangala and Swarnamangala, while Mohitnagar, Kahikuchi, Madhuramangala, Nalbari and Shatamangala were released under indigenous accessions.

Though tall varieties possess high yield potential, they are prone to wind damage. The tall nature of palm hinders various operations like spraying and harvesting which are quite labour intensive and cumbersome. Therefore, arecanut breeding programme in addition to yield improvement are also aimed at development of dwarf varieties/hybrids. Hirehalli Dwarf (HD) a natural mutant identified for its short stature, is a good genetic source for arecanut improvement. Dwarf varieties with high yield potential will directly benefit the growers by way of enhanced returns and reduced cost of various cultural operations like harvesting, spraying, in addition to minimising damages to palms due to heavy wind owing to greater mechanical support of stem. Therefore, exploitation of dwarfing genes in breeding dwarf hybrid varieties with high yield potential was initiated. Hybrids involving Hirehalli Dwarf and released high-yielding tall/ semi-tall varieties (Mangala, Sumangala, Sreemangala and Mohitnagar) as parents, were developed and evaluated for yield performance and dwarfness. Among hybrids, HD x Sumangala and HD x Mohitnagar were identified as superior for yield with relatively dwarf plant habit and recommended for cultivation as, VTLAH-1 (Vittal Areca Hybrid-1) and VTLAH-2 (Vittal Areca Hybrid-2), respectively.

Varieties and Hybrids

Mangala

This variety is a selection from VTL-3, an accession introduced from China and possesses a number of desirable characters such as earliness in bearing, more number of female flowers per inflorescence, higher nut set, initial and cumulative higher yield, quicker stabilization of production and lesser height in comparison with South Kanara Local (S.K. Local). The average yield of this variety is 2.90 kg chali/palm/year. The variety was released for commercial cultivation in the coastal areas of Karnataka and Kerala during 1972 as Mangala and is characterized by partially drooping crown with well spread leaves and more number of leaflets as compared to S.K. Local. The leaflets are dark green in colour with characteristic crinkling at the tip.

Sumangala

This variety is developed from VTL-11 accession obtained from Indonesia and evaluation for yield traits showed an increase in yield over South Kanara Local. In view of the substantial increase in yield, the variety was released for all areca-growing areas in general and coastal Karnataka and Kerala in particular during 1985. It is a tall type with partially drooping crown. Under good management palms flowers in 4-5 years. The colour of the ripe nuts is deep yellow to orange and oblong to round in shape. The variety recorded an average yield of 3.28 kg chali/palm/year.

Sreemangala

The variety was developed from the accession VTL-17 introduced from Singapore and showed high yield potential with increase in yield over S.K. Local. The palm is tall with partially drooping crown with longer internodes and sturdy stem. It starts flowering in 4-5 years. It is high yielder with an average chali yield of 3.18 kg per palm per year. Ripe nuts are usually bold in size and round in shape with deep yellow colour. This cultivar was released Sreemangala during 1985 for coastal areas of Karnataka and Kerala.

Mohitnagar

It is a selection from the indigenous arecanut accession VTL-60 collected from West Bengal, with high yield potential and was released for commercial cultivation during 1991. The important feature of this variety is its greater uniformity. The bunches are well placed and nuts are loosely arranged on spikes which help in their uniform development and also enable efficient plant protection measures. Early stabilization in yield has been observed as compared to Sumangala and Sreemangala. The variety is a consistent high yielder with an average chali yield of 3.67 kg/palm/year. This variety is recommended for cultivation in arecanut growing areas of West Bengal and coastal areas of Karnataka and Kerala.

Swarnamangala

It is a high-yielding selection from the exotic accession VTL-12 collected from Vietnam. It is a tall high yielding variety with medium thick stem and partially drooping crown and homogeneous population. The palms flowers 4-5 years after planting. Nuts are bigger and heavier with high recovery of chali or dry kernel. Average yield of this palm is 3.88 kg chali/ palm/year. It is recommended for cultivation in Karnataka and Kerala and was released during 2005.

Kahikuchi Tall

It is a selection from indigenous accession VTL-64 from Assam, possess high yielding nature with medium thick stem, longer internodes, partially drooping crown, homogeneous population and regular bearing. The fruits are orange in colour, bold and round shaped, with high recovery of chali from fresh nuts. The palms come to bearing by 5th year. The variety is a consistent high yielder with an average yield of 3.70 kg dry kernel/palm/year. The variety is recommended for commercial cultivation in Assam and Meghalaya during 2008.

Madhuramangala

It is a selection from indigenous accession VTL-62 collected from Maharashtra. The yield performance of the variety is higher than the released varieties, *viz.* Mangala, Sumangala, Sreemangala and also traditional local types. The variety is suitable for both tender nut and ripe nut processing and also fetches more price in the market because of its quality and marble appearance of the split nut. The average yield is 2.95 kg dry tender processed nuts/palm/year or 3.54 kg dry kernel/ palm/year. The variety was notified for commercial cultivation in Karnataka and Maharashtra during 2014.

Nalbari

It is a selection from indigenous germplasm (VTL-75) collected from Assam. The yield performance of the variety is higher as compared to all the earlier released varieties (Mangala, Sumangala, Sreemangala, Mohitnagar, Swarnamangala, Kahikuchi etc.) and found suitable for ripe nut processing. The variety possesses high yielding nature, with medium thick stem, longer internodes and partially drooping crown. The variety is a regular bearer with consistent yield and bunches are well placed on the stem. The fruits are round shaped, yellow in colour with high recovery of dry kernel from fresh nuts. The palms come to bearing by 5th year. The average yield is 4.15 kg dry kernel/palm/year. The variety has been notified for cultivation in Karnataka, West Bengal and North Eastern region during 2014.

Shatamangala

It is a high yielding arecanut variety developed from VTL-146, the indigenous accession collected from Gujarat and was released during 2016 to commemorate the centenary year of ICAR-CPCRI and notified during 2021. The yield performance of the variety is higher than the released varieties, *viz.* Mangala, Sumangala, Sreemangala and traditional local types, and is suitable for both tender nut and ripe nut processing. The average yield is 3.26 kg dry tender processed nuts/palm/ year or 3.96 kg dry kernel/palm/year. The variety is recommended for commercial cultivation in Karnataka and Gujarat.

VTLAH -1: Vittal Arecanut Hybrid-1

It is hybrid between hybrid between Hirehalli Dwarf (female parent). It is hybrid between . The palms have sturdy stem with super imposed nodes, reduced canopy size, well spread leaves, partial drooping crown. The fruits are medium sized, oval to round in shape and yellow-orange in color. Early stabilization in yield and high recovery of chali are the other striking features of this hybrid. The average chali yield of this hybrid is 2.54 kg/palm/year. This variety is recommended for commercial cultivation in arecanut growing tracts of Karnataka.

VTLAH -2: Vittal Arecanut Hybrid-2

It is involving Hirehalli Dwarf (female parent) and Mohitnagar (male parent). It is dwarf in growth habit and was released in 2007. Medium thick stem with super imposed nodes, reduced canopy size, well spread leaves, drooping crown, medium sized oval nuts, early stabilization in yield and high recovery of chali are the striking features of this hybrid. The average chali yield of this hybrid is 2.64 kg/ palm/year. This variety is recommended for commercial cultivation in Karnataka.

References

Abraham, A, Mathew, P.M., and Ninan, C.A. 1961. Cytologia 26: 327-32.

Ananda, K.S. 2007. Arecanut Descriptors, Part-I, Central Plantation Crops Research Institute, Kasaragod, Kerala, p. 172.

Ananda, K.S. 2010. Collection, evaluation and utilization of genetic resources in Arecanut (Areca catechu L.). In: National Conference on Horticultural Biodiversity for Livelihood, Economic Development and Health Care (*Swadesh Prem Jagriti Sangosthi*-2010) UHS, Bangalore, 28-31 May, 2010.p.11-12.

Bavappa, K.V.A., Nair, M.K. and Premkumar, T. (1982). The Areca Palm (Areca catechu Linn.) Central Plantation Crops Research Institute, Kasaragod, Kerala, India.

Bavappa, K.V.A. and Nair, M.K. 1982. Cytogenetics and Breeding. In: The Arecanut Palm. Bavappa, K.V.A., Nair, M.K. and Prem Kumar, T (Eds.) Central Plantation Crops Research Institute, Kasaragod, pp 51-96.

Bavappa, K.VA., and Ramach*ander, P.R. 1*967. Indian Journal of Genetics and plant Breeding 27: 93-100.

Bavappa, KVA, and Raman, VS. 1965. Journal ofindian Botanical Society 44: 495- 505.

CPCRI, Kasarogod 2017. Annual Report 2016-17. Central Plantation Crops Research Institute, Kasaragod, Kerala, India.

Darlington, CD., and]anaki Aml11al, E.K. 1945. Chromosome Atlas of Cultivated Plants. London: George Allen and Unwin Ltd. P. 397.

FAO STAT. 2020. https://www.fao.org/faostat/en/#data/QCL accessed on 10.08.2022.

Mohan Rao, M. 1982. In: The Arecanut Palm, edited by Bavappa, K.V.A., Nair, M .K., and Premkumar, T., Central Plantation Crops Research Institute, Kasaragod, India, pp. 1-9.

Multhy, K.N., and Bavappa, K. VA. 1960. Arecanut Journal 11: 51-55.

Nagaraja, N.R., Elain Apshara, S. and Niral, V. 2022. Improved varieties of coconut, arecanut and cocoa for higher productivity. Indian Horticulture 67(6): 31-37.

Nagaraja, N.R. and Ananda, K.S. 2021. Improved high yielding varieties and hybrids of arecanut. Extension Folder no. 291. ICAR-CPCRI, Kasaragod, Kerala.

Nair, M.K., and Ratnambal, M.J. 1978. Current Science 47: 172-73.

Raghavan, Y., and Baruah, H.K. 1958. Economic Botany 12: 315-45.

Shanna, AX., and Sarkar, S.K. 1956. Genetics 28: 361-488.

Venkatasubban, K.R. J945. Proceeding of/he Indian Academy of Science 22:193-207.

Watt, G. 1889. Vol. Periodicals Experts, Delhi P. 559.

Section 05: Crop Protection

21

Ecological Engineering A New Approach for Pest Management in Horticultural Crops

K. Rolania[1], A. Kumar, S. Pilania and S.M. Haldhar[2]

[1]*Department of Entomology, COA, CCS Haryana Agricultural University Hisar, Haryana*

[2]*Department of Entomology, COA, Central Agricultural University, Imphal*

Ecological engineering has recently emerged as a paradigm for considering pest management approaches that rely on the use of cultural techniques to manipulate farm habitat and to enhance biological control. In contrast to the intensive use of ecosystems, these days to enhance productivity can affect agro-ecosystems through soil and water depletion, biodiversity loss, challenging pest problems and disruption in flow of ecosystem services. Ecological engineering for pest management mainly focuses on increasing the abundance, diversity and function of natural enemies in horticultural crops habitats by providing refuges and alternate or supplementary food resources and also attracts different kinds of pollinators like honeybees. The aim of ecological engineering in horticulture ecosystem is to integrate soil and pest management strategies with regular practices of farmers for the benefit of environment and farming community. It involves knowledge of horticulture, ecology and farm economics, for restoration and construction of healthy and sustainable horticulture ecosystems. By redesigning the horticulture ecosystem, farmers can enhance biodiversity on their farms through adopting polyculture, cover crops, corridors, crop rotations and various habitats. This is a key strategy in sustainable horticulture to enhance biodiversity at the landscape and field level. For example, the planting of buckwheat, *Fagopyrum esculentum* as a cover crop in vineyards and alyssum, *Lobularia maritima* between rows of vegetables provide resources for predators and parasitoids resulting in reduced herbivore damage. The main approach in ecologically-based pest management is to increase agro-ecosystem diversity and complexity as a foundation for establishing beneficial interactions that keep pest populations in control.

Ecosystems, rich in biodiversity support economic growth, sustainable development and human wellbeing. The continuous loss of biodiversity, bringing about in drastic reductions in ecosystem goods and services, contrarily affecting economic wealth and environmental sustainability is disturbing the ecological equilibrium. Earth's ecosystems are degrading as a result of damage, unsustainable development and a failure to invest and reinvest in their productivity, health and sustainability and overexploitation of natural resources. Horticulture is impacted by a variety of biotic and abiotic factors where a strategic science based approach is needed to deal with the problems of plant health and productivity. The integrity and conservation of horti-ecosystem is the key for sustainable agriculture. Enhanced productivity based intensive horti-ecosystems has created a multitude of problems including soil erosion, water depletion/contamination, biodiversity loss, challenging pest problems and disruption in flow of ecosystem services, which will directly affect the plant health and safe food production (Edpuganti and Rajan, 2018).

In the early 1970s, Integrated Pest Management (IPM) concept arose in response to concerns about effects of pesticides on the environment and has proved to be an incredible strategy for reducing the dependence of crop protection on chemical control methods (Prokopy and Kogan, 2003; Kogan, 1998). IPM requires a well-planned methodology which includes the combined use of various pest-control methods, *viz.* mechanical, physical and cultural control; host plant resistance; biological control *etc*; informed by monitoring of pest densities. Each of the specific methodological tactics used in IPM has tended to become a specialised area of research with sometimes only a restricted correspondence between scientists across regions. This has prompted calls for more noteworthy collaboration and exchange of ideas between different sub-disciplines. On account of biological control, Gurr and Wratten (1999) proposed the idea of "integrated biological control", which utilizes conservation biological control techniques to support classical, inoculation and inundation biological control. According to Eilenberg *et al.* (2001), conservation biological control (CBC) has been defined as "modification of the environment or existing practices to protect and enhance specific natural enemies of other organisms to reduce the effect of pests".

There are two approaches that affect CBC they are:

- To reduce the factors causing mortality of natural enemies *i.e.* better targeting of pesticides in time and space, reducing rates of application or using narrower spectrum compounds
- To improve natural enemy fitness and effectiveness by habitat manipulation.
    - With the help of second approach the species diversity and structural complexity of horti-ecosystems can be increased. In regard to CBC, habitat manipulation aims to provide natural enemies with resources such as nectar (Baggen and Gurr, 1998), pollen (Hickman and Wratten, 1996),

physical refugia (Halaji *et al.*, 2000), alternative prey (Abou-Awad, 1998), alternative hosts (Viggiani, 2003) and liking sites (Sutherland *et al.*, 2001). Ecological engineering involves the habitat manipulation approaches which enhance the natural enemies by providing these resources and operate to reduce pest densities.

Ecological Engineering

Dr Howard T. Odum (1962) was the first to coin the term, "ecological engineering," which was defined as "environmental manipulation by man using small amounts of supplementary energy to control systems in which the main energy drives are still coming from natural sources". As per Ahmad and Pathania (2017), ecological engineering can also be termed as "habitat manipulation". In later years, ecological engineering have defined as "the design of human society with its natural environment for the benefit of both" by Mitsch and Jorgensen (1989). Ecological engineering for pest management has recently emerged as a paradigm for considering pest management approaches that rely on the use of cultural techniques to manipulate farm habitat and to enhance biological control (Gurr *et al.*, 2004) and informed by ecological knowledge rather than on high technology approaches such as synthetic pesticides and genetically engineered crops.

Ecological Engineering vs. Ecosystem Engineering

Ecological engineering differs from the more newly recognized concept "ecosystem engineering" in that it is an intentional human endeavour. This refers to how other animals shape habitats via their innate biology rather than through deliberate construction.Termites, for example, modify the structural properties of soils (Dangerfield *et al.*, 1998), so affecting the availability of nutrients to other organisms (Thomas *et al.*, 1999).

The term "eco technology" is a synonym for ecological engineering. Restoration ecology, sustainable agro ecology, habitat reconstruction, ecosystem rehabilitation, river and wetland restoration, and reclamation ecology are all allied to ecological engineering (Mitsch, 2004). These sub-sets show the variety of domains where ecological engineering has been used, such as wetlands restoration, waste water treatment and utilisation, integrated fish culture systems, mining technology (Mitsch and Jorgensen, 1989) and wildlife conservation (Morris *et al.*, 1994).

Pest Management vs. Ecological Engineering

It might be contradicted that all pest control measures are forms of ecological engineering, regardless of whether they affect the physical environment (*i.e.*, through tillage), the chemical environment (*i.e.*, through pesticide use), or the biotic environment (*i.e.*, via the use of novel crop varieties). The adoption of cultural practices to manipulate habitat and improve biological control, on the other hand, fits the idea of ecological engineering the best. These cultural practises usually:

- Need very low energy or material inputs;
- Depends on natural processes (for example, natural enemies or herbivore responses to vegetation diversification);
- Have evolved in accordance with ecological principles;
- Are fine-tuned by ecological experimentation;
- Contribute to theoretical and practical ecological knowledge.

The primary goal of ecological engineering for pest management is to increase the number, diversity and function of natural enemies in horticultural ecosystems by offering refuges and alternate or supplemental food supplies (Gurr *et al.*, 2012). Ecological engineering is a horti-ecological component that promotes accuracy (a trait of engineering) in the outcome of an intervention. While the idea has been implemented with some assurance of efficacy in ecosystem restoration and landscape productivity, utilising functional plants that frequently operate as bio-filters or provide nutrients to the system (Lu *et al.*, 2015; Mitsch, 2004), it is still in its infancy in pest management, notably in horticulture crops. In practice, planting buckwheat, *Fagopyrum esculentum* Moench, as a cover crop in vineyards (Berndt*et al.*, 2006), and Alyssum, *Lobularia maritima* (L) Desv., between rows of vegetables (Gillespie *et al.*, 2011), provides resources for predators and parasitoids, resulting in the less herbivore damage. The basic strategy in ecological pest control is to improve agro-ecosystem diversity and complexity as a basis for generating positive interactions that maintain pest population under control (Gurr *et al.*, 2004).

Evolution of Ecological Engineering

Habitat modification has its roots in tactics that have been employed in agricultural systems for ages to increase generalist predators (Sweetman, 1958). The use of straw shelters to create temporary spider refugia and overwintering places during cyclic agricultural disturbances is an example of an early habitat modification method that has been employed by Chinese farmers for over 2000 years and is still in use today (Dong and Xu, 1984). Another method, established in Burma in the 1770s, utilised bamboo canes to connect citrus trees, allowing predatory ants to migrate between them and remove caterpillar pests (van Emden, 1991).

Ecological Engineering for Pest Management – Above Ground

- Raise the flowering plants/ compatible cash crops along the orchard border by arranging shorter plants towards main crop and taller plants towards the border to attract natural enemies as well as to avoid immigrating pest population

- Grow flowering plants on the internal bunds inside the orchard
- Not to uproot weed plants those are growing naturally like *Tridax procumbens*, *Ageratum* sp, *Alternanthera* sp etc. which act as nectar source for natural enemies,
- Not to apply broad spectrum chemical pesticides, when the P: D ratio is favourable. The plant compensation ability should also be considered before applying chemical pesticides.

Ecological Engineering for Pest Management – Below Ground

- Keep soils covered year-round with living vegetation and/or crop residue.
- Add organic matter in the form of farm yard manure (FYM), vermicompost, crop residue which enhance below ground biodiversity.
- Reduce tillage intensity so that hibernating natural enemies can be saved.
- Apply balanced dose of nutrients using biofertilizers.
- Apply mycorrhiza and plant growth promoting rhizobacteria (PGPR)
- Apply *Trichoderma* spp. and Pseudomonas fluorescens as seed/seedling/ planting material, nursery treatment and soil application (if commercial products are used, check for label claim. However, biopesticides produced by farmers for own consumption in their fields, registration is not required).

Habitat Manipulation

Top down control: Here, natural bio-agents (third trophic level) inhibit herbivores (second trophic level), and this sort of strategy is shown in "Augmentative biological control." Pest suppression as a result of this phenomenon, according to Root (1973), supports the "enemy" hypothesis.

Bottom up control: Pests will be immediately affected by manipulations inside the crop, such as green mulches and cover crops (first trophic level) under this approach. This strategy is used in "Conservation biological control" as habitat alteration. Root (1973) used the phrase "resource concentration hypothesis" to describe pest suppression caused by non-natural enemy impacts, implying that the resource (crop) was successfully "diluted" by signals from other plant species.

Horti-ecosystem Management

Many strategies for managing the Horti-ecosystem can work independently of natural enemies (Fig.1) and, as discussed below, constitute a type of ecological engineering.

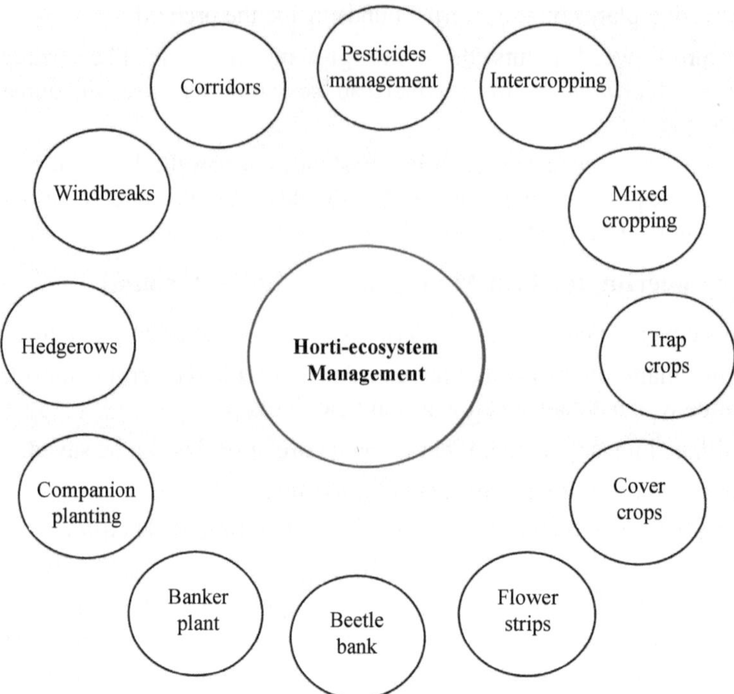

Fig. 1: Horti-ecosystem management strategies to enhance farm biodiversity

Intercropping

Intercropping is a key cultural strategy in pest management, based on the principle of minimising insect pests, by enhancing the diversity of horti-ecosystems. The practise of cultivating two or more crops (typically from different families) in area and time is known as intercropping. Strip cropping is a type of intercropping in which two or more crops are grown in alternating strips (Rao *et al.,* 2012). Both approaches boost biodiversity while making the environment less conducive to pest development. When planted in a monoculture rather than a mixed crop, pests have an easier job locating host plants. This is based on the "Resource Concentration Hypothesis," which states that plant-feeding insects are more likely to identify and stay in patches of their host plants that are dense and less diversified (Root, 1973). Maqsood *et al.* (2022) evaluated impact of ecological engineer-ing on the flea beetles, *Phyllotreta striolata* and *Altica himensis* and their natural enemies on brinjal crop. It showed that Treatment I caused maximum increase in mean number of natural enemies (1.11/ 10 plants) which in turn brought maximum mean pest reduction. Treatment II caused second maximum increase in mean number of natural enemies (0.92/ 10 plants). Treatment III caused minimum increase in mean number of natural enemies (0.68/ 10 plants). The diversity of predators was documented in different treatments. Simpson's diversity index, Shannon-

Weiner index and Evenness index were found higher in Treatment I followed by Treatment II and Treatment III. The maximum mean percent increase of natural enemies in main crop over control (250.52 %) with maximum mean % reduction of target pest (63.46 %) was observed in Treatment I. The non-host plants when raised with host plants were found to reduce pests and pro-mote natural enemies. Border crop and intercrop along with main crop hinders insect pest development and favours natural enemies by providing supplementary food and refugee (Staver *et al.*, 2001). Marigold increases longevity and fecundity of natural enemies by providing nectar and pollen (Baggen *et al.*, 1999). Results also agree with finding that insect pest number was reduced in brinjal crop when brinjal was intercropped with coriander (Khorsheduzzaman *et al.*, 1997).

Other interesting examples of ecofriendly pest management by intercropping in various horit-ecosystems, such as cabbage with mustard and beans with corn, can be evaluated.

Table 1: Cropping pattern in different treatments of the experiment

	Main crop	Trap crop	Attractant	Repellent	Perimeter
Treatment 1	Brinjal {72) Sofanum meiongena L.	Sunflower as border (21) Helianthus annuus L.	Coriander as Intercrop (24) Coriandrum sativum	Onion as buffer (32) Allium cepa	Marigold (56) Tagetes sp.
Treatment 2	Brinjal {72) Solanum meiongena L	Turnip as buffer (32) Brassica rapa	Buckwheat as border (21) Fagopyrum escuientum	Mint as intercrop (24) Mentha sp.	Marigold (56) Tagetes sp.
Treatment 3	Brinjal {72} Solanum metongena L.	Marigold as buffer (32) Tagetes sp.	Buckwheat as intercrop (24) Fagopyrum escuientum	Nettle as border (21) Urtico dioica	**Dill (56)** Anethum graveo- lens

Source: Maqsood *et al.*, 2022

Mixed Cropping (push pull system)

In this cropping system, specifically chosen companion plants are grown in between the main crop to act as a push component and on the border to act as a pull component. Using semio-chemicals, the 'push' component repels insect pests while the 'pull' component (trap crop) attracts insect pests away from the main crop (Cook *et al.*, 2007). A thorough understanding of the chemical ecology of plant–insect interactions on various crops is required for such a system. Candidate crops should be tested extensively in field trials. In smallholder vegetable production in Kenya, a push–pull technique was designed at the International Centre of Insect Physiology and Ecology (ICIPE) especially for the management of vegetable pests. It was revealed that some intercrops (silver leaf, *Desmodium uncinatum*, and green leaf, *Desmodium intortum*) provided additional benefits in terms of Striga weed reduction. For a higher yield, this impact is equally as significant as pest control in vegetables. However, because intercrop root exudates (non-volatile) reduce Striga germination, this underlying mechanism has an allelopathic impact and hence only needs the intercrop component of the push–pull system (Khan *et al.*, 2010).

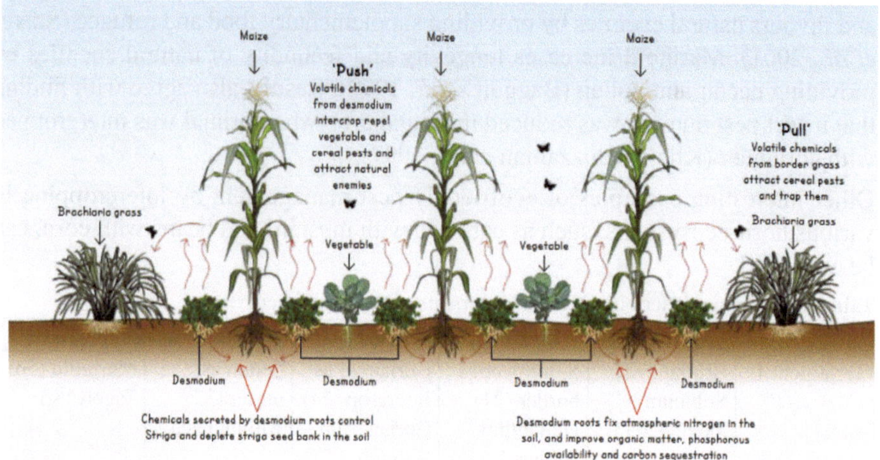

Fig. 2: Push-pull systems in vegetable crops for pest management (source: https://upscale-h2020.eu/wp-content/uploads/2022/05/Push%E2%80%93Pull-System-Intensification-With-Vegetables-Upscale.pdf)

***Trap crops*:** A trap crop is a crop that is grown to entice insect pests away from the cash crop (main crop). The trap crop can be a different plant species, variety, or growth stage of the same species, as long as it is more appealing to the pests while they are present. The trap crop must be more attractive to pests than the cash crop, and precautions must be made to prevent pests in the trap crop from migrating to the cash crop later. Therefore, successful trap crop utilisation is a challenging task (Sreedhar and Krishnamurthy, 2009). Pests that are poor fliers and/or use the wind to disseminate are ineffective against trap crops (e.g., aphids, spider mites). Trap crops were developed to be used in traditional pest control systems where pesticides may be used to kill the pests trapped. In organic systems, approved insecticides can be used, but pests can also be eradicated by destroying the crop. The timing is pivotal; if the trap crop is destroyed too early or too late, it can eliminate the advantages of traps or possibly cause the mass movement of insect pests into the crop. Trap crops are therefore important in reducing insect populations in many horti-ecosystems. According to the study of Nafziger and Fadamiro (2011) on *Microplitis croceipes* wasp, the greatest longevity was recorded for honey-fed wasps followed bythe longevity observed on buckwheat, licorice mint and sweet alyssum.

Table 2: Important trap crop for pest management in horticultural crops

Trap crop	Main crop	Method of planting	Pest controlled
Basil and marigold	Garlic	Border crop	Thrips
Castor plant	Cotton	Border crop	Heliotis sp.
Chervil	Vegetables, ornamentals	Among plants	Slugs
Chinese cabbage, mustard, and radish	Cabbage	Planted in every 15 rows of cabbage	Cabbage webworm, flea hopper, mustard aphid
Collards	Cabbage	Border crop	Diamondback moth
Dill and lovage	Tomato	Row intercrop	Tomato hornworm
Horse radish	Potato	Intercrop	Colorado potato beetle
Hot cherry pepper	Bell pepper	Border crop	Pepper maggot
Indian mustard	Cabbage	Strip intercrop in between cabbage plots	Cabbage head caterpillar
Marigold (French and African marigold)	Solanaceous, crucifers, legumes, cucurbits	Row/strip intercrop	Nematodes
Medick (Medicago littoralis)	Carrot	Strip intercrop in between carrot plots	Carrot root fly
Onion and garlic	Carrot	Border crops or barrier crops in between plots	Carrot root fly Thrips
Radish	Cabbage family	Row intercrop	Flea beetle, root maggot
Nasturtium	Cabbage	Row intercrop	Aphids flea beetle, cucumber beetle, squash vine borer
Tansy	Potato	Intercrop	Colorado potato beetle
Tomato	Cabbage	Intercrop (tomato is planted 2 weeks ahead at the plots' borders)	Diamondback moth

Sources: http://agropedia.iitk.ac.in/content/trap-crops-tool-managing-insect-pests-damage

Cover Crops

Cover crops include living mulches, green manures, and catch crops. Some cover crops like Dhaincha are incorporated into the soil to serve as nitrogen source and organic matter to the subsequent crop, while typically legumes or grasses, are grown to minimise soil erosion and weed suppression. When compared to bare tilled soil, cover crops were found to lower soil temperature, increase relative humidity (RH), and make easy availability of more free water. During winters, they can also serve as a shelter for the beneficial species such as spiders and ground beetles. A more intensive strategy is to plant insectary strips of temporary nectar and pollen plants such as buckwheat, sweet clover, white clover and some brassicas (mustards, oilseed radish) within the crop field itself. Beneficial organisms will migrate from the cover crop to the crops once the cover crops have ceased flowering, however mowing the cover crops when beneficial organisms are required on the main crop will stimulate an earlier migration (Meena *et al.*, 2017). Another unique strategy is the use of brassicas (Mustard, oilseed, canola and radish) as cover crops which then ploughed into the soil and the substances released during decomposition helps in nematode management (Dana, 2002).

As per Nicholls *et al.* (2001), vineyard systems with flowering cover crops were characterized by lower densities of leafhopper nymphs and adults than monoculture vineyards in Hopland. Despite the fact that *Anagrus epos*, the most important leafhopper parasitoid, found enough food in the cover crops, few travelled to the grapevines to look for leafhopper eggs.As a result, every other row of cover crops was mowed to push *Anagrus* wasps and predators onto the vines.Leafhopper nymphal densities on vines were similar in the cover-cropped rows before mowing. One week after mowing, numbers of nymphs declined on vines where the cover crop was mowed, coinciding with an increase in *Anagrus* densities in mowed cover crop rows. During the second week, such nymphal decline was even more pronounced, coinciding with an increase in numbers of *Anagrus* wasps in the foliage.

Flower Strips

Tshumi *et al.* (2016) observed the effect of flower strips on the natural enemies of aphid on potato by growing the seed mixture of the annual flower strip to provide continuous and high levels of floral and extra-floral nutrients that are attractive and accessible to important natural enemies of aphids.After a thorough review of literature, flowering plants such as *Anethum graveolens* L. (Apiaceae), *Bellis perennis* L. (Asteraceae), *Calendula arvensis* L. (Asteraceae), *Anthriscus cerefolium* Hoffm. (Apiaceae), *Camelina sativa* (L.) Crantz (Brassicaceae), *Centaurea cyanus* L. were selected. These flower strips improved biological control of aphid in neighbouring potato fields while also providing benefits to natural enemy biodiversity and the use of insecticides was also reduced. Berndt and Wratten (2005) observed that the spatial distribution of natural enemies in and around blooming buckwheat and alyssum crops enhances the fertility and lifespan of *Dolichogenidea tasmanica*, an important parasitoid of leaf roller (*Epiphyaspost vittana*) in vineyards, where parasitism rates rise and leaf-roller densities fall. The abundance of predators (*Orius* spp., Coccinellids and Syrphids) and thrips was studied in insectary and at 30 and 60 m away from it by Nicholls *et al.* (2001). *Orius* was found in low abundance in vines away from the insectary, a trend that was negatively correlated to thrips densities. A farmer's decision to establish flower strips, on the other hand, will be based on benefit cost ratio. As an extra economic incentive for farmers to invest in this management tool, flower strips may need to be pushed through horti-environmental subsidies or incentives (Tshumi *et al.*, 2016).

Beetle Bank

Due to loss of hedgerows and other natural habitats, under current horticultural practices, edges of fields serve as the overwintering habitat for the predatory insects. These predators that survived the winter in the perennial grasses near the field boundary or bunds would take substantial amount of time to penetrate the crop or may not travel or reach to the centre, lowering the efficacy of natural control. So the beetle bankconcept emerged, with a goal to control insect pest by increasing the predator density of the area, to provide refuges or "islands" of diversity within fields to create ideal environment for predatory insects in non-crop period. Beetle banks are permanent strips of natural vegetation within the crop, created by making ridges of suitable height (0.5 m – 0.75 m) and width (2 m) along the full length of fields. One beetle bank is required for every 20 hectare field in European countries such as the United Kingdom. With this density of beetle banks, spread of predatory beetles from beetle banks in the spring resulted in a uniform distribution throughout horticulture crops after the bank had been formed for three years (Thomas *et al.*, 1991).

Banker Plant

Pest and natural enemy-infested plants that are brought to a region as a source of natural enemies. By providing an alternate food source, banker plants serve as a site for raising bio-control agents in the greenhouse. The banker plants can be the same crop/crop pest as the one you're seeking to eradicate, or they can be a different host and prey. Examples include: green peach aphid control with bird cherry aphid on wheat; whitefly management in greenhouses with greenhouse whitefly on eggplant (Osborne *et al.,* 2005).

Companion Planting

This is a broad term that refers to the addition of specific plants to help nearby crops grow and produce better. Companion plants are commonly used in pest management to prevent or repel pests. The African marigold, for example, produces thiopene, a nematode repellent that makes it an excellent companion plant for a variety of garden crops (Parker*et al.*, 2013).

Hedgerows

A hedgerow is a long, linear row of flowering shrubs and small trees, occasionally with understory wildflowers and grasses.Hedgerow plants are often planted via transplants or live-stake cuttings. Hedgerows are commonly found along property lines, fence lines, roadways, and as physical barriers between agriculture fields. As per the results of Morandin *et al.* (2014), except for leafminers, fewer pests (aphids, cucumber beetle, flea beetle, lygus bug, stink bug, weevil *etc.*) were found in the hedgerow locations than in the weedy control plots. Some hedgerows consisting of dense evergreen trees can provide wind screening and reduce pesticide drift from neighbouring farms, allowing natural enemies to thrive in the field.

Windbreaks

Windbreaks are a source of alternate hosts and provide shelter for many natural enemies. Windbreaks are similar to hedgerows; however they are often planted with taller-growing species to reduce wind velocity more effectively than hedgerows. Carabids and staphylinids (Coleoptera) that feed on agricultural pests were more common towards the edge of multi-row windbreaks than in the interior in North Dakota (Katayama, 1980). Because the edge of a single-row elm windbreak makes up the majority of the windbreak, carabid and staphylinid abundance should be rather stable across the windbreak (Frye *et al.*, 1988). Insectivorous birds maintain large territories and prefer windbreaks that are larger and wider. When feeding in fields along the edge, other species may benefit from curved or undulating windbreak designs that provide more edge and less exposure.

Corridors

Individuals, genes, and ecological processes can travel freely via corridors throughout the landscape. Corridors are frequently utilised to link animal habitat pieces. Conservation biologists have historically utilized corridors to maintain biological diversity because they provide various paths for species to circulate and disperse across the environment (Rosenberg *et al.*, 1997). Nicholls et al. (2001) found that predators on the flowers of dominant corridor plants such as fennel (*Foeniculum vulgare*), yarrow (*Achillea millefolium*), *Erigeron annuus*, and *Buddleja* spp. included *Chrysoperla carnea*, *Orius* spp., *Nabis* spp., *Geocoris* spp., several members of the families Coccinellidae, Syrphidae, Mordellidae, and some species of thomisid spiders. Certain predator species were often discovered to be associated with certain flowering plants.

Pesticides Management

Despite the fact that pesticides are the most extensively used pest management method, the majority of pesticides kill not only the target insect but also many of its natural invertebrate foes.They can also injure vertebrate natural enemies and other non-target animals, and most insect pests can become resistant to pesticides over time.Pesticides can be used wisely to reduce the negative effects of these chemicals on natural enemies. Pesticides should only be used when absolutely essential, such as spot treatments wherever feasible and when natural enemies are least sensitive to pesticide application, which can assist to minimise natural enemy losses (Van Emden, 1988). Pesticide management includes not only the minimum dosage ofpesticides, but also the use of selected pesticides.Pesticide usage in Indonesian rice fields has been reduced without affecting crop yield. Rice output in Indonesia was on the verge of collapse in 1985-86 due to enormous outbreaks of brown planthopper. As a result of indiscriminate use of chemicals, the insect gained pesticide resistance swiftly, and its principal predators (such as spiders) were wiped off. Rice yields were so low that for the first time in several years, rice had to be imported. Following consultation with scientists, the elimination of pesticide subsidies and the prohibition of broad spectrum insecticides resulted in a 67% reduction in pesticide use.It was a great example of pesticide management that resulted in greater pest control by decreasing the usage of pesticides that harm the natural enemy.

Constraints and Future Prospects

- Integration of the conservation and manipulation techniques in the IPM modules should be done and be tested for proper pest management practices for different crop pests.
- There is basic require to strengthen on research defining the role of the tritrophic interactions, cultural practices and other practices in improving the competence of natural enemies.
- A concerted research is necessary to develop viable technologies with consideration to the conserving or increasing the efficiency of natural enemies.
- Removing the extension gap between researcher and farmer is pivotal for success of the conservation and manipulation techniques.
- Most of the experiments especially on the use of semio-chemicals were conducted in smaller area, so studies should be conducted in larger areas so as to generate good amount of data on the use of the semio-chemicals.

Conclusion

Enhancing biodiversity at the landscape and field level through the use of cover crops, corridors, and other habitats is a fundamental technique in sustainable horticulture. The basic goal of ecological pest management is to improve horti-ecosystem diversity and complexity as a basis for generating positive interactions that keep pest populations under control. Motivating farmers to follow these eco-friendly techniques on a community level is essential for the effective deployment of this technology. It is necessary to avoid using pesticides during the first 40 days of crop growth in order to allow beneficial insects to establish themselves and effectively manage pests. A cooperative effort by the horticultural community will not only reduce insect populations, but will also improve soil microflora and provide organic matter to the soil.

References

Abou-Awad, B.A., El-Sherif, A.A. et al., 1998. Studies on development, longevity, fecundity and predation of AmblyseiusoliviNasr and Abou-Awad (Acari: Phytoseiidae) on various kinds of prey and diets. Zeitschrift fur Pflanzenkrankheiten und Pflanzenschutz 105 (5): 538-44.

Ahmad, M. and Pathania, S. S. 2017. Ecological engineering for pest management in agro ecosystem-a review. International journal of current microbiology and applied sciences, 6(7): 1476-85.

Baggen, L. R., Gurr, G. M. and Meats, A. 1999. Flowers in tri-trophic systems: mechanisms allowing selective exploitation by insect natural enemies for conservation biological control. Entomol Exp Appl. 91:155

Baggen, L. R. and Gurr, G. M. 1998. The influence of food on Copidosoma koehleri, and the use of flowering plants as a habitat management tool to enhance biological control of potato moth, Phthorimaea operculella. Biological Control, 11(1):9-17.

Berndt, L. A., Wratten, S. D. and Scarratt, S. L. 2006. The influence of floral resource subsidies on parasitism rates of leaf rollers (Lepidoptera: Tortricidae) in New Zealand vineyards. Biological Control, 37(1): 50-55.

Berndt, L. A. and Wratten, S. D. 2005.Effects of alyssum flowers on the longevity, fecundity, and sex ratio of the leaf roller parasitoid Dolichogenidea tasmanica. Biological Control, 32:65-69.

Cook, S. M., Khan, Z. R. and Pickett, J. A. 2007. The use of 'push–pull'strategies in integrated pest management. Annual Review of Entomology, 52:375-400.

Dana, J. A. (2002). Report on The Farm as Natural Habitat: Reconnecting Food Systems with Ecosystems.

Dangerfield, J. M., McCarthy, T. S. and Ellery, W. N. 1998. The mound-building termite Macrotermes michaelseni as an ecosystem engineer. Journal of Tropical Ecology, 507-20.

Dong, C.X. and Xu, C.E. 1984. Spiders in cotton fields and their protection and utilization. China Cotton 3: 45–47.

Edpuganti, L. S. And Rajan, S. J. 2018. Ecological engineering for sustainable agriculture: simple concept with greater impact. IJSRP, 8(2): 123-25.

Eilenberg, J., Hajek, A. and Lomer, C. 2001. Suggestions for unifying the terminology in biological control.Biocontrol, 46: 387–400.

Frye, R. D., Dix, M. E. and Carey, D. R. 1988. Effect of two insecticides on abundance of insect families associated with Siberian elm windbreaks. Journal of the Kansas Entomological Society 278-284.

Gillespie, M., Wratten, S., Sedcole, R. and Colfer, R. 2011. Manipulating floral resources dispersion for hoverflies (Diptera: Syrphidae) in a California lettuce agro-ecosystem. Biological Control, 59 (2): 215-20

Gurr, G., Wratten, S. D. and Altieri, M. A. (Eds.) 2004. Ecological Engineering for Pest Management: Advances in Habitat Manipulation for Arthropods. Csiro Publishing.

Gurr, G. M., Wratten, S. D. and Snyder, W. E. (Eds.). 2012. Biodiversity and Insect Pests: Key Issues for Sustainable Management. John Wiley & Sons.

Gurr, G. M. And Wratten, S. D. 1999. FORUM 'Integrated biological control': A proposal for enhancing success in biological control. International Journal of pest management, 45(2): 81-84.

Halaji, J., Cady, A. B. and Uetz, G. W. 2000. Modular habitat refugia enhance generalist predators and lower plant damage in soybeans.Environmental Entomology 29:383-93.

Hickman, J. M. and Wratten, S. D. 1996. Use of Phacelia tanacetifolias trips to enhance biological control of aphids by hoverfly larvae in cereal fields.Journal of Economic Entomology 89:832-40.

Katayama, R. W. 1980. 'Seasonal abundance of insects in shelterbelts' (Doctoral dissertation, PhD Thesis. Department of Entomology, North Dakota State University, Fargo)

Khan, Z. R., Midega, C. A., Bruce, T. J., Hooper, A. M. and Pickett, J. A. 2010. Exploiting phytochemicals for developing a 'push–pull'crop protection strategy for cereal farmers in Africa. Journal of Experimental Botany, 61(15): 4185-96.

Khorsheduzzaman, A. K. M., Ali, M. I., Mannan, M. A. and Ahmed, A. 1997. Brinjal-coriander intercropping an effective IPM component against brinjal shoot and fruit borer, Leucinodes orbonalis Guen (*Pyralidae: Lepidoptera*). Bangladesh Journal of Entomology 7: 85-91.

Kogan, M. 1998. Integrated pest management: historical perspectives and contemporary developments. Annu Rev Entomol 43: 243–70.

Lu, Z., Zhu, P., Gurr, G. M., Zheng, X., Chen, G. And Heong, K. L. 2015. Rice pest management by ecological engineering: a pioneering attempt in China. In: Rice Planthoppers (pp. 161-178). Springer, Dordrecht.

Maqsood, S., Buhroo, A. A., Sherwani, A. and Mukhtar M. 2022. Optimizing control of flea beetles through ecological engineering of vegetable agroeco-system in Kashmir. Plant Science Today; https://doi.org/10.14719/pst.1450.

Meena, A., Meena, M., Kumar, R. And Meena, B. M. 2017. Farmscaping: An ecological approach to insect pest management in agroecosystem. Journal of Entomology and Zoology Studies 5(3): 598-603.

Mitsch, W. J. 2004. Ecological Engineering and Ecosystem Restoration. New Jersey: John Wiley and Sons.

Mitsch, W. J. And Jorgensen, S. E. 1989. Ecological engineering: an introduction to ecotechnology.

Morandin, L. A., Long, R. F. And Kremen, C. 2014. Hedgerows enhance beneficial insects on adjacent tomato fields in an intensive agricultural landscape. Agriculture, Ecosystems & Environment 189: 164-70.

Morris, M. G., Thomas, J. A., Ward, L. K., Snazell, R. G., Pywell, R. F., Stevenson, M. J. and Webb, N. R. 1994. Re-creation of early-successional stages for threatened butterflies—an ecological engineering approach. Journal of Environmental Management, 42 (2): 119-35.

NafzigerJr, T. D. and Fadamiro, H. Y. 2011. Suitability of some farmscaping plants as nectar sources for the parasitoid wasp, Microplitis croceipes (Hymenoptera: Braconidae): effects on longevity and body nutrients. Biological Control, 56 (3): 225-229.

Nicholls, C. I., Parrella, M., Altieri, M. A. 2001.The effects of a vegetational corridor on the abundance and dispersal of insect biodiversity within a northern California organic vineyard. Landscape Ecology 16(2):133-146.

Osborne, L. S., Landa, Z., Taylor, D. J., Tyson, R. V. 2005. Using banker plants to control insects in greenhouse vegetables. Proceedings of the Florida State Horticulture Society 118:127-28.

Parker, J. E., Snyder, W. E., Hamilton, G. C. and Rodriguez-Saona, C. 2013. Companion planting and insect pest control, (pp. 1-30). Rijeka, Croatia: InTech.

Prokopy, R. and Kogan, M. 2003.Integrated pest management, in Encyclopedia of Insects, ed. by Resh VH and Carde RT. Academic Press, San Diego, pp. 589–95.

Rao, M. S., Rao, C. R., Srinivas, K., Pratibha, G., Sekhar, S. M., Vani, G. S. 2012. Intercropping for management of insect pests of castor, Ricinus communis, in the semi–arid tropics of India. Journal of Insect Science 12:1536-2442.

Root, R. B. 1973. Organization of a plant-arthropod association in simple and diverse habitats: the fauna of collards (Brassica oleracea). Ecological monographs 43 (1): 95-124.

Rosenberg, D. K., Noon, B. R. and Meslow, E. C. 1997. Biologicalcorridors: form, function and efficacy. BioScience 47(10): 677–87.

Sreedhar, U. and Krishnamurthy, V. 2009. Eco-friendly strategy for management of budworm H. armigera in Virginia tobacco. In: 5th International Conference on bio-pesticides: TERI and Society for Promotion and Innovation of Bio-pesticides held at New Delhi from 26-30 April, 2009

Staver, C., Guharay, F., Monterroso, D. and Muschler, R.G. 2001. Designing pest-suppressive multi strata perennial crop system: Shade-grown coffee in Central America. Agrofor Syst. 53 (2): 151-70.

Sutherland, J. P., Sullivan, M. S. and Poppy, G. M. 2001. Distribution and abundance of aphidophagous hoverflies (Diptera: Syrphidae) in wildflower patches and field margin habitats.Agricultural and Forest Entomology 3:57-64.

Sweetman, H. L. 1958. The Principles of Biological Control.Interrelation of Hosts and Pests and Utilization in Regulation of Animal and Plant Populations. The Principles of Biological Control.Interrelation of Hosts and Pests and Utilization in Regulation of Animal and Plant Populations (revd. edn. 11 1/4× 8 1/2).

Thomas, F., Poulin, R., de Meeus, T., Guegan, J. F. and Renaud, F. 1999. Parasites and ecosystem engineering: what roles could they play?. Oikos: 167-71.

Thomas, M. B., Wratten, S. D., &Sotherton, N. W. (1991). Creation of 'island' habitats in farmland to manipulate populations of beneficial arthropods: predator densities and emigration. Journal of applied Ecology, pp. 906-17.

Tschumi, M., Albrecht, M., Collatz, J., Dubsky, V., Entling, M. H., Rodriguez, A. J. N.et al. 2016.Tailored flower strips promote natural enemy biodiversity and pest control in potato crops. Journal of Applied Ecology 1365:2653-2664.

Van Emden, H. F. 1988. The potential for managing indigenous natural enemies of aphids on field crops. Philosophical Transactions of the Royal Society of London. B, Biological Sciences, 318(1189): 183-201.

Van Emden, H. F. 1991. Pest control. Cambridge University Press.

Viggiani, G. 2003. Functional biodiversity for the vineyard agroecosystem: aspects of the farm and landscape management in Southern Italy. Bulletin Oilb/Srop 26(4):197-202.

22

Augmentation of Immunity in Human Beings in Challenging Times

Hare Krishna, S N S Chaurasia, Anant Bahadur, Swati Sharma, Rajeev Kumar, Manoj Kumar Singh, R B Yadava and T K Behera

Division of Vegetable Production, ICAR-Indian Institute of Vegetable Research, Varanasi-221 305, Uttar Pradesh, India

Nutrition affects the development of human body immune system, playing a vital role in modifying the threat of exposure and susceptibility to infectious pathogens, acuteness of morbidity and success of treatment. Keeping in view the present pandemic and emerging viruses such Eastern Equine Encephalitis, Hantaviruses, Nipah viruses for which no effective preventive and curative treatment is in hand, a healthy immune system could be one of the most important defences to fight off (El-Gamal *et al*., 2011). Vegetables are rich source of essential vitamins, minerals, antioxidants and dietary fibres; however, certain vegetables hold more significance in view of boosting immunity. Daily consumption of vegetables with immunomodulation properties (modification of immune response or functioning of the immune system) helps secure a good health and is a critical step towards a salubrious life style. Therefore, consuming vegetables, especially those which can regulate immunity with their immunomodulatory properties, could be an important step towards building up a sound immune system.

Medicinal Importance of Vegetables

The old adage 'you become what you eat' has its deep roots in civilizational history. Since ages, vegetables have been traditionally used as medicine or advised to be ingested during specific illness in developing world. Ancient Indian medicine system (*Ayurveda*) describes 'Do's and Don'ts of diet', which also include consumption of specific vegetables in a particular illness. Last few decades have witnessed researches validating numerous traditional medicinal uses of vegetables. About 80 % of population in developing countries of Asia and Africa rely on use of traditional medicines to cater their primary health care needs. The WHO traditional medicine strategy also aims to support countries to integrate traditional medicine into their own health systems to keep their populations healthy. The plant parts such as roots, young shoots, leaves, stems, fruits are used for treating the illness like cold and cough, toothache, muscle pain, dysentry, skin disorders, renal and cardial problems, inflammation of spleen, hyperglycemia etc. Traditional medicinal uses of amaranthus, drumsticks, purslane, curry leaves, bottle gourd, ridged gourd, chayote, garlic, welsh onion, parsley, celery, coriander, bitter gourd, pumpkin, turnip, cabbage etc. are well documented. Tribal and local people in Western India have belief that consumption of some seasonal wild vegetables improves immunity during the rains, when susceptibility to infection increases. The clue for further research on immunomodulatory properties of unexplored vegetables can be drawn from their ethno-botanical uses through indigenous traditional knowledge existing among the local culture and inhabitants.

Immune System and Immunomodulatory

The immune system consists of two distinguished mechanisms to fight off infection: i) antibody-mediated defense system or humoral immunity and ii) the cell-mediated defense system or cellular immunity. The humoral immune system deals with extracellular antigens, while the cellular immunity occurs within the infected cells and is mediated by T lymphocytes. Further based on the function, immune system is categorized as innate immune (non-specific) and adaptive (specific or acquired) immune systems. Each immunity system consists of humoral and cellular responses, and to realize full advantage of immunity, both responses needs to function in coordination (Kumar *et al.*, 2012). The primary defense system is innate immunity, which entails physical (such as skin and epithelial lining of the organs) and biochemical barriers (like secretions, bile, mucus and gastric acid) with a non-specific, leukocyte-mediated cellular response to defend against invading foreign antigens (Maggini *et al.*, 2018). The adaptive immune system distinguishes particular individual foreign protein (antigens) and memorizes them through immunological memory.

Lymphocytes (B cells and T cells) are crucial components of the adaptive immune systems as they play pivotal roles by employing specific receptors to distinguish specific foreign antigens. Further, B cells synthesize antibodies, immunoglobulins (including immunoglobulin A (IgA), IgG, and IgE,), in response to invading foreign antigens e.g. pathogens and allergens, thereby causing hypersensitivity reactions (Nair *et al.*, 2019). On the other hand, the T cells express T-cell receptors and can be grouped into two subpopulations CD4 or CD8 molecules. CD4+ T cells are primarily supporting T cells aiding and managing the activities of other immune cells. CD8+ T cells are cytotoxic as they kill cells infected by viruses or mutated cells with the potential to cause cancer (Hachimura *et al.*, 2018).

The healthy immune system upholds homeostasis within the body. The function and efficiency of the immune system are affected by several exogenous and endogenous factors leading to immunostimulation or immunosuppression. Agents (biological or synthetic substances) having capabilities to stimulate, suppress or modulate any aspect of the immune system including both adaptive and innate immune system are termed as immunomodulators. If they work to increase the immune response, they are called as immunostimulants else if they diminish the immune response, they are termed as immunosuppressants.

Vegetables *vis-a-vis* Immunomodulation

Despite immense nutrition and health value, the role of vegetables in preventing infection by exerting immunomodulatory effect on immune system has been discussed seldom. Consumption of such vegetables can improve the body's immune-responsiveness against invading foreign antigen through activation of the immune system including innate and adaptive immune arms. Different plant parts of such vegetable crops have been used to evaluate their immunomodulatory properties and isolate and characterize novel compounds for their potential clinical applications. These vegetables modulate immune system via cytokine modulation, release of nitric oxide, enhancement of immune cells, enzymatic and phagocytic activities, alteration of composition of the gut microbiota etc. The immunomodulatory activities of such vegetables are attributed to presence of compounds like lectins, isothiocynate 'Sulforaphane', triterpenoid 'Bryonolic acid', glycoproteins 'Lagenin', 'Luffaculin', 'Momorchochin', novel polysaccharides etc. Likewise, carrots, celery, coriander and parsley of Umbelliferae family are rich in certain coumarins and flavonoids like quercetin, isopimpinellin, rutin, bergapten and xanthotoxin. These compounds enhanced lymphocyte activation including including CD8+ T cells enhancement. Consumption of natural foods rich in vitamin C, like vegetable and fruits, had most positive impact in the course of infections and the decreased incidence of bacterial, fungal and viral infections (Polak *et al*., 2021).

Owing to undesirable immune responses (during autoimmune disease like arthritis; allergy; severe cell rupture in blood transfusion and organ transplant rejection), use of immunosuppressive drugs has risen in last decade. However very often, serious side effects such as hepatotoxicity, nephrotoxicity, neurotoxicity, hypertension and risk of diabetes have been noticed with the use of certain immunosuppressive drugs. The growing concerns over the use of synthetic immunosuppressive drugs have drawn attention towards natural alternatives. Parsley essential oil has shown immunosuppressive activities by modulating both cellular and humoral responses. Immunosuppressive activities have also been reported in drumsticks (Tang *et al*., 2018) and purple sweet potato (Mahajan and Mehta, 2010). Similarly, polyacetylene, falcarinol, which is exclusively found in vegetable crops of Umbelliferae family, has shown immunosuppressive activities by reducing intestinal inflammation (Stefanson and Bakovic, 2020).

Apart from exerting immunomodulatory properties, such vegetables are also rich sources of several vitamins, essential minerals and dietary fibres. It is a well-established fact that different vitamins such as A, B-complex, C, D, E and folate and minerals like iron, copper, zinc & selenium are vital for functioning of immune system owing to their synergistic roles during various phases of the immune response (Gombart *et al.*, 2020). Watzl *et al.* (1999) reported modulation of human T-lymphocyte functions by the consumption of carotenoid-rich vegetables such as tomato, carrot and spinach. Reduced risk of various kinds of cancers and, to some extent, coronary heart disease is being consistently associated with the routine intake of tomato and its products (Riso *et al.*, 2006).

Immunomodulatory Vegetables *vis-à-vis* non-communicable Diseases

Non-communicable diseases (NCDs) like stroke, heart disease, diabetes, cancer and chronic lung disease, together cause almost 71% (41 million people) of total deaths across the globe (WHO, 2021). One of the major risk factors, which drive rise of NCDs is unhealthy diet. Therefore for reducing deaths from NCDs, it is essential to keep a check on unhealthy lifestyle choices, which lead to their development. These comprise upholding an active lifestyle and following a healthy diet. Eating vegetables every day is crucial for maintaining a healthy life style. Apart from providing protection against infection, the immunomodulatory properties of vegetables help circumventing the risk of NCDs as well. For instances, bitter melon demonstrated the potential clinical benefits in preventing and delaying the advancement of oral cancer (Rao, 2018) and Head and neck squamous cell carcinoma (Bhattacharya *et al.*, 2016) by enhancing the natural killer-cell (major components of innate immunity which has the ability to mediate antitumor activity) killing activity against cancerous cells (Bhattacharya *et al.*, 2017).

'Sulforaphane', a naturally occurring isothiocyanate from broccoli have been shown to prevent colon cancer (Bessler and, Djaldetti, 2018). Similarly, pectic oligosaccharides, Rhamno galacturonan I arabino galactan (RGI-AG) from tomato exhibited anticancer potential on gastric cancer (Kapoor and Dharmesh (2017). Lectin binding rhamno galacturonan-I containing pectic polysaccharide have also been isolated from pumpkin has potential application in pharmaceutical industries (Zhao *et al.*, 2017). Taro has displayed a gamut of health-benefitting effects, such as antimutagenic, anti-inflammatory, antitumoral, immunomodulatory, anti-hyperglycemic, and anti-hyperlipidemic activities, owing to presence of several bioactive compounds, like tarin, A-1/B-2 α-amylase inhibitors, taro polysaccharides 1 and 2 (TPS-1 and TPS-2), taro-4-I polysaccharide, digalactosyldiacylglycerols, monogalactosyldiacylglycerols and antioxidants (Pereiera *et al.*, 2021).

Vegetables like beet root and curry leaf had also inhibited the progression of tumor via immunostimulatory effect and inflammatory reaction. Besides, curry leaves and okra (Liu *et al.* 2018) has hypoglycemic property, which helps managing diabetes. Another perennial vegetable, drumsticks has been found effective in treatment of colitis and wound healing, which might be attributed to presence of phenols and flavonoids. Recently, Vezza *et al.* (2019) reported that the 'propyl-propane thiosulfonate', from garlic, prevents colitis anti-inflammatory activity. Furthermore, diallyl thiosulfinate (allicin), a compound derived from garlic, delays the progression of diabetic nephropathy (a serious kidney-related complication of type 1 diabetes and type 2 diabetes) through antioxidant and anti-inflammatory mechanisms (Arellano-Buendía *et al.* 2018). Likewise, allicin also enhances antimicrobial activity of macrophages during TB infection and fungal infection. In addition, garlic is reported to have cardio-protective and antiparasitic effects (Vathsala and Krishna Murthy, 2020). Common purslane has shown potential for treatment of cancer and asthama due to its anti-inflammatory and immunomodulatory properties. Purslane is also used for treating cough in children. The protective effect against cough could be attributed to the decreased white blood cell counts, neutrophils and eosinophils counts of blood. Further, the protective effects are also extended on antibody titer, serum immunoglobulin and cytokines (Moghaddam *et al.* 2020). While, cruciferous vegetables have also shown to exert protection against pathogenic infection. 'Ascorbigen', an indole-containing derivative of l-ascorbic acid showed immunomodulatory properties. One of the most active ascorbigen has been found to be *N*-methylascorbigen, which is found in *Brassica* sp. It has exhibited therapeutic effects such as inhibition of tumor growth and protection from bacterial & viral infections. The immunomodulatory activity of natural ascorbigen highlights the importance of cruciferous-vegetable diet owing to their anticarcinogenic properties. 'Emodin', a natural anthraquonoid compound, found in several vegetables such as cabbage, rhubarb, lettuce and peas is reported to regulate blood pressure in hypertension through modulation of cytokines (Vasdev *et al.* 2011).

Certain vegetable have been reported to source of some compounds which has potential application in pharmaceutical industries as adjuvants e.g. polysaccharides from bitter melon & rhamnogalacturonan-I-type polysaccharides from okra and *Moringa* lectin, a protein isolated from seeds (Coriolano *et al.* 2018). Adjuvants are a vital constituent of cancer vaccines as they help eliciting a potent immune response against the vaccine antigens. Okra polysaccharides have unique dual-functions as immunosuppressant and immunostimulant, which can be a exploited for the application of the same in the nutraceutical and pharmaceutical industries (Hayaza *et al.* 2021).

Vegetables, Source of Edible Vaccines

A vaccine is defined as a biological preparation, which enhances immunity by encouraging the antibodies' production in human body system. Likewise, edible plant vaccines are immunogenic preparations, which have antigenic proteins in lieu of microbes. This precludes any looming threat resulting from the resurgence of diseases due to use of the attenuated pathogen. Edible vaccines are developed through the introduction of the selected gene(s) into the host plant to produce the desired encoded protein. The coat protein of a specific microbe (virus or bacteria), which has no pathogenicity is exploited for transformation (Gunasekaran and Gothandam 2020). A plant-based edible vaccine has several advantages like avoidance of necessity of specific animal for experimentation, purification of product and refrigeration; higher production involving low expenditure; avoidance of requirement for trained medical personnel for vaccination; easy and effective oral deliverance, especially, among the children; exclusion of possible hazard of reuse of needles etc. (Sohrab *et al.* 2020).

Epidemic-prone diseases, including emerging and re-emerging diseases constitute the greatest threat to public health security. Vaccination lowers the chances of catching diseases by strengthening immunity. Selected vegetables have been exploited to produce the edible vaccines against life-threatening diseases like Hepatitis B, rabies, *Escherichia coli* infection, SARS *etc*. The *Agrobacterium*-mediated transformation of carrot for the gene of urease B (UreB) an effective and common immunogen against *Helicobacter pylori* (stomach infecting bacteria) resulted into expression of 25 µg/g roots of UreB protein (Zhang *et al.* 2010). The results suggested that the UreB transgenic carrot could potentially be used as an edible vaccine for controlling *H. pylori*. The first plant based approved drug by US Food and Drug Administration was Elelyso (taliglucerase alfa), an enzyme produced in genetically engineered carrot cells, for treating type 1 Gaucher's disease (Fox 2012). However, many other vegetable-based edible vaccines are at different phase of evaluation (Concha *et al.*, 2017)

Conclusion

Some vegetables exert immunomodulatory activity in experimental models at a particular dose both in vivo and in vitro. The immunomodulators, from vegetables, have the ability to modify the immune responses (humoral and cellular) through innate or adaptive mechanisms. Including selected vegetables with immunomodulatory properties as part of the daily diet may reduce the risk of infection and NCDs. In future, efforts should be made to explore the immunomodulatory properties of more vegetables, particularly, underutilized vegetables as such vegetables are largely used by the rural inhabitants and tribal people, who have very enriched indigenous traditional knowledge and are well aware of their ethno-botanical uses. Further, breeding for enhancement of immunomodulating compounds may also be taken up to identify or develop elite vegetable genotypes with higher contents of immunomodulator compounds. Immunomodulator compounds, like lectins, pectic polysaccharides, triterpenoids, ribosomes inactivating proteins etc., isolated from such vegetables have potential application in pharmaceutical industries as well. The use of vegetable-derived natural immunomodulators and adjuvants will instill confidence among the users due to lack of any likely side-effects. Promotion of health benefits of vegetables consumption could be very crucial for encouraging adoption of healthy life style in the society and an important step towards realization of sustainable development goals set by UN by 2030.

References

Arellano-Buendía AS, Tostado González M, Sánchez Reyes O, García Arroyo FE, Argüello García R, Tapia E, Sánchez Lozada LG and Osorio Alonso H. 2018. Immunomodulatory Effects of the Nutraceutical Garlic Derivative Allicin in the Progression of Diabetic Nephropathy. International Journal of Molecular Sciences 19(10): 3107.

Bessler H and Djaldetti M. 2018. Broccoli and human health: immunomodulatory effect of sulforaphane in a model of colon cancer. International Journal of Food Sciences and Nutrition. 69(8): 946-53.

Bhattacharya S, Muhammad N, Steel R, Peng G and Ray RB. 2016. Immunomodulatory role of bitter melon extract in inhibition of head and neck squamous cell carcinoma growth. Oncotarget 7(22): 33202-9.

Bhattacharya S, Muhammad N, Steel R, Kornbluth J and Ray R B. 2017. Bitter Melon Enhances Natural Killer-Mediated Toxicity against Head and Neck Cancer Cells. Cancer Prevention Research (Philadelphia) 10(6): 337-44.

Concha C, Cañas R, Macuer J, Torres MJ, Herrada AA, Jamett F, Ibáñez C. 2017. Disease Prevention: An Opportunity to Expand Edible Plant-Based Vaccines? Vaccines (Basel) 5(2): 14.

Coriolano MC, Santana Brito J de, Siqueira Patriota LL de, Araujo Soares AK de, Lorena de VMB, Paiva PMG, Napoleao TH, Coelho LCBB and de Melo CML. 2018. Immunomodulatory Effects of the Water-soluble Lectin from Moringa oleifera Seeds (WSMoL) on Human Peripheral Blood Mononuclear Cells (PBMC). Protein & Peptide Letters 25(3): 295-301.

El-Gamal, Y.M., O.A. Elmasry and D.H. El-Ghoneimy. 2011. Immunomodulatory effects of food. Egypt J Pediatr Allergy Immunol. 9: 3–13.

Fox J. 2012. First plant-made biologic approved. Nature Biotechnology 30: 472.

Gunasekaran B and Gothandam KM. 2020. A review on edible vaccines and their prospects. Brazilian Journal of Medical and Biological Research 53(2): e8749.

Hachimura S, Totsuka M and Hosono A. 2018. Immunomodulation by food: impact on gut immunity and immune cell function. Bioscience, Biotechnology and Biochemistry 82(4): 584-599.

Hayaza S, Wahyuningsih SPA, Susilo RJK, Husen SA, Winarni D, Doong RA and Darmanto W. 2021. Dual role of immunomodulation by crude polysaccharide from okra against carcinogenic liver injury in mice. Heliyon 7(2): e06183.

Kapoor S., and Dharmesh, S.M. 2017. Pectic Oligosaccharide from tomato exhibiting anticancer potential on a gastric cancer cell line: Structure-function relationship. Carbohydrate Polymers 160:52–61. 10.1016/j.carbpol.2016.12.046

Kumar D, Arya V, Kaur R, Bhat ZA, Gupta VK and Kumar V. 2012. A review of immunomodulators in the Indian traditional health care system. Journal of Microbiology, Immunology and Infection 45(3): 165-84.

Liu J., Zhao Y, Wu Q, John A, Jiang Y, Yang J, Liu H and Yang B. 2018. Structure characterisation of polysaccharides in vegetable "okra" and evaluation of hypoglycemic activity. Food Chemistry 242:211–6.

Maggini S, Pierre A and Calder PC. 2018. Immune Function and Micronutrient Requirements Change over the Life Course. Nutrients 10(10): 1531.

Mahajan SG and Mehta AA. 2010. Immunosuppressive activity of ethanolic extract of seeds of Moringa oleifera Lam. in experimental immune inflammation. Journal of Ethnopharmacology 130(1):183-6.

Mortazavi Moghaddam SG, Kianmehr M, Khazdair MR. 2020. The Possible Therapeutic Effects of Some Medicinal Plants for Chronic Cough in Children. Evid Based Complement Alternative Medicines 2020: 2149328.

Nair A, Chattopadhyay D and Saha B. 2019. Plant-derived immunomodulators. In: New Look to Phytomedicine, 1st edn, Khan AMS, Ahmad I, Chattopadhyay D (Eds.). Academic Press: Cambridge, MA, USA, pp. 435–99.

Polak E, Stępień AE, Gol O and Tabarkiewicz J. 2021. Potential Immunomodulatory Effects from Consumption of Nutrients in Whole Foods and Supplements on the Frequency and Course of Infection: Preliminary Results. Nutrients 13(4).

Rao CV. 2018. Immunomodulatory Effects of Momordica charantia Extract in the Prevention of Oral Cancer. Cancer Prevention Research (Philadelphia) 11(4): 185-86.

Sohrab SS. 2020. An edible vaccine development for coronavirus disease 2019: the concept. Clinical and Experimental Vaccine Research 9(2):164-68.

Stefanson A, Bakovic M. 2020. Dietary polyacetylene falcarinol upregulated intestinal heme oxygenase-1 and modified plasma cytokine profile in late phase lipopolysaccharide-induced acute inflammation in CB57BL/6 mice. Nutrition Research 80: 89-105.

Tang C, Sun J, Zhou B, Jin C, Liu J, Gou Y, Chen H, Kan J, Qian C and Zhang N. 2018. Immunomodulatory effects of polysaccharides from purple sweet potato on lipopolysaccharide treated RAW 264.7 macrophages. Journal of Food Biochemistry 42: e12535.

Vasdev S, Stuckless J and Richardson V. 2011. Role of the immune system in hypertension: modulation by dietary antioxidants. International Journal of Angiology 20(4): 189-212.

Vathsala PG and Krishna Murthy P. 2020. Immunomodulatory and antiparasitic effects of garlic-arteether combination via nitric oxide pathway in Plasmodium berghei-infected mice. Journal of Parasitic Diseases 44(1): 49-61.

Vezza T, Algieri F, Garrido-Mesa J, Utrilla MP, Rodríguez-Cabezas ME, Baños A, Guillamón E, García F, Rodríguez-Nogales A and Gálvez J. 2019. The Immunomodulatory Properties of Propyl-Propane Thiosulfonate Contribute to its Intestinal Anti-Inflammatory Effect in Experimental Colitis. Molecular Nutrition and Food Research 63(5): e1800653.

WHO. 2021. Noncommunicable diseases. https://www.who.int/health-topics/noncommunicable-diseases#tab=tab_1 August 01, 2021

Zhang H, Liu M, Li Y, Zhao Y, He H, Yang G and Zheng C. 2010. Oral immunogenicity and protective efficacy in mice of a carrot-derived vaccine candidate expressing UreB subunit against Helicobacter pylori. Protein Expression and Purification 69(2): 127–31

Zhao J, Zhang F, Liu X, St Ange K, Zhang A, Li Q and Linhardt RJ. 2017. Isolation of a lectin binding rhamnogalacturonan-I containing pectic polysaccharide from pumpkin. Carbohydrate Polymers 163: 330–36.

23

Functional Properties of Secondary Metabolites in Medicinal Plants

Manish Das

Director, ICAR-Directorate of Medicinal & Aromatic Plants Research, Indian Council of Agricultural Research, Boriavi-387310, Anand, Gujarat, India

Plants have the ability to synthesize a wide variety of chemical compounds that are used to perform important biological functions, and to defend against attack from predators such as insects, fungi and herbivorous mammals. Many of these phyto-chemicals have beneficial effects on long-term health when consumed by humans, and can be used to effectively treat human diseases. At least 12,000 such compounds have been isolated so far; a number estimated to be less than 10% of the total (Tapsell *et al*., 2006; Lai and Roy, 2004). These phyto-chemicals are divided into: (1) primary metabolites such as sugars and fats, which are found in all plants; and (2) secondary metabolites (SM) – compounds which are found in a smaller range of plants, serving a more specific function (Meskin, 2002). For example, some secondary metabolites are toxins used to deter predation and others are pheromones used to attract insects for pollination. It is these secondary metabolites and pigments that can have therapeutic actions in humans and which can be refined to produce drugs—examples are inulin from the roots of dahlias, quinine from the cinchona, morphine and codeine from the poppy, and digoxin from the foxglove (Meskin, 2002). Chemical compounds in plants mediate their effects on the human body through processes identical to those already well understood for the chemical compounds in conventional drugs; thus herbal medicines do not differ greatly from conventional drugs in terms of how they work. This enables herbal medicines to be as effective as conventional medicines, but also gives them the same potential to cause harmful side effects (Tapsell *et al*., 2006; Lai and Roy, 2004).

Most cultures have a tradition of using plants medicinally. In Europe, apothecaries stocked herbal ingredients for their medicines. In the Latin names for plants created by Linnaeus, the word *officinalis* indicates that a plant was used in this way. For example, the marsh mallow has the classification *Althaea officinalis*, as it was traditionally used as an emollient to soothe ulcers (William, 2004). Ayurvedic medicine, herbal medicine and traditional Chinese medicine are other examples of medical practices that incorporate medical uses of plants. Pharmacognosy is the branch of modern medicine about medicines from plant sources.

Modern medicine now tends to use the active ingredients of plants rather than the whole plants. The photochemicals may be synthesized, compounded or otherwise transformed to make pharmaceuticals. Examples of such derivatives include Digoxin, from digitalis; capsaicine, from chili; and aspirin, which is chemically related to the salicylic acid found in white willow (William, 2004). The opium poppy continues to be a major industrial source of opiates, including morphine. Few traditional remedies, however, have translated into modern drugs, although there is continuing research into the efficacy and possible adaptation of traditional herbal treatments.

Secondary Metabolites (SM)

Secondary metabolites are organic compounds that are not directly involved in the normal growth, development, or reproduction of an organism (Fraenkel, 1959). Unlike primary metabolites, absence of secondary metabolites does not result in immediate death, but rather in long-term impairment of the organism's survivability, fecundity, or aesthetics, or perhaps in no significant change at all. Secondary metabolites are often restricted to a narrow set of species within a phylogenetic group (Stamp, 2003). Secondary metabolites often play an important role in plant defense against herbivory (Samuni-Blank *et al.*, 2012) and other interspecies defenses (Chizzali *et al.*, 2012). Humans use secondary metabolites as medicines, flavorings, and recreational drugs.

Studies on plant secondary metabolites have been increasing over the last 50 years. These molecules are known to play a major role in the adaptation of plants to their environment, but also represent an important source of active pharmaceuticals. Plant cell culture technologies were introduced at the end of the 1960s as a possible tool for both studying and producing plant secondary metabolites (Bourgaud *et al.*, 2001). Different strategies, using *in vitro* systems, have been extensively studied with the objective of improving the production of secondary plant compounds. Undifferentiated cell cultures have been mainly studied, but a large interest has also been shown in hairy roots and other organ cultures. Specific processes have been designed to meet the requirements of plant cell and organ cultures in bioreactors. Despite all of these efforts of the last 30 years, plant biotechnologies have led to very few commercial successes for the production of valuable secondary compounds. Compared to other biotechnological fields such as microorganisms or mammalian cell cultures, this can be explained by a lack of basic knowledge about biosynthetic pathways, or insufficiently adapted reactor facilities (Harborne, 1999). More recently, the emergence of recombinant DNA technology has opened a new field with the possibility of directly modifying the expression of genes related to biosyntheses. It is now possible to manipulate the pathways that lead to secondary plant compounds. Many research projects are now currently being carried out and should give a promising future for plant metabolic engineering.

General Modes of Action of Secondary Metabolites

A number of plants are well-known for their toxic or hallucinogenic properties. Very often, these plants contain certain alkaloids, terpenoids or other SM which specifically modulate a corresponding molecular target in animals or humans. Such targets are often neuroreceptors, enzymes which degrade neurotransmitters, ion channels, ion pumps, or elements of the cytoskeleton (mostly tubulin or microtubules). Quite a number of these SM are presently been extracted from plants and are used in modern medicine as chemical entities with established applications (Wink, 2015) (Table 1). These SM appear to be quite specific for a given target. It is speculated that their shape had been formed during evolution and selection by a process of "evolutionary molecular modeling" in analogy to chemical molecular modeling in medicinal chemistry (Wink, 2015).

Plant Secondary Compounds

Plant secondary compounds are usually classified according to their biosynthetic pathways (Harborne, 1999). Three large molecule families are generally considered: phenolics, terpenes and steroids, and alkaloids. A good example of a widespread metabolite family is given by phenolics: because these molecules are involved in lignin synthesis, they are common to all higher plants. However, other compounds such as alkaloids are sparsely distributed in the plant kingdom and are much more specific to defined plant genus and species. This narrower distribution of secondary compounds constitutes the basis for chemotaxonomy and chemical ecology (Bourgaud *et al.*, 2001).

Due to their large biological activities, plant secondary metabolites have been used for centuries in traditional medicine. Nowadays, they correspond to valuable compounds such as pharmaceutics, cosmetics, fine chemicals, or more recently nutraceutics. Recent surveys have established that in western countries, where chemistry is the backbone of the pharmaceutical industry, 25% of the molecules used are of natural plant origin (Payne *et al.*, 1991). A good example could be aspirin (acetylsalicylate) which derives from salicylate. The genuine molecule can be isolated in large quantities from many plants (*Spiraea ulmaria*, *Betula lenta*), but the chemical is synthesized as an acetyl-derivative in order to lower secondary effects (stomachaches). Chemical synthesis apart, the production of plant secondary metabolites has, for a long time, been achieved through the field cultivation of medicinal plants (Bourgaud *et al.*, 2001). However, plants originating from particular biotopes can be hard to grow outside their local ecosystems. It also happens that common plants do not withstand large field cultures due to pathogen sensitiveness (anthracnose on *Hypericum perforatum* or *Arnica montana*). This has led scientists and biotechnologists to consider plant cell, tissue and organ cultures as an alternative way to produce the corresponding secondary metabolites (Payne *et al.*, 1991).

Production of Secondary Compounds

Industrial interest emerged quite early in the story of plant vitro cultures and secondary metabolites. Indeed, at least six large-scale industrial productions were engaged between 1976 and 1986 elsewhere in the world (Zryd, 1988). However, since that time, this technology has led to only a few applications for the production of commercial compounds. This lack of success can be attributed to several bottlenecks (Steward *et al.*, 1999a).

First of all, it is obvious that industrial processes were started at a time when basic knowledge was severely lacking on both plant vitro cultures and secondary compounds. It may be surprising to realize that a key-problem like cell viability, and its assessment, has been studied only recently with plant cells (Steward *et al.*, 1999a; Steward *et al.*, 1999b), whereas, this topic has been considered for a long time with animal cell cultures (Cook and Mitchell, 1989). Also, secondary metabolites are produced following long biosynthetic pathways that can involve dozens of enzymes. This synthesis is much more complex than for recombinant proteins produced with mammalian or prokaryotic biotechnologies, which usually involve one or two genes. This can partially explain the lack of success of plant cell and tissue cultures compared to other expression systems (Cook and Mitchell, 1989).

A second bottleneck has to do with the economic feasibility of plant cell and organ cultures. Indeed, this technology requires high-cost bioreactors associated with aseptic conditions that are expensive to maintain (Curtis, 1999). Besides, unlike mammalian cell cultures which produce high-value molecules, plant vitro cultures have also been considered for the production of low or moderate-value molecules such as food ingredients. As a consequence, economic feasibility is even more difficult to reach (Cook and Mitchell, 1989).

Strategies to use Plant *vitro* Culture Technologies

There have been many attempts in the last decade to address the problem of cost-effectiveness of plant vitro culture technologies. Several new routes have been investigated. The first one is the design of low-cost bioreactors associated with a reduction of online controls and probes, and a simplified sterilization process (Curtis, 1999). Another possible strategy could be the use of plastic bags as already in use for the cultivation of microalgae (Borowitzka, 1999). These plastic bags are obviously much cheaper to use than culture reactors. However, they do not address all the problems encountered with cell cultures such as cell autotrophy or sedimentation. Others have tried to change more radically the concept of plant cell tissue and organ cultures in bioreactors without the use of a traditional culture chamber. A new system has been proposed by Borisjuk and co-workers (Borisjuk *et al.*, 1999) although it was designed for the production of recombinant proteins from genetically modified plants. The basic concept is to grow entire plants in a system where roots are maintained in sterile conditions and separated from the aerial parts. The recombinant proteins are recovered in the nutrient solution that surrounds the roots with the help of the rhizospheric excretion from the plant. This concept is interesting because it uses whole plants. It addresses the problem of biomass production encountered with cell suspension cultures that are hard to develop.

Another system has been recently proposed by Gontier and co-workers (Gontier *et al.*, 2000). This system has been designed specifically for the production of secondary compounds. The basic principle is to get rid of sterility but to keep a culture medium that allows the manipulation of secondary compound production (elicitation, addition of precursors, etc.). Sterility is essential because a traditional *in vitro* culture medium is constituted of organic compounds, especially sugars that would be altered by microorganisms if maintained in non-sterile conditions. If we use whole plants that are photosynthetically active, it is possible to avoid the use of organic compounds in a mineral-based medium. In this case, the medium is less amenable to contamination in open conditions. The other basic concept of this device is to yield secondary compounds in a non-destructive manner for the plants, therefore allowing repeated and regular harvests from the same biomass. In practice, this is achieved by growing the plants in hydroponic or aeroponic non-sterile conditions and by treating the medium in order to elicit the synthesis and to allow the excretion of the desired compounds, out of the roots. The last step is the recovery of the compounds from the medium with the help of purification processes. As a conclusion, this system uses the plants for what they can do best in natural conditions: they grow quite quickly when they are photosynthetically active. It also uses the concept of culture medium for what it is most useful in reactor culture technology: it allows a close control of secondary compound production (Gontier *et al.* 2000). This technology has been successfully used at pilot scale on three different model plants: *Datura innoxia* for tropane alkaloids, *Ruta graveolens* for furocoumarins, and *Taxus baccata* for paclitaxel (Borisjuk *et al.*, 1999).

Plant Metabolic Engineering and Pharmaceutical Compounds

Plant metabolic engineering has been successfully applied to the production of pharmaceutically useful secondary metabolites. Many pathways are currently being investigated but alkaloids have probably received more attention because of their pharmaceutical relevance (Yun *et al.*, 1992). Pioneer investigations on medicinal plants and metabolic engineering were conducted by Yun *et al.* (1992) on tropane alkaloids. These authors used a hyoscyamine 6β-hydroxylase (*H6H*) gene from *Hyoscyamus niger*, controlled by a 35s promoter, and succeeded in overexpressing h6h activity in *Atropa belladonna*. Transgenic *Atropa* plants displayed an enhanced conversion of hyoscyamine into scopolamine, which is a more pharmaceutically useful compound. These results clearly demonstrate that it is possible to considerably modify secondary metabolite patterns, playing on enzymes located downstream to the synthesis (Bourgaud *et al.*, 2001). Other experiments based on single gene transformations have been carried out by the same group on various alkaloid producing plants (Yun *et al.*, 1992). Overexpression of tobacco putrescine-*N*-methyl-transferase (PMT) in *Atropa belladonna* led to unmodified alkaloid profiles in the transformants whereas the same transformation carried out on *Nicotiana sylvestris* gave a 40% increase in leaf nicotine. Unlike H6H, PMT is placed upstream in alkaloid synthesis. PMT experiments demonstrate that it is possible to increase the metabolite flux so that it can benefit a downstream compound (nicotine). However, this strategy is not always successful, as in the case of *Atropa belladonna*, as other enzymes possibly limit the synthesis. Another strategy has been put forward by Sato and co-workers (Sato *et al.*, 2001) with experiments on scoulerine 9-*O*-methyltransferase (SMT). It consists in playing with pathways that derive from the same branching point. SMT converts scoulerine in successive compounds that lead to berberine, but scoulerine can also be transformed into sanguinarine by a parallel pathway (Bourgaud *et al.*, 2001). SMT overexpression allowed an increase in the berberine content of *Coptis* transgenic cells (20%) to the detriment of the competitive pathway.

Many attempts have been made to invent new devices that could lead to breakthroughs for the production of secondary compounds. However, the major changes in the area of plant secondary metabolites have probably been achieved thanks to the rise of molecular genetics in recent years, through the so-called metabolic engineering approach.

Bioactive Natural Products

The discovery of new bioactive natural products is still a fascinating field in organic chemistry as demonstrated by the recent paradigms of the anticancer drug epothilon, the immunosuppressant rapamycin, or the proteasome inhibitor salinosporamide, to name but a few of hundreds of possible examples (Dickschat, 2011). Finding new secondary metabolites is a prerequisite for the development of novel pharmaceuticals, and this is an especially urgent task in the case of antibiotics due to the rapid spreading of bacterial resistances and the emergence of multiresistant pathogenic strains, which poses severe clinical problems in the treatment of infectious diseases (Dickschat, 2011). Biosynthetic aspects are closely related to functional investigations, because a deep understanding of metabolic pathways to natural products, not only on a chemical, but also on a genetic and enzymatic level, allows for the expression of whole biosynthetic gene clusters in heterologous hosts (Ravishankar and Rao, 2000). This technique can make interesting, new secondary metabolites available from unculturable microorganisms, or may be used to optimise their availability by fermentation, for further research and also for production in the pharmaceutical industry.

Role of Light in Secondary Metabolites

The content and distribution of SMs in medicinal plants are closely related to environmental factors, especially light. In recent years, artificial light sources have been used in controlled environments for the production and conservation of medicinal germplasm (Zhang *et al.*, 2021). Therefore, it is essential to elucidate how light affects the accumulation of SMs in different plant species. There is need to systematically collate and understand recent advances in regulatory roles of light quality, light intensity, and photoperiod in the biosynthesis of three main types of SMs (polyphenols, alkaloids, and terpenoids), and the underlying mechanisms. Generally, when plants are exposed to UV-B radiation, the mechanism of light signaling pathway regulates expression of genes encoding transcription factors or key enzymes involved in the biosynthesis of SMs, to affect the accumulation of SMs under UV-B. Phytochromes (phyA and phyB) and cryptochromes (cry1 and cry2) regulate the accumulation of SMs in a different way (Zhang *et al.*, 2021). Under blue, red, and far-red light conditions, HY5 functions its roles normally and promotes the production of SMs. The transcription factors stimulate the expression of related biosynthetic genes and enhance the content of SM like artemisinin in *Artemisia annua.* Phytochrome Interacting Factors (PIFs) also mediate the biosynthesis of SMs in many plant species (Zhang *et al.*, 2021).

Secondary Metabolites Tissue Cultures

Plants are a tremendous source for the discovery of new products of medicinal value for drug development. Today several distinct chemicals derived from plants are important drugs currently used in one or more countries in the world. Many of the drugs sold today are simple synthetic modifications or copies of the naturally obtained substances. The evolving commercial importance of secondary metabolites has in recent years resulted in a great interest in secondary metabolism, particularly in the possibility of altering the production of bioactive plant metabolites by means of tissue culture technology (Vanisree *et al.*, 2004). Plant cell culture technologies were introduced at the end of the 1960's as a possible tool for both studying and producing plant secondary metabolites. Different strategies, using an *in vitro* system, have been extensively studied to improve the production of plant chemicals. The focus is now on the application of tissue culture technology for the production of some important plant pharmaceuticals.

Many higher plants are major sources of natural products used as pharmaceuticals, agrochemicals, flavor and fragrance ingredients, food additives, and pesticides (Balandrin and Klocke, 1988). The search for new plant derived chemicals should thus be a priority in current and future efforts toward sustainable conservation and rational utilization of biodiversity (Phillipson, 1990). In the search for alternatives to production of desirable medicinal compounds from plants, biotechnological approaches, specifically, plant tissue cultures, are found to have potential as a supplement to traditional agriculture in the industrial production of bioactive plant metabolites (Rao and Ravishankar, 2002). Cell suspension culture systems could be used for large scale culturing of plant cells from which secondary metabolites could be extracted. The advantage of this method is that it can ultimately provide a continuous, reliable source of natural products (Vanisree *et al.*, 2004).

Discoveries of cell cultures capable of producing specific medicinal compounds at a rate similar or superior to that of intact plants have accelerated in the last few years. New physiologically active substances of medicinal interest have been found by bioassay. It has been demonstrated that the biosynthetic activity of cultured cells can be enhanced by regulating environmental factors, as well as by artificial selection or the induction of variant clones (Vanisree *et al.*, 2004). Some of the medicinal compounds localized in morphologically specialized tissues or organs of native plants have been produced in culture systems not only by inducing specific organized cultures, but also by undifferentiated cell cultures. The possible use of plant cell cultures for the specific biotransformations of natural compounds has been demonstrated (Cheetham, 1995; Scragg, 1997; Krings and Berger, 1998; Ravishankar and Rao, 2000). Due to these advances, research in the area of tissue culture technology for production of plant chemicals has bloomed beyond expectations.

Biosynthetic Activities of Cultured Cells

In order to obtain high yields suitable for commercial exploitation, efforts have focused on isolating the biosynthetic activities of cultured cells, achieved by optimizing the cultural conditions, selecting high-producing strains, and employing precursor feeding, transformation methods, and immobilization techniques (Dicosmo and Misawa, 1995). Transgenic hairy root cultures have revolutionized the role of plant tissue culture in secondary metabolite production. They are unique in their genetic and biosynthetic stability, faster in growth, and more easily maintained. Using this methodology a wide range of chemical compounds have been synthesized (Shanks and Morgan, 1999; Giri and Narasu, 2000). Advances in tissue culture, combined with improvement in genetic engineering, specifically transformation technology, has opened new avenues for high volume production of pharmaceuticals, nutraceuticals, and other beneficial substances (Hansen and Wright, 1999). Recent advances in the molecular biology, enzymology, and fermentation technology of plant cell cultures suggest that these systems will become a viable source of important secondary metabolites.

Genome manipulation is resulting in relatively large amounts of desired compounds produced by plants infected with an engineered virus, whereas transgenic plants can maintain constant levels of production of proteins without additional intervention (Sajc *et al.*, 2000). Large-scale plant tissue culture is found to be an attractive alternative approach to traditional methods of plantation as it offers a controlled supply of biochemicals independent of plant availability (Sajc *et al.*, 2000). Kieran *et al.* (1997) detailed the impact of specific engineering-related factors on cell suspension cultures. Current developments in tissue culture technology indicate that transcription factors are efficient new molecular tools for plant metabolic engineering to increase the production of valuable compounds (Gantet and Memelink, 2002). *In vitro* cell culture offers an intrinsic advantage for foreign protein synthesis in certain situations since they can be designed to produce therapeutic proteins, including monoclonal antibodies, antigenic proteins that act as immunogenes, human serum albumin, interferon, immuno-contraceptive protein, ribosome unactivator trichosantin, antihypersensitive drug angiotensin, leu-enkephalin neuropeptide, and human hemoglobin (Hiatt *et al.*, 1989; Manson and Arntzen, 1995; Wahl *et al.*, 1995; Arntzen, 1997; Hahn *et al.*, 1997; La Count *et al.*, 1997; Marden *et al.*, 1997; Wongsamuth and Doran, 1997; Doran, 2000).

Future Perspectives

In vitro propagation of medicinal plants with enriched bioactive principles and cell culture methodologies for selective metabolite production is found to be highly useful for commercial production of medicinally important compounds (Vanisree *et al.*, 2004). The increased use of plant cell culture systems in recent years is perhaps due to an improved understanding of the secondary metabolite pathway in economically important plants. Advances in plant cell cultures could provide new means for the cost-effective, commercial production of even rare or exotic plants, their cells, and the chemicals that they will produce. Knowledge of the biosynthetic pathways of desired compounds in plants as well as of cultures is often still rudimentary, and strategies are consequently needed to develop information based on a cellular and molecular level. Because of the complex and incompletely understood nature of plant cells in *in vitro* cultures, case-by-case studies have used to explain the problems occurring in the production of secondary metabolites from cultured plant cells. A key to the evaluation of strategies to improve productivity is the realization that all the problems must be seen in a holistic context. At any rate, substantial progress in improving secondary metabolite production from plant cell cultures has been made within last few years. These new technologies will serve to extend and enhance the continued usefulness of higher plants as renewable sources of chemicals, especially medicinal compounds. It is hoped that a continuation and intensification efforts in this field will lead to controllable and successful biotechnological production of specific, valuable, and as yet unknown plant chemicals.

Conclusion

Using the new tools of molecular cell biology, genetics, immunology and NGS, many bioactivities of SM can be studied in more detail and precision. It is important that phytochemists not only isolate SM and describe their chemical structures but that they also study their biological activities alone or in combinations. In order to translate the findings from various laboratories, we need clinical trials to corroborate the efficacy of herbal drugs and to market them as evidence based medicines.

Table 1: Use and bioactivty of a few selective secondary metabolites which are applied as isolated compounds in medicine; alkaloid (A), terpenoids (T)

Plant species	Substance (class)	Mode of action	Properties/applications
Aconitum napellus	aconitine (A)	activates Na^+ channels	analgesic
Atropa belladonna	L-hyoscyamine (A)	antagonist of mAChR	parasympathomimetic
Camptotheca acuminate	camptothecin (A)	inhibitor of DNA topoisomerase	tumour therapy
Cannabis sativa	tetrahydrocannabinol (T)	activates THC receptor	analgesic
Catharanthus roseus	dimeric Vinca alkaloids (A)	inhibit microtubule assembly	tumor therapy
Chondrodendrontomentoum	tubocurarine (A)	inhibits nAChR	muscle relaxant
Cinchona pubescens	quinidine (A)	inhibits Na^+ channels	antiarrhythmic
Coffea arabica	caffeine (A)	inhibits phosphodiesterase and adenosine receptors	stimulant
Colchicum autumnale	colchicine (A)	inhibits microtubule assembly	gout treatment
Cytisus scoparius	sparteine (A)	inhibits Na^+ channels	antiarrhythmic
Digitalis lanata	digitoxin, digoxin (T)	inhibits Na^+,K^+-ATPase	heart insufficiency
Erythroxylum coca	cocaine (A)	inhibits Na^+ channels and reuptake of noradrenaline and dopamine	analgesic; stimulant
Galanthus woronowii	galanthamine (A)	inhibits AChE	alzheimer treatment
Lycopodium clavatum	huperzine A (A)	inhibits AChE	alzheimer treatment
Papaver somniferum	morphine (A)	agonist of endorphine receptors	analgesic, hallucinogen
Physostigma venenosum	physostigmine (A)	inhibits AChE	alzheimer treatment
Pilocarpus joborandi	pilocarpine (A)	agonist of mAChR	glaucoma treatment
Psychotria ipecacuanha	emetine (A)	protein biosynthesis inhibitor inhibits the uptake of noradrenalin	treatment of amebae infections; emetic
Rauvolfia reserpina	reserpine (A)	into postsynaptic vesicles	hypertonia treatment
Sanguinaria canadensis	sanguinarine (A)	DNA intercalator	antibacterial, antiviral
Strophantus gratus	ouabain (T)	inhibits Na^+, K^+-ATPase	heart insufficiency
Taxus brevifolia	paclitaxel (taxol) (A)	inhibits microtubule disasssembly	tumer therapy

References

Arntzen C J. 1997. High tech herbal medicine: plant based vaccines. Nature Biotechnol. 15(3): 221-22.

Balandrin M J and J A Klocke. 1988. Medicinal, aromatic and industrial materials from plants. In Y.P.S. Bajaj (ed.), Biotechnology in Agriculture and Forestry. Medicinal and Aromatic Plant, vol. 4. Springer-Verlag, Berlin, Heidelberg, pp. 1-36.

Borisjuk N V, Borisjuk L G, Logendra S, Petersen F, Gleba Y and Raskin I. 1999. Production of recombinant proteins in plant root exudates, Nature Biotechnology, 17:466–69.

Borowitzka M A. 1999. Commercial production of microalgae: ponds, tanks, tubes and fermenters, Journal of Biotechnology, 70:313–21.

Bourgaud F, Gravot A, Milesi S and Gontier E. 2001. Production of plant secondary metabolites: a historical perspective, Plant Science, 161:839-51.

Cheetham P S J. 1995. Biotransformations: new routes to food ingredients. Chem Ind., pp. 265-268.

Chizzali L, Cornelia M and Beerhues L. 2012. Phytoalexins of the Pyrinae: Biphenyls and dibenzofurans. Beilstein J. Org. Chem. 8:613–20.

Cook J A and Mitchell J B. 1989. Viability measurements in mammalian cell systems, Analytical Biochemistry 179:1–7.

Curtis W R. 1999. Achieving economic feasibility for moderate-value food and flavor additives: a perspective on productivity and proposal for production technology cost reduction, In: Plant Cell and Tissue Culture for the Production of Food Ingredients, Fun, T.J., Singh, G., Curtis, W.R. (Eds.) Kluwer Academic, Plenum Publisher pp. 225–236.

Dicosmo F and Misawa M. 1995. Plant cell and tissue culture: A lternatives for metabolite production. Biotechnol. Adv. 13(3): 425-53.

Dickschat J S. 2011. Biosynthesis and function of secondary metabolites, Beilstein J. Org. Chem, 7:1620–21.

Doran P M. 2000. Foreign protein production in plant tissue cultures. Curr. Opin. Biotechnol. 11: 199-204.

Fraenkel, G S. 1959. The raison d'Etre of secondary plant substances. Science, 129 (3361):1466–70.

Gantet P and Memelink J. 2002. Transcription factors: tools to engineer the production of pharmacologically active plant metabolites. Trends Pharmacol. Sci. 23:563-69.

Giri A and Narasu M L. 2000. Transgenic hairy roots: recent trends and applications. Biotechnol. Adv. 18:1-22.

Gontier E, Clement A, Bourgaud F and Guckert A. 2000. Plant metabolite production by hydroponic or aeroponic cultures. Plant Cell Report, 20:177–90.

Hahn J J, Eschenlauer C A, Narrol H M, Somers A D and Srienc F. 1997. Growth kinetics, nutrient uptake, and expression of the Alcaligenes eutrophus poly(-hydroxybutarate) synthesis pathway in transgenic maize cell suspension cultures, Biotechnol. Prog. 13(4):347-54.

Hansen G and Wright M S. 1999. Recent advances in the transformation of plants, Trends Plant Sci. 4: 226-31.

Harborne J B. 1999. Classes and functions of secondary products, In: Chemicals from Plants, Perspectives on Secondary Plant Products (Eds.), N.J. Walton, D.E. Brown, Imperial College Press, pp. 1–25.

Hiatt A, Cafferkey R and Boedish K. 1989. Production of antibodies in transgenic plants. Nature, 342: 76-78.

Kieran P M, MacLoughlin P F and Malone D M. 1997. Plant cell suspension cultures: some engineering considerations, J. Biotechnol. 59: 39-52.

Krings U and Berger R G. 1998. Biotechnological production of flavours and fragrances. Appl. Microb. Biotechnol, 49:1-8.

La Count W, An G and Lee J M. 1997. The effect of PVP on the heavy chain monoclonal antibody production from plant suspension cultures. Biotechnol. Let. 19(1):93-96.

Lai PK and Roy J. 2004. Antimicrobial and chemo preventive properties of herbs and spices. Curr. Med. Chem. 11(11):1451–60.

Manson H S and Arntzen C J. 1995. Transgenic plants as vaccine production system, Trends Biotechnol, 3:388-92.

Marden M C, Dieryck W, Pagnier J, Poyart C, Gruber V, Bournat P, Baudino S and Merot B. 1997. Human hemoglobin from transgenic tobacco, Nature 342:29-30.

Meskin Mark S. 2002. Phytochemicals in Nutrition and Health. CRC Press. p. 123.

Payne G F, Bringi V, Prince C and Shuler M L. 1991. The quest for commercial production of chemicals from plant cell culture, In: Plant Cell and Tissue Culture in Liquid Systems, Hanser, Payne, G.F., Bringi, V., Prince, C. Shuler, M.L. (Eds.) pp. 1–10.

Phillipson J D. 1990. Plants as source of valuable products. In: B.V. Charlwood, and M.J.C. Rhodes (Eds.), Secondary Products from Plant Tissue Culture. Oxford: Clarendon Press, pp. 1-21.

Rao R S and Ravishankar G A. 2000. Biotransformation of protocatechuic aldehyde and caffeic acid to vanillin and capsaicin in freely suspended and immobilized cell cultures of Capsicum frutescens. J. Biotechnol 76:137-46.

Ravishankar G A and Rao R S. 2000. Biotechnological production of phyto-pharmaceuticals. J. Biochem. Mol. Biol. Biophys 4:73-102.

Sajc L, Grubisic D and Vunjak-Novakovic G. 2000. Bioreactors for plant engineering: an outlook for further research, Biochem. Eng, J. 4:89-99.

Samuni-Blank M, Izhaki I, Dearing MD, Gerchman Y, Trabelcy B, Lotan A, Karasov WH, Arad, Z 2012. Intraspecific directed deterrence by the mustard oil bomb in a desert plant. Current Biology. 22:1-3.

Sato F, Hashimoto T, Hachiya A, Tamura K I, Choi K B and Morishige T. 2001. Metabolic engineering of plant alkaloid biosynthesis, In: Proceedings of the National Academy Science, USA, 98:367–72.

Scragg A H. 1997. The production of aromas by plant cell cultures. In: Adv Biochem. Eng. Biotechnol. T. Schepier (ed.) Vol. 55. Berlin: Springer-Verlag, pp. 239-63.

Shanks J V and Morgan J. 1999. Plant hairy root culture. Curr. Opin. Biotechnol. 10:151-55.

Tapsell LC, Hemphill I and Cobiac L. 2006. Health benefits of herbs and spices: the past, the present, the future. Med. J. Aust. 185:4–24.

Stamp N. 2003. Out of the quagmire of plant defense hypotheses. The Quarterly Review of Biology, 78 (1): 23–55.

Steward N, Martin R, Engasser J M and Goergen J L. 1999a. A new methodology for plant cell viability assessment using intracellular esterase activity, Plant Cell Report, 19:171–76.

Steward N, Martin R, Engasser J M and Goergen J L. 1999b. Determination of growth and lysis kinetics in plant cell suspension cultures from the measurement of esterase release Biotechnology and Bioengineering, 66:114–21.

Vanisree M, Lee Chen-Yue, Lo Shu-Fung, Nalawade S M, Lin C Y, and Tsay Hsin-Sheng. 2004. Studies on the production of some important secondary metabolites from medicinal plants by plant tissue cultures. Bot. Bull. Acad. Sin., 45:1-22.

Wahl M F, An G and Lee J M. 1995. Effects of DMSO on heavy chain monoclonal antibody production from plant cell culture. Biotechnol, Lett. 17(5):463-68.

William S H. 2004. Officina. Medical meanings: a glossary of word origins. p.162.

Wink M. 2015. Modes of action of herbal medicines and plant secondary metabolites. Medicines, 2(3): 251-86.

Wongsamuth R and Doran M P. 1997. Production of monoclonal antibodies by tobacco hairy roots. Biotechnol. Bioeng. 54(5): 401-15.

Yun D J, Hashimoto T and Yamada Y. 1992. Metabolic engineering of medicinal plants: transgenic Atropa belladonna with an improved alkaloid composition, In: Proceedings of the National Academy Science, USA, 89:11799–803.

Zhang S, Zhang L, Zou H, Qiu L, Zheng Y, Yang D and Wang Y. 2021. Effects of light on secondary metabolite biosynthesis in medicinal plants. Front. Plant Sci. 12- https://doi.org/10.3389/fpls.2021.781236.

Zrÿd J P. 1988. Cultures in vitro et production de metabolites secondaires, In: Cultures de Cellules, Tissus, et Organes Végétaux, Fondements Théoriques et Utilisations Pratiques, Zrÿd, J.P. (Ed.), Polytechniques Romandes Presses, pp. 228–34.

Section 06: Institution

24

Dr YSR Horticulture University, Heralding Horticultural Revolution in Andhra Pradesh

T. Janakiram, B. Srinivasulu, E Keruna Sree and V. Deepthi

Dr.Y.S.R. Horticultural University, Venkataramannagudem – 534 101
West Godavari District, Andhra Pradesh

Government of Andhra Pradesh has identified Horticulture sector as one of the growth engine with a focus to increase the production, productivity and quality all horticultural crops coupled with value chain development and marketing linkages for better price realization. Total area under horticultural crops is 17.95 lakh ha with a production of ` 314.78 lakh tonnes. Horticulture sector has recorded 10.17 % growth over previous year and contributing ` 49,189 crores during 2020-21 to GVA of Andhra Pradesh (DH, Govt. of Andhra Pradesh, 2021).

The state ranks first in productivity of oil palm, papaya, lime, cocoa, tomato, coconut and chilli, second in mango, sweet orange and turmeric in India. Andhra Pradesh registration of Horticulture Nurseries (Regulation) Act, 2010, covers only fruit nurseries, Government of Andhra Pradesh amended the Act and the said Act came into force from 18 January 2022 to include shadenet / poly house nurseries, raising chilli, tomato, vegetable seedlings etc. and tissue culture nurseries for ensuring production and supply of quality plant material to farmers (A.P Registration Act, 2010).

The Compound Annual Growth Rate (CAGR) of horticultural production for 2020-21 was 4.7%, this showing increasing trend of horticulture production. Thus, Andhra Pradesh has emerged as the largest producer of fruits in our country, contributing to 15.6 % of the total production. As per EXIM Bank report, Andhra Pradesh contributes 7.8 % of vegetables and vegetable products (Reddy, 2016; DH, Govt of 2020).

New crops like dragon fruit, Taiwan guava, Mauritius variety of pineapple, strawberry, cinnamon, nutmeg, Japanese mint, improved variety of black pepper, turmeric and ginger have been introduced for generating higher income to the farmers by the Department of Horticulture, Government of Andhra Pradesh. Andhra Pradesh is the largest producer of cocoa. Every year new plantations are being taken up in 500 ha and processing industries for value-addition are promoted creating employment opportunities.

Andhra Pradesh covered 10% of the total area under microirrigation and 15 % of the total area under drip in our country. Among top 10 districts under area coverage in the country, 2 districts (Anantapuramu and YSR Kadapa) are from Andhra Pradesh During 2021-22, an additional area of 62,271 ha was brought under Horticulture crops. Organic farming has been promoted in 11,250 ha of mango, cashew, turmeric, chilli and vegetables by Department of Horticulture, Government of A.P. Andhra Pradesh climate is quite suitable for vegetables and flowers on off season / year round. Major crops grown under protected cultivation are capsicum, English cucumber, chrysanthemum, gerbera, carnation, orchids, roses and hybrid tomato. Andhra Pradesh has emerged as the biggest exporter of banana. During 2020-21 94,571 MT of banana have been exported from Andhra Pradesh.29195 crop specific "Thotabadi" programmes(Farm School Training) were organized in various horticultural crops covering 8,75,850 farmers for productivity and quality enhancement in Horticultural crops (Anon, 2019 & Anon 2021).

Role of Dr. YSR Horticulture University

Dr.YSR Horticultural University (Dr.YSRHU), a state university, 2nd of its kind in the country was established by the Government of Andhra Pradesh on 26th June, 2007. This university is functioning with a mandate to develop human resources through Education, Research and Extension in Horticulture and allied sectors through forty two institutes i.e constituent Colleges of Horticulture (4), constituent Horticultural Polytechnics (4), Horticultural Research Stations (19), Krishi Vigyan Kendras (4), affiliated Colleges of Horticulture (4) and affiliated Horticultural Polytechnics (7)

HRD through Education, Research and Extension

Dr.YSRHU is offering B.Sc (Hons.) Horticulture, M.Sc (Horticulture) and Ph.D in horticulture with specialization in Fruit Science, Vegetable Science, Floriculture & Landscape Architecture, Plantation, spices, Medicinal & Aromatic Crops, Post Harvest Technology, Plant Pathology and Entomology, besides two year Diploma in Horticulture.It is one and only university in Andhra Pradesh conducting research on horticulture crops, evolving varieties and technologies in horticulture crops suitable to Andhra Pradesh through coordinated research services of nineteen Research Stations established throughout the state.

This university is concentrating on transfer of latest technologies in Horticulture through KrishiVigyanKendras (KVKs), Horticultural Research Stations (HRSs), Colleges of Horticulture and Horticultural Polytechnics for the benefit of farmers, rural youth and women directly and through RythuBharosaKendrams (RBKs), District Resource Centres (DRCs) and Department of Horticulture, Government of Andhra Pradesh.

Towards Higher Productivity of Horticultural Crops

University has released 27 varieties in various horticultural crops and 13 released varieties were notified at National Level Vide Gazette No.1369, dt:7 April, 2021of Ministry of Agriculture and Farmers Welfare, Government of India namely Turmeric (Lavanya), Coriander (Suruchi), Chilli (LCA-620, LCA-625, LCH-111), Colacasia (Godavari Chema), Fenugreek (Lam Methi-2), Banana (Godavari Bontha), Cassava (PDP CMR-1) and Coconut (Vasista Ganga, Gauthami Ganga, Abhaya Ganga, Vynateya Ganga)

Termeric Lavanya (KTS-3) | Coriander: Suruchi (LCC-234) | Chilli: LCA-620

Chilli: LCA-625 | Chilli: LCH-111 | Colacasia: Godavari Chema (KCS-3)

Cassava: TCMS-5/PDP CMR-1 | Fenugreek: Lam Methi-2 | Coconut: Vasista Ganga

Coconut: Gauthami Ganga | Coconut: Abhaya Ganga | Coconut: Vynateya Ganga | Banna: Godavari Bontha

Technology Support

The university has developed important technologies in production, protection and value-addition in horticultural crops

- Developed Biocontrol management practices for major pests and diseases.
- Organic farming protocols developed for turmeric, ginger, colocasia, diascoria, sweet orange, acid lime, elephant-foot yam, etc.,

- Fertigation schedules standardized in acid lime, sweet orange, oil palm, banana, etc.
- Standardized protocols for production of bio-fertlizers in solid and liquid forms.
- Tissue culture protocols standardized for banana (local varieties), tuber crops.
- Established cocoa model nursery funded by DCCD at HRS, Vijayarai.
- Established Model Nursery at Horticultural Research Station, Chintapalli for black pepper – funded by DASD, Kozhikode.
- Standardized IPM/IDM technologies in sweet orange, banana, mango, vegetables, spices crops and chilli.
- Identified biocontrol agents, i.e. Dichocrysa astur a predator against Rugose Spirilling White fly and standardized protocols for mass multiplication for supply to farmers.
- Standardized protocols for value added products (palmyrah neera, syrup, jaggery, jaggery powder, sugar, crystal, chocolate, fruit bar, fruit flour, tuber flour and spongy endosperm flour) in Palmyrah with FSSAI certification with license authorized to manufacturing/ repack/ re-label from Govt. of Andhra Pradesh.
- Biotechnology labs established at CRS, Tirupati, HRS, Kovvur and HRS, Lam
- DNA fingerprinting of varieties released by DrYSRHU was done in respect of acid lime (Balaji, Petlur Selection -1), coconut (Gouthami Ganga, Vasista Ganga, Vynateya Ganga, Abhaya Ganga), turmeric (Lavanya), Colocasia (Godavari Chema), banana (Godavari Bontha) and tapioca (PDP CMR-1)
- Released varieties of different crops were registered with NBPGR and got IC numbers.

Dr YSRHU Success Stories

The University has successful transferred technologies to farmers and played much role in doubling farmers' income. Success stories are:

- Assured Income Generation Through Integrated Farming Systems (IFS) For Livelihood of Tribal Folks: Recorded an assured net income of ` 90,000 to 1,20,000/ha in Integrated Farming System where as ` 40,000 - ` 50,000/ha before implementation of IFS. Dr.YSRHU – KVK, Venkataramannagudem, intervened tribal women farmer Smt. T. Ramana of Rekulakunta village received ICAR award "Pandit DeenDayal Upadhyay Anthyodaya Krishi Puraskar-2019" at zonal level for her excellent contribution towards IFS model adoption.

- Management of cashew orchards for sustainable livelihood of tribal farmers: A total of 57.8 tonnes of cashew nuts were marketed by 223 tribal farmers with an average price of ` 123.5/kg. Twelve per cent yield increase was recorded over the last year. This activity was very much accepted by the farmers and were happy with an average increased income of ` 31,250/ acre.
- Improved vegetables cultivation in tribal areas of West Godavari District: The technology is spreading in tribal areas of West Godavari District and now at present the area under improved vegetable cultivation is about 200 acres with a net income of ` 40,000/- to ` 1,20,000/- per acre
- Cultivation of multiple crops in organic farming: organic method of cultivation has reduced dependence on external inputs and created high market demand and motivated nearly 35-40 farmers to adopt organic farming.
- Apiculture - a successful skill based intervention: Honey was extracted @ 20-25 Kg from each box from December to June. Honey extracted from these colonies was sold @ Rs. 300/ - 500/ kg, depending on the season and demand.
- Millet processing as an income generating activity: The tribals are empowered to market various products in open and online markets. Business of an amount of ` 2,00,000/ month is done by the groups through this activity, with a regular income of ` 5,000-6,000/month for each member
- Livelihood Improvement of Rubber Growers in Tribal Area of East Godavari District, A.P.: On an average, each tribal rubber grower earned ` 35,295/- (rupees thirty five thousand two hundred and ninty five only) over a period of 160 days with intervention of Dr. YSRHU-KVK, Pandirimamidi.
- Introduction of colacasia variety KCS-3 for higher yields: Net income for one hectare was ` 6,90,000/- in demo plot where as in farmers practice it was Rs.4,25,000/- Percentage increase in the yield was 11.76% cost benefit ratio was 1:5.3.
- Integrated Farming System Module in Tribal Area of East Godavari district: horti based Integrated Farming System has doubled the income of tribal farmers and motivated to shift from mono cropping of paddy to IFS.
- University has intervened strawberry growing tribal farmers of Lambasingi agency area and trained them to prepare strawberry jam and dehydrated strawberry fruits, resulted into fetching income to tribal strawberry farmers.

Outreach Programmes to Farmers

The university has developed good agricultural practices to major horticultural crops of Andhra Pradesh. University is concentrating on transfer of good agricultural practices to farmers by way of Farmers Advisory Cell (FAC), Plant Protection Advisory Cell (PPAC), OFTs, FLDs, Trainings and Skill trainings besides mass communication strategies, viz. Dr. YSRHU Udyana Mitra, Farmers Awareness Creation Centre (FACC), Electronic Wing, Community Radio Station (Dr. YSRHU Udyana Vani), Vice-Chancellor to Village Programme, Our University – Our Organic Village programme, Dr.YSRHU year of Coconut 2020-21, Dr. YSRHU year of Citrus 2021-22, Dr. YSRHU Phone-in programmes, TV Phone-in-Live programmes, Year round Training Calendar, Dr.YSRHU you tube channel, Dr. YSRHU – Dr. YSR RBKs : Dr. YSRHU Udyana Bandhu and Dr. YSRHU Udyana Sandarshana etc.,

Graduate Readiness Programme and Curriculum Improvement

Dr.YSRHU has awarded horticultural degrees so far to 2654 students (2125 in B.Sc (Hon.) Horticulture, 467 in M.Sc (Horticulture) and 62 in Ph.D in Horticulture), besides 2281 in Diploma in Horticulture. University has organized Coconut & Citrus Graduate Readiness programmes for the benefit of students at National level. Certain suggestions for improvement in the curriculum are i) Early implementation on National Educational Policy (NEP) in the universities ii) Introduction of courses/curriculum on Organic Horticulture, Tribal Horticulture, Integrated Farming Systems, Marketing Strategies Market Intelligence, Farm Mechanization / Artificial Intelligence and Digital Extension.

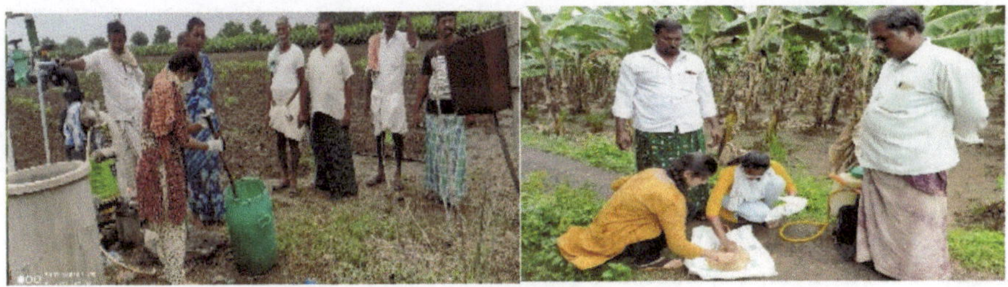

Remunerative Crops and Technologies

University with available state of art facilities and in collaboration with National institutes and Department of Horticulture, Government of Andhra Pradesh making stride in introducing new Horticultural crops in Andhra Pradesh viz., Nutmeg, Dragon fruit, Moringa, Black Pepper, Strawberry, Avacado, Annonamuricata and processing varieties in horticultural crops are future remunerative crops. University also working on crop diversification, organic farming, farm mechanization, validation of farmers technologies and creation of awareness on exotic crops in co-ordination with Department of Horticulture, Government of Andhra Pradesh.

References

Andhra Pradesh Registration of Horticulture Nurseries (Regulation) Act, 2010 (Act No. 13 of 2010)

DH Govt. of AP. 2020. Annual Report, AP Micro Irrigation Project, Department of Horticulture, Government of Andhra Pradesh.

Export-Import Bank, India 2019. Potential for enhancing exports from Andhra Pradesh, 2019. Export-Import Bank of India, occasional Paper No. 182.

DH Govt. of AP. 2021. Department of Horticulture, 2021, Government of Andhra Pradesh (Source : http:horticulture.ap.nic.in).

Reddy, K. Y. 2016. Micro-Irrigation in Participatory Mode Pays Huge Dividends-Apmip Experiences, India. Irrigation and Drainage 65:72–78. doi:10.1002/ird.2039

Index